Protein Engineering Protocols

METHODS IN MOLECULAR BIOLOGY™

John M. Walker, SERIES EDITOR

386. **Peptide Characterization and Application Protocols**, edited by *Gregg B. Fields, 2007*
385. **Microchip-Based Assay Systems:** *Methods and Applications*, edited by *Pierre N. Floriana, 2007*
384. **Capillary Electrophoresis:** *Methods and Protocols*, edited by *Philippe Schmitt-Kopplin, 2007*
383. **Cancer Genomics and Proteomics:** *Methods and Protocols*, edited by *Paul B. Fisher, 2007*
382. **Microarrays, Second Edition:** *Volume 2, Applications and Data Analysis*, edited by *Jang B. Rampal, 2007*
381. **Microarrays, Second Edition:** *Volume 1, Synthesis Methods*, edited by *Jang B. Rampal, 2007*
380. **Immunological Tolerance:** *Methods and Protocols*, edited by *Paul J. Fairchild, 2007*
379. **Glycovirology Protocols**, edited by *Richard J. Sugrue, 2007*
378. **Monoclonal Antibodies:** *Methods and Protocols*, edited by *Maher Albitar, 2007*
377. **Microarray Data Analysis:** *Methods and Applications*, edited by *Michael J. Korenberg, 2007*
376. **Linkage Disequilibrium and Association Mapping:** *Analysis and Application*, edited by *Andrew R. Collins, 2007*
375. **In Vitro Transcription and Translation Protocols:** *Second Edition*, edited by *Guido Grandi, 2007*
374. **Quantum Dots:** *Methods and Protocols*, edited by *Charles Z. Hotz and Marcel Bruchez, 2007*
373. **Pyrosequencing® Protocols**, edited by *Sharon Marsh, 2007*
372. **Mitochondrial Genomics and Proteomics Protocols**, edited by *Dario Leister and Johannes Herrmann, 2007*
371. **Biological Aging:** *Methods and Protocols*, edited by *Trygve O. Tollefsbol, 2007*
370. **Adhesion Protein Protocols**, *Second Edition*, edited by *Amanda S. Coutts, 2007*
369. **Electron Microscopy**: *Methods and Protocols, Second Edition*, edited by *John Kuo, 2007*
368. **Cryopreservation and Freeze-Drying Protocols**, *Second Edition*, edited by *John G. Day and Glyn Stacey, 2007*
367. **Mass Spectrometry Data Analysis in Proteomics**, edited by *Rune Matthiesen, 2007*
366. **Cardiac Gene Expression:** *Methods and Protocols*, edited by *Jun Zhang and Gregg Rokosh, 2007*
365. **Protein Phosphatase Protocols:** edited by *Greg Moorhead, 2007*
364. **Macromolecular Crystallography Protocols:** *Volume 2, Structure Determination*, edited by *Sylvie Doublié, 2007*
363. **Macromolecular Crystallography Protocols:** *Volume 1, Preparation and Crystallization of Macromolecules*, edited by *Sylvie Doublié, 2007*
362. **Circadian Rhythms:** *Methods and Protocols*, edited by *Ezio Rosato, 2007*
361. **Target Discovery and Validation Reviews and Protocols:** *Emerging Molecular Targets and Treatment Options, Volume 2*, edited by *Mouldy Sioud, 2007*
360. **Target Discovery and Validation Reviews and Protocols:** *Emerging Strategies for Targets and Biomarker Discovery, Volume 1*, edited by *Mouldy Sioud, 2007*
359. **Quantitative Proteomics by Mass Spectrometry**, edited by *Salvatore Sechi, 2007*
358. **Metabolomics:** *Methods and Protocols*, edited by *Wolfram Weckwerth, 2007*
357. **Cardiovascular Proteomics:** *Methods and Protocols*, edited by *Fernando Vivanco, 2006*
356. **High-Content Screening:** *A Powerful Approach to Systems Cell Biology and Drug Discovery*, edited by *D. Lansing Taylor, Jeffrey Haskins, and Ken Guiliano, and 2007*
355. **Plant Proteomics:** *Methods and Protocols*, edited by *Hervé Thiellement, Michel Zivy, Catherine Damerval, and Valerie Mechin, 2006*
354. **Plant–Pathogen Interactions:** *Methods and Protocols*, edited by *Pamela C. Ronald, 2006*
353. **Protocols for Nucleic Acid Analysis by Non-radioactive Probes**, *Second Edition*, edited by *Elena Hilario and John Mackay, 2006*
352. **Protein Engineering Protocols**, edited by *Katja M. Arndt and Kristian M. Müller, 2006*
351. ***C. elegans:*** *Methods and Applications*, edited by *Kevin Strange, 2006*
350. **Protein Folding Protocols**, edited by *Yawen Bai and Ruth Nussinov 2007*
349. **YAC Protocols**, *Second Edition*, edited by *Alasdair MacKenzie, 2006*
348. **Nuclear Transfer Protocols:** *Cell Reprogramming and Transgenesis*, edited by *Paul J. Verma and Alan Trounson, 2006*
347. **Glycobiology Protocols**, edited by *Inka Brockhausen-Schutzbach, 2006*
346. ***Dictyostelium discoideum*** **Protocols**, edited by *Ludwig Eichinger and Francisco Rivero, 2006*
345. **Diagnostic Bacteriology Protocols**, *Second Edition*, edited by *Louise O'Connor, 2006*
344. **Agrobacterium Protocols**, *Second Edition: Volume 2*, edited by *Kan Wang, 2006*
343. **Agrobacterium Protocols**, *Second Edition: Volume 1*, edited by *Kan Wang, 2006*
342. **MicroRNA Protocols**, edited by *Shao-Yao Ying, 2006*
341. **Cell–Cell Interactions:** *Methods and Protocols*, edited by *Sean P. Colgan, 2006*
340. **Protein Design:** *Methods and Applications*, edited by *Raphael Guerois and Manuela López de la Paz, 2006*
339. **Microchip Capillary Electrophoresis:** *Methods and Protocols*, edited by *Charles S. Henry, 2006*
338. **Gene Mapping, Discovery, and Expression:** *Methods and Protocols*, edited by *M. Bina, 2006*

METHODS IN MOLECULAR BIOLOGY™

Protein Engineering Protocols

Edited by

Katja M. Arndt

Kristian M. Müller

Institut für Biologie III, Universität Freiburg, Freiburg, Germany

HUMANA PRESS TOTOWA, NEW JERSEY

Softcover reprint of the hardcover 1st edition 2007
999 Riverview Drive, Suite 208
Totowa, New Jersey 07512

www.humanapress.com

This publication is printed on acid-free paper. ∞
ANSI Z39.48-1984 (American Standards Institute)

Permanence of Paper for Printed Library Materials.

Production Editor: Melissa Caravella

Cover design by Patricia F. Cleary

Cover illustration: From Fig. 1, Chapter 16, "A General Method of Terminal Truncation, Evolution, and Re-Elongation to Generate Enzymes of Enhanced Stability," by Jochen Hecky, Jody M. Mason, Katja M. Arndt, and Kristian M. Müller

For additional copies, pricing for bulk purchases, and/or information about other Humana titles, contact Humana at the above address or at any of the following numbers: Tel.: 973-256-1699; Fax: 973-256-8341; E-mail: orders@humanapr.com; or visit our Website: www.humanapress.com

10 9 8 7 6 5 4 3 2 1

eISBN 978-1-59745-187-1

Library of Congress Cataloging in Publication Data

Protein engineering protocols / edited by Kristian M. Müller, Katja M. Arndt.
p. cm. -- (Methods in molecular biology ; v. 352)
Includes bibliographical references and index.
ISBN 978-1-61737-338-1
1. Protein engineering. I. Arndt, Katja M. II. Müller, Kristian M. III. Series: Methods in molecular biology (Clifton, N.J.) ; v. 352.
TP248.65.P76P746 2006
660.6'3--dc22

2006041110

Preface

Protein engineering is a fascinating mixture of molecular biology, protein structure analysis, computation, and biochemistry, with the goal of developing useful or valuable proteins. *Protein Engineering Protocols* will consider the two general, but not mutually exclusive, strategies for protein engineering. The first is known as rational design, in which the scientist uses detailed knowledge of the structure and function of the protein to make desired changes. The second strategy is known as directed evolution. In this case, random mutagenesis is applied to a protein, and selection or screening is used to pick out variants that have the desired qualities. By several rounds of mutation and selection, this method mimics natural evolution. An additional technique known as DNA shuffling mixes and matches pieces of successful variants to produce better results. This process mimics recombination that occurs naturally during sexual reproduction.

The first section of *Protein Engineering Protocols* describes rational protein design strategies, including computational methods, the use of non-natural amino acids to expand the biological alphabet, as well as impressive examples for the generation of proteins with novel characteristics. Although procedures for the introduction of mutations have become routine, predicting and understanding the effects of these mutations can be very challenging and requires profound knowledge of the system as well as protein structures in general. Consequently, this section focuses on the question of how to design a protein with the desired properties, and examples are chosen to cover a wide range of engineering techniques, such as protein–protein interactions, DNA binding, antibody mimics, and enzymatic activity.

The second section of *Protein Engineering Protocols* deals with evolutionary techniques. In contrast to rational design, directed evolution strategies do not require prior structural knowledge of a protein, nor is it necessary to be able to predict what effect a given mutation will have. Indeed, the results of directed evolution experiments are often surprising in that desired changes are often caused by unexpected mutations. Several factors determine the success of such a strategy: the library design and quality, the choice of the method for evolution and/or DNA recombination, and the selection or screening method. Consequently, this second section of *Protein Engineering Protocols* provides instructions to each of these steps, starting from general ideas of library design and statistical assessment of library quality. New methods for DNA shuffling as well as different selection strategies are presented. Examples are given for the evolution of different characteristics, such as protein folds, folding, thermostability, and enzyme activity.

This volume provides a comprehensive guide to the methods used at every stage of the engineering process. It combines a thorough theoretical foundation with detailed protocols and will be invaluable to all research workers in the area, from graduate students to senior investigators. We would like to thank all authors for their excellent contributions and Prof. John M. Walker for his editorial guidance, patience, and assistance throughout the editorial process.

Katja M. Arndt
Kristian M. Müller

Contents

Contributors

Katja M. Arndt • *Institut für Biologie III, Universität Freiburg, Freiburg, Germany*

Jamie M. Bacher • *The Skaggs Institute for Chemical Biology, The Scripps Research Institute, La Jolla, CA*

Peter L. Bergquist • *Biotechnology Research Institute, Macquarie University, Sydney, NSW, Australia; Department of Molecular Medicine and Pathology, Auckland University Medical School, Auckland, New Zealand*

Luke H. Bradley • *Department of Chemistry, Princeton University, Princeton, NJ*

François-Xavier Campbell-Valois • *Département de Biochimie, Université de Montréal, Montréal, Québec, Canada*

Michel Denault • *Department of Quantitative Methods, HEC Montréal, Montréal, Québec, Canada*

Andrew D. Ellington • *Institute for Cellular and Molecular Biology and Department of Biochemistry, University of Texas, Austin, TX*

Birte K. Feld • *Department of Chemistry and the Institute for Genomics and Bioinformatics, University of California, Irvine, CA*

Peter Friedhoff • *Institut für Biochemie, Justus-Liebig-Universität, Giessen, Germany*

Satoshi Fujita • *Department of Chemistry and Biotechnology, School of Engineering, The University of Tokyo, Hongo, Tokyo, Japan; Research Institute for Cell Engineering, National Institute of Advanced Industrial Science and Technology (AIST), Tokyo, Japan*

Farid J. Ghadessy • *MRC Laboratory of Molecular Biology, Cambridge, United Kingdom; Department of Oncology, University College Medical School, London, United Kingdom*

Moreland D. Gibbs • *Biotechnology Research Institute, Macquarie University, Sydney, NSW, Australia*

Michael H. Hecht • *Department of Chemistry, Princeton University, Princeton, NJ*

Jochen Hecky • *Institut für Biology III, Universität Freiburg, Freiburg, Germany*

Philipp Holliger • *MRC Laboratory of Molecular Biology, Cambridge, United Kingdom*

Mitsuhiko Ikura • *Division of Molecular and Structural Biology, Ontario Cancer Institute and Department of Medical Biophysics, University of Toronto, Toronto, Ontario, Canada*

Akiko Koide • *Department of Biochemistry and Molecular Biology, The University of Chicago, Chicago, IL*

Shohei Koide • *Department of Biochemistry and Molecular Biology, The University of Chicago, Chicago, IL*

Hidetoshi Kono • *Computational Biology Group, Neutron Science Research Center, Quantum Beam Science Directorate, Japan Atomic Energy Agency, Kyoto, Japan*

Jody M. Mason • *Institut für Biologie III, Universität Freiburg, Freiburg, Germany*

Stephen W. Michnick • *Département de Biochimie, Université de Montréal, Montréal, Québec, Canada*

Kristian M. Müller • *Institut für Biologie III, Universität Freiburg, Freiburg, Germany*

Atsushi Miyawaki • *Laboratory for Cell Function and Dynamics, Advanced Technology Development Center, Brain Science Institute, RIKEN, Wako City, Saitama, Japan*

Wataru Nomura • *Institute for Chemical Research, Kyoto University, Uji, Kyoto, Japan*

Joelle N. Pelletier • *Département de Chimie, Université de Montréal, Montréal, Québec, Canada*

Alfred Pingoud • *Institut für Biochemie, Justus-Liebig-Universität, Giessen, Germany*

Jeffery G. Saven • *Makineni Theoretical Laboratories, Department of Chemistry, University of Pennsylvania, Philadelphia, PA*

Asako Sawano • *Laboratory for Cell Function and Dynamics, Advanced Technology Development Center, Brain Science Institute, RIKEN, Wako City, Saitama, Japan; Brain Science Research Division, Brain Science and Life Technology Research Foundation, Itabashi, Tokyo, Japan*

Sachdev S. Sidhu • *Department of Protein Engineering, Genentech Inc., South San Francisco, CA*

Sabine C. Stebel • *Institut für Biologie III, Universität Freiburg, Freiburg, Germany*

Yukio Sugiura • *Faculty of Pharmaceutical Sciences, Doshisha Women's University, Koudo, Kyotanabe, Japan*

KAZUNARI TAIRA • *Department of Chemistry and Biotechnology, School of Engineering, The University of Tokyo, Hongo, Tokyo, Japan; Gene Function Research Laboratory, National Institute of Advanced Industrial Science and Technology (AIST), Tsukuba Science City, Japan*

PETER THUMFORT • *Department of Chemistry, Princeton University, Princeton, NJ*

KEVIN TRUONG • *Institute of Biomaterials and Biomedical Engineering, Department of Electrical and Computer Engineering, University of Toronto, Toronto, Ontario, Canada*

WEI WANG • *Makineni Theoretical Laboratories, Department of Chemistry, University of Pennsylvania, Philadelphia, PA*

YINAN WEI • *Department of Chemistry, Princeton University, Princeton, NJ*

GREGORY A. WEISS • *Department of Chemistry and the Institute for Genomics and Bioinformatics, University of California, Irvine, CA*

CHRISTINE WURTH • *Department of Chemistry, Princeton University, Princeton, NJ*

JING-MIN ZHOU • *Department of Chemistry and Biotechnology, School of Engineering, The University of Tokyo, Hongo, Tokyo, Japan; Gene Function Research Laboratory, National Institute of Advanced Industrial Science and Technology (AIST), Tsukuba Science City, Japan*

I

Design and Computational Strategies for Protein Engineering

1

Combinatorial Protein Design Strategies Using Computational Methods

Hidetoshi Kono, Wei Wang, and Jeffery G. Saven

Summary

Computational methods continue to facilitate efforts in protein design. Most of this work has focused on searching sequence space to identify one or a few sequences compatible with a given structure and functionality. Probabilistic computational methods provide information regarding the range of amino acid variability permitted by desired functional and structural constraints. Such methods may be used to guide the construction of both individual sequences and combinatorial libraries of proteins.

Key Words: *De novo* protein design; combinatorial libraries; computational protein design; biased codons.

1. Introduction

1.1. Protein Design

Through attempts to design protein structures, including those having particular functions, researchers can refine the understanding we have of the forces and effects that specify the properties of the folded state. In addition, control over the design of particular folded state structures will likely lead to new synthetic proteins having the efficiency and specificity of biological proteins. Such applications include therapeutics, sensors, catalysts, and materials. The successful design of proteins is possible even without a complete quantitative understanding of all the forces involved in specifying their structures.

Designing proteins is nontrivial, however, because of both their complexity and the subtlety of the interactions that specify the folded state. Proteins are large (tens to hundreds of amino acid residues), and many structural variables specify the folded state, including sequence, backbone topology, and side-chain conformations. Each residue may have multiple conformations, even if the backbone

From: *Methods in Molecular Biology, vol. 352: Protein Engineering Protocols*
Edited by: K. M. Arndt and K. M. Müller © Humana Press Inc., Totowa, NJ

structure is specified. In addition to this structural complexity, there is also sequence complexity. Design involves identifying folding sequences from the enormous ensemble of possible sequences. This search is guided by the large degree of "consistency" observed in folded proteins ***(1)***. On average, a folded protein is atomically well-packed with favorable van der Waals interactions, hydrophobic residues are sequestered from solvents, and most hydrogen-bonding interactions are satisfied. However, this consistency is often complex and may have little simplifying symmetry. In addition, such noncovalent interactions are some of the most difficult to accurately quantify, and estimating free energies associated with mutation or structural ordering remains a subtle area of computational research ***(2,3)***. Despite their predictive power, presently we cannot expect to determine the relative stability changes of large numbers of sequences using detailed simulation methods for estimating free energy differences. Nonetheless, molecular potentials derived from small molecules and from the protein structure database do contain partial information regarding the interactions and forces known to be important for specifying and stabilizing protein structures. In some cases, the optimization of such potentials has lead to striking successes in protein design ***(4)***. Such potentials are necessarily approximate, and any sequence so designed is likely sensitive to the particular potential and target structure used. Alternatively, the partial information contained in these potentials may be used in a probabilistic manner, to yield the likelihoods of the amino acids. A probabilistic approach is also appropriate for characterizing the full variability of sequences that may fold to a common structure, because there are likely to be an enormous number of such sequences—far more than can be addressed via sequence search or enumeration.

Such probabilistic approaches are also particularly appropriate for *de novo* protein design in the context of combinatorial protein experiments, which create and rapidly assay many sequences. Although combinatorial methods can address very large numbers of sequences (10^4–10^{12}), these numbers are still infinitesimal compared with the numbers of possible protein sequences, e.g., $20^{100} \approx 10^{130}$ for a 100-residue protein. Thus, even with combinatorial methods, we still must focus on selected regions of sequence space. This is most often accomplished by preselecting a few residue sites within the protein by inspection and allowing full or partial variability at these sites. Recently, computational methods have been developed that can keep track of a much wider range of sequence variability and provide quantitative methods for winnowing and focusing the sequence space. Herein, we discuss computational methods for sequence design with an emphasis on probabilistic methods that address the site-specific amino acid variability for a given structure.

1.2. Directed Methods of Protein Design

Here, "directed protein design" refers to the identification of a sequence (or a set of sequences) likely to fold to a predetermined backbone structure. Each such sequence can then be experimentally realized using peptide synthesis or gene expression, to confirm its folded structure and other molecular properties. Early efforts in design were guided by trends observed among naturally occurring structures and identified protein sequences that were compact, with substantial secondary structure but not necessarily well-defined tertiary structures ***(5)***. With their abilities to quantify and tabulate interresidue interactions, computational methods have dramatically accelerated successful protein design. Typically, such methods cast the sequence search as an optimization process, in which amino acid identity and side-chain conformation are varied to optimize a scoring function that quantifies sequence structure compatibility. Exhaustive searching of all m^N possible sequences is feasible only if a small number of residues N are allowed to vary or the number of allowed amino acids m is substantially reduced, e.g., from $m = 20$ to $m = 2$. To arrive at sequences with well-packed interiors having favorable interatomic interactions on average, the search must also include variation in the different side-chain conformations (rotamer states) of each amino acid (*see* **ref. *6***). As a result, the complexity of the search increases, because m, the number of possible states for a residue, increases by a factor of 10 or more, depending on the number of rotamers associated with each amino acid. If only a few residues are allowed to vary and the conformations of the remaining residues are constrained, complete enumeration of all possible combinations can be performed to identify low-energy sequence–rotamer combinations. Such complete enumeration is typically not feasible, because of the exponential dependence on chain length and numbers of rotamers. For such cases, the sequence space can be sampled in a directed manner to move progressively toward optimal (or nearly optimal) sequences. Stochastic methods, such as genetic algorithms and simulated annealing, involve searching sequence space in a partially random fashion, in which, on average, the search progressively moves toward better scoring (lower energy) sequences ***(7–10)***. Such searches have sufficient "noise" or recombination to permit escape from local minima in the sequence–rotamer landscape. When applied to atomically detailed representations, the stochastic methods focus primarily on repacking the interior of a structure with hydrophobic residues ***(9)*** and have been applied to the wild-type structures of 434 Cro ***(10)***, ubiquitin ***(11)***, the B1 domain of protein G ***(12)***, the WW domain ***(4)***, and helical bundles ***(13,14)***. Although, in many cases, these methods have aided in identifying experimentally viable sequences ***(4,15)***, stochastic search methods need not identify global optima ***(16)***. For potentials comprising only site and pair interactions, elimination methods, such as "dead-end elimination" can find the global optimum ***(16–20)***. Such methods successively remove individual amino acid–rotamer states that can not be part of the global optimum until no further states can be eliminated. The Mayo group has applied such methods to automate the full sequence design of a 28-residue zinc finger mimic ***(21)*** and, after patterning hydrophobic and polar sites, a 51-residue homeodomain motif *(22)*. The group has also redesigned residue subsets within portions of a variety of proteins *(23–25)*. Functional properties, such as metal binding or catalysis, may also be included as elements of the design process *(26–28)*. The elements and algorithms of directed protein design has been the subject of several recent reviews *(4,29,30)*.

Despite some striking successes, computational methods for the directed design of sequences have limitations with respect to characterizing the features of protein sequences folding to a particular structure. Stochastic methods can be applied to large proteins and permit many sites to be varied simultaneously, but the computational times and resources for such calculations are extensive, even for small proteins. Directed methods will necessarily be sensitive to the energy or scoring function used, because they identify the optimum of a particular energy function. All such energy functions, however, are necessarily approximate, and uncertainties in the energy function may not merit the search for global optima. Many naturally occurring proteins are not optimized. In fact, most proteins are only marginally stable, e.g., $\Delta G^o < 10$ kcal/mol for folding *(31)*. In addition, sequences that function, e.g., those that bind another molecule, need not be the global optimum with respect to structural stability. It is important to develop methods complementary to those used for directed protein design, methods that reveal the features of sequences likely to fold to a particular structure but which may not be structurally optimal. Such techniques may be used in designing protein sequences. In addition, such computational methods may be applied to a new class of protein design studies, combinatorial experiments, in which large numbers of proteins may be simultaneously synthesized and screened.

1.3. Probabilistic Approaches to Protein Design

In the context of protein design, we refer to the use of site-specific amino acid probabilities rather than specific sequences as "probabilistic protein design." Probabilistic approaches, as opposed to directed or deterministic approaches, are often used in the quantitative sciences for cases in which there is only partial information regarding a problem. For protein design, such a probabilistic approach is motivated by the complexities and uncertainties associated with the folding process. Protein folding is a complicated kinetic process, with myriad interactions specifying the folded state. Each of the stabilizing noncovalent interactions is of comparable magnitude and there seems to be no one overriding determinant of folding. The means of quantifying these interactions are necessarily approximate (*see* **Note 1**). Probabilistic design methods also directly provide very useful sequence information, particularly regarding structurally important amino acids. The amino acid probabilities can guide the design of specific sequences and can also highlight sites likely to tolerate mutation with minimal impact on structure; such sites can be targets of variation after multiple rounds of protein design.

Probabilistic methods may be used in several ways to guide protein design. Sequences should be generated in a manner consistent with the calculated probabilities. First, the most straightforward choice is a consensus sequence or the sequence comprising the most probable amino acid at each position. If necessary, repeated calculations may be performed with successive iterations determining an increasing fraction of the residues in the protein. Such an approach has been used to arrive at a 114-residue, dinuclear metalloprotein *(32)* and a solubilized variant of an integral membrane protein *(33)*. Second, the calculated probabilities may be used to guide a search for sequences. A Monte Carlo-based method has been presented, wherein the calculated amino acid probabilities are used to bias the selection of trial sequences that are either accepted or rejected at each point in the Monte Carlo Markov trajectory *(34)*. Such methods address correlated amino acid identities, but at the cost of the computational overhead associated with the search, although this overhead is diminished if information is used to guide the search. Last, probabilistic methods may be used to quantitatively guide the design of combinatorial libraries of proteins *(35)*.

1.4. Combinatorial Experiments

Combinatorial protein experiments may be used to investigate sequence structure compatibility and to discover novel sequences folding to a specific structure. In protein combinatorial design experiments, large numbers of sequences (libraries) are screened for evidence of folding to a predetermined structure. Depending on how the sequence diversity is generated and assayed, experiments of this type can explore a large number of sequences, up to 10^{12} sequences *(36)*. A library can then be screened using a selection assay, such as ligand binding or enzymatic activity. Such experiments can go "beyond the protein sequence database," in a manner in which the diversity of the sequences is at the control of the researcher. Features important to folding (and other biological properties) may be explored in a manner decoupled from the evolutionary pressures of nature's proteins. Combinatorial methods have been used to identify helical proteins *(37–39)*; ubiquitin variants *(40)*; self-assembled protein monolayers *(41)*; proteins with amyloid-like properties *(41)*; metal-binding peptides *(42)*; and stable interhelical oligomers *(43)*. Several excellent reviews of combinatorial experiments and methodology have appeared recently *(44–47)*.

2. Methods

Probabilistic methods for protein design provide estimates of the site-specific probabilities of the amino acids within a particular protein structure. Here, we discuss several methods for estimating these probabilities and focus on an entropy-based self-consistent formalism that solves for these probabilities directly.

2.1. Alignment of Related Sequences

The sequence variability of a protein structure can be considered using sequence and structural databases. Sequences known to fold to very similar structures can be identified from the Protein Data Bank or a database of structural alignments ***(48)***. If the structure of a sequence is known, other proteins having sufficient sequence similarity (e.g., >40% sequence identity) may be assumed to share the same structure. Multiple sequence alignments of such structurally similar proteins can used to estimate the site-specific probabilities of the amino acids as simply the frequencies of each amino acid at each position in the alignment ***(49)***. Such a set of probabilities is often termed the sequence profile. If the number of sequences is insufficient, such that some amino acids are never observed at particular sites, pseudocount and other methods may be used to "regularize" the frequencies, such that they are more representative sequences folding to the chosen structure ***(50)***. Nonetheless, the probabilities from such a profile will be heavily biased toward the properties of sequences in the database. We may wish to obtain a broader understanding of the full range of sequence variability, because there are many natural examples of sequences folding to similar structures with little sequence similarity. These database-derived profiles are also not appropriate for novel protein structures, for which we have no sequences in the database. More general computational methods permit the amino acid probabilities to be determined *ab initio* using a given backbone structure as a template.

2.2. Directed Search Methods to Build Profiles

Repeated use of directed search methods can estimate the properties of an ensemble of sequences. For such calculations, a target structure is chosen by specifying the coordinates of the main chain (backbone) atoms. Several recent directed design studies have yielded sequences with substantial similarity to the wild-type sequence if a single protein structure is used ***(51–54)***. For a given structure, multiple sequence search calculations may be run independently, resulting in a set of sequences whose alignment yields the site-specific probabilities. Desjarlais and co-workers have used independent runs of their sequence prediction algorithm for each member of an ensemble of closely related structures consistent with a particular fold ***(55)***. For each structure, an optimal "nucleating" sequence is identified and, subsequently, the sequence/rotamer variability is explored throughout the structure. The method has been used to identify sequences consistent with the fold of a WW domain, a small β-sheet protein ***(4,55)***. By designing sequences for each of 100 minor structural variants (1 Å root mean square deviation) of a particular fold using sequence prediction algorithm, Larson et al. have built computational profiles exhibiting much more diversity than those obtained using a single structure ***(56)***. Workers at Xencor, Inc. have recently used Monte Carlo sampling of sequences near an "optimal" sequence, in which residues near the active site of β-lactamase were mutated ***(57)***. Sequences with more than 1000-fold increases in resistance to an antibiotic were identified. These types of approaches to building profiles are computationally intensive, however, because repeated directed searches are performed to build the site-specific frequencies of the amino acids.

2.3. Statistical Theory of Sequence Ensembles

A statistical, entropy-based formalism has been developed for identifying the set of site-specific amino acid probabilities (the sequence profile) for a given backbone structure, rather than just identifying the optimum sequence ***(58,59)***. Ideas from statistical mechanics are used to address the number and composition of sequences that are consistent with a particular backbone structure. The theory also addresses the whole space of available compositions, not just the small fraction that is accessible to experiment and to computational enumeration and sampling. The properties of suboptimal sequences may be readily examined. Large protein structures (>100 residues) may be readily considered in the calculations. Here the "entropy" is the number of sequences consistent with the target structure. Concepts from thermodynamics are used to reduce the number of possible sequences: constraints on the sequences reduce the entropy, and the entropy should decrease with decreasing energy.

The input for the method is a target backbone structure and an energy function that quantifies sequence–structure compatibility. For a target backbone structure, the method yields the probabilities of each of the amino acids at each residue site (*see* **Note 2**). Both global (e.g., the overall energy of the sequences in the target) and local features (e.g., the allowed amino acids at a particular site) can be included as constraints in the theory. Many sets of amino acid probabilities are possible. This method determines the "most probable" such set by maximizing an effective entropy, whereby this maximization is subject to constraints. Using such constraints effectively, the method provides a systematic means to reduce the size of sequence space to be searched to an experimentally feasible size.

Among sequences with desired properties specified by constraint functions, let $w_i[\alpha,r_k(\alpha)]$ denote the probability that amino acid α is present at residue position i and that its side chain is any one of a discrete set of conformations, $r_k(\alpha)$ (rotamer states; **refs. *6*** and ***60***). The total sequence–conformational entropy, S_c (here, simply referred to as "conformational entropy") can be defined as:

$$S_c = -\sum_{i,\alpha,k} w_i\,[\alpha, r_k(\alpha)] \ln w_i\,[\alpha, r_k(\alpha)]$$

The sum extends over each sequence position, i, and all available amino acids, α, at each position. For each amino acid, the sum also extends over each of the k possible rotamer states, $r_k(\alpha)$. The $w_i[\alpha,r_k(\alpha)]$ are determined as those that maximize S_c subject to any constraints, f_i. This maximization is done using the method of Lagrange multipliers ***(61)***. A variational functional V of the $w_i[\alpha,r_k(\alpha)]$ is defined as:

$$V = S - \beta_1 f_1 - \beta_2 f_2 - \ldots$$

In general, the constraints f_i are also functions of the probabilities $w_i[\alpha,r_k(\alpha)]$. In identifying the state probabilities consistent with particular values of constraints, the mth constraint function, f_m, is constrained to have a particular value, f^o_m. The set of equations that determine the probabilities and the Lagrange multipliers then take the form (*see* **Note 3**):

$$0 = \partial V / \partial w_i[\alpha, r_k(\alpha)]$$

$$f^o_m = f_m\{w_i[\alpha, r_k(\alpha)]\}$$

This large set of coupled, nonlinear equations is solved using root-finding methods. Although there are many choices for such methods, we find a globally convergent method to be broadly applicable ***(62)***.

2.3.1. Energy Functions

In the calculation, two energies, the conformational energy, E_c, and the environmental energy, E_{env}, are considered and used as constraints in maximizing a conformational entropy, S_c.

The conformational energy, E_c, is calculated using an atom-based potential, the AMBER force field ***(63)***. E_c includes van der Waals interactions, electrostatic interactions with a distant dependent dielectric ($4\varepsilon r_{ij}$) and a modified hydrogen bond term ***(64)***. For a particular sequence ($\alpha_1,\ldots,\alpha_N$) in which the conformational states of these amino acids are $[r_1(\alpha_1),\ldots,r_N(\alpha_N)]$, E_c is:

$$E_c = \sum_i \varepsilon_i\,[\alpha, r_k(\alpha)] + \sum_{i,j>i} \varepsilon_{i,j}[\alpha, r_k(\alpha); \alpha', r_{k'}(\alpha')]$$

In the context of protein energy functions, the one-body term $\varepsilon_i[\alpha,r_k(\alpha)]$ includes the interaction energies between backbone and amino acid side chains, as well as a reference energy (*see* **Subheading 2.3.3.**) of amino acid. The two-body term, $\varepsilon_{i,j}[\alpha,r_k(\alpha);\alpha',r_{k'}(\alpha')]$, is a sum over the interaction energies between two rotamers at two different sites in the structure. For a large number of sequences sharing common energetic properties, we assume that the fluctuation in E_c about its mean value caused by variation of sequence is small. We can then write:

$$E_c \approx \overline{E}_c = \sum_{i,\alpha,k} \varepsilon_i[\alpha, r_k(\alpha)]w_i[\alpha, r_k(\alpha)] + \sum_{\substack{i,j>i \\ \alpha,\alpha' \\ k,k'}} \varepsilon_{i,j}[\alpha, r_k(\alpha);\alpha', r_{k'}(\alpha')]w_i[\alpha, r_k(\alpha)]w_j[\alpha', r_{k'}(\alpha')]$$

As another constraint, an environmental energy, E_{env} is introduced to account for the hydrophobic effect in a way that is efficient within the statistical theory *(59)*. This potential takes into account the surface-exposure preferences of the amino acids. Regarding E_c, we can write E_{env} using amino acid probabilities as:

$$E_{env} \approx \overline{E}_{env} = \sum_{i,\alpha,k} \varepsilon_{env}[\alpha, r_k(\alpha)]w_i[\alpha, r_k(\alpha)]$$

where ε_{env} is a local environmental energy defined in **Subheading 2.3.2**. Note that this energy contains no two-body interactions and is dependent only on the amino acid and rotamer state at each position.

2.3.2. Solvation and Hydrophobic Energy

An important input to any protein design method is some means of quantifying the hydrophobic effect and other solvation properties. Explicit representation of solvent is impractical for calculations that examine large variation in sequence, and even calculating solvent-accessible surface areas—which often correlate well with hydrophobic tendencies—can be computationally intensive. In an effort to consider solvation effects in a practical manner consistent with the statistical calculations, an environmental energy has been introduced that is a function of the density ρ of C_β atoms in the vicinity of each side chain *(59)*. On average, hydrophobic residues tend to be located in buried regions of proteins, whereas hydrophilic residues tend to be located at the surface. Thus, hydrophobic residues are likely to have a higher C_β density than hydrophilic ones. Using 500 different globular proteins of known structure (the training set), we derived effective potentials for the amino acids using the usual equation for "statistical" potentials:

$$\varepsilon_{env}(\alpha, \rho) = -T_e \ln \frac{p(\alpha, \rho)}{p(\alpha)p(\rho)}$$

where $p(\alpha,\rho)$ is the fraction of times a local C_β density of ρ is observed for amino acid α; $p(\alpha)$ is the fraction of times amino acid α is observed in the training set;

and $p(\rho)$ is the fraction of times a local density of ρ is observed, regardless of amino acid type. T_e is the effective temperature. The density, ρ, is defined as the density of C_β atoms within the "free volume" within a sphere centered at a particular orientation of the side chain. The free volume refers to an average volume not excluded by the side chain:

$$\rho(\alpha) = \frac{n_\beta}{\frac{4}{3}\pi R^3 - \langle V_{access}(\alpha)\rangle}$$

Here, n_β is the number of C_β atoms within a distance R (set to be 8 Å) from the side chain center of mass, and $\langle V_{access}(\alpha)\rangle$ is the average excluded volume of amino acid α, and is calculated as an average volume over all possible rotamer states of α. We note that the local density is dependent on the rotamer state of the amino acid, so $\varepsilon_{env}\{\alpha, \rho[r_k(\alpha)]\} \equiv \varepsilon_{env}\,[\alpha, r_k(\alpha)]$. This C_β density-based potential has a good correlation with other amino acid hydrophobicity scales ***(59)***. For the sequence probability calculations, E_{env} is constrained to have a value of a known sequence with that structure (if one is known), or a value representative of proteins of that particular size or chain length.

2.3.3. Reference Energy

In protein design, we seek to optimize the energy of a particular sequence when in the target structure relative to that of the ensemble of unfolded states. To address this issue regarding unfolded states, a reference energy, $\gamma_{ref}(\alpha)$, for each amino acid is introduced into the energy, E_c, to mimic the effects of the denatured state ***(51,65)***. The energy is calculated as a "free energy" of each amino acid α in the form of *N*-acetyl-α-*N′*-methylamide, with averaging over multiple backbone states. This crudely approximates an average over extended unfolded states. The reference energy involves a sum over possible rotamers and possible backbone configurations, approximated by varying each of the backbone ϕ and φ angles in increments of 10°. The reference energies of each amino acid may then be estimated using:

$$\gamma_{ref}(\alpha,\beta_{ref}) = -\beta_{ref}^{-1}\ln\left[\frac{z_{ref}(\alpha,\beta_{ref})}{z_{ref}(G,\beta_{ref})}\right]$$

$$z_{ref}(\alpha,\beta_{ref}) = \sum_{\phi,\varphi,k}\exp\{-\beta_{ref}\varepsilon_{ref}[\phi,\varphi,r_k(\alpha)]\}$$

where ε_{ref} is an conformational energy in a particular conformation of amino acid α in the *N*-acetyl-*N′*-methylamide amino acid, as determined using a molecular potential. Here, $\beta_{ref} = 1/(k_B T)$, where k_B is Boltzmann's constant and T is a temperature appropriate for the conformation sampling of side chain and backbone conformations (e.g., T = 300 K). Here, a backbone-dependent rotamer library is used throughout ***(60)***. Reference energies are measured with respect to that of glycine (G), which has no side chain. The energy constraint on the sequences involving interatomic interactions then takes the form:

$$E_c \approx \overline{E}_c = \sum_{i,\alpha,k} \{\varepsilon_i[\alpha, r_k(\alpha)] - \gamma_{ref}[\alpha, \beta_{ref}] w_i[\alpha, r_k(\alpha)]\} + \sum_{\substack{i,j>i \\ \alpha,\alpha' \\ k,k'}} \varepsilon_{i,j}[\alpha, r_k(\alpha); \alpha', r_{k'}(\alpha')] w_i[\alpha, r_k(\alpha)] w_j[\alpha', r_{k'}(\alpha')]$$

2.3.4. Rotamer and Identity Probabilities

The theory maximizes the total conformational entropy, S_c, yielding a probability $w_i[\alpha, r(\alpha)]$ that a particular amino acid is present at site *i* and is in side-chain conformation *k*. The amino acid probability, $w_i(\alpha)$, can then be determined using:

$$w_i(\alpha) = \sum_k w_i[\alpha, r_k(\alpha)]$$

Using an analogy to statistical thermodynamics, the Lagrange multiplier that arises from constraining the conformational energy, β_c, may be considered an effective inverse temperature, $1/\beta_c = T_c$. The corresponding "heat capacity," C_v, is defined as:

$$C_v = \frac{\partial E_c}{\partial T_c} = \beta_c^2 \sum_i \left(\langle \varepsilon^2{}_i \rangle - \langle \varepsilon_i \rangle^2 \right)$$

$$\langle \varepsilon_i \rangle = \sum_{\alpha,k} \varepsilon_i^{loc}[\alpha, r_k(\alpha)] w_i[\alpha, r_k(\alpha)]$$

$$\varepsilon_i^{loc}[\alpha, r_k(\alpha)] = \varepsilon_i[\alpha, r_k(\alpha)] - \gamma_{ref}(\alpha, \beta_c) + \sum_{j,\alpha',k'} \varepsilon_{i,j}[\alpha, r_k(\alpha); \alpha', r_{k'}(\alpha')] w_j[\alpha', r_{k'}(\alpha')]$$

where $\langle \varepsilon_i \rangle$ is termed a local mean field energy, which denotes the average local field around a particular amino acid side, *i*. The effective heat capacity, C_v, provides a quantitative measure of the fluctuations in the sequence–rotamer identities as values of the constraint conditions, such as the overall energy, are modulated during a calculation.

By way of example, the theory is applied to a particular protein, an SH3 domain. The conformational entropy decreases with decreasing the effective temperature T_c (i.e., decreasing E_c; **Fig. 1**). At high energies (high T_c, low β_c), there are many unfavorable (high-energy) interactions between residues and a broad distribution of sequence–rotamer states at each site. On average, the number of probable

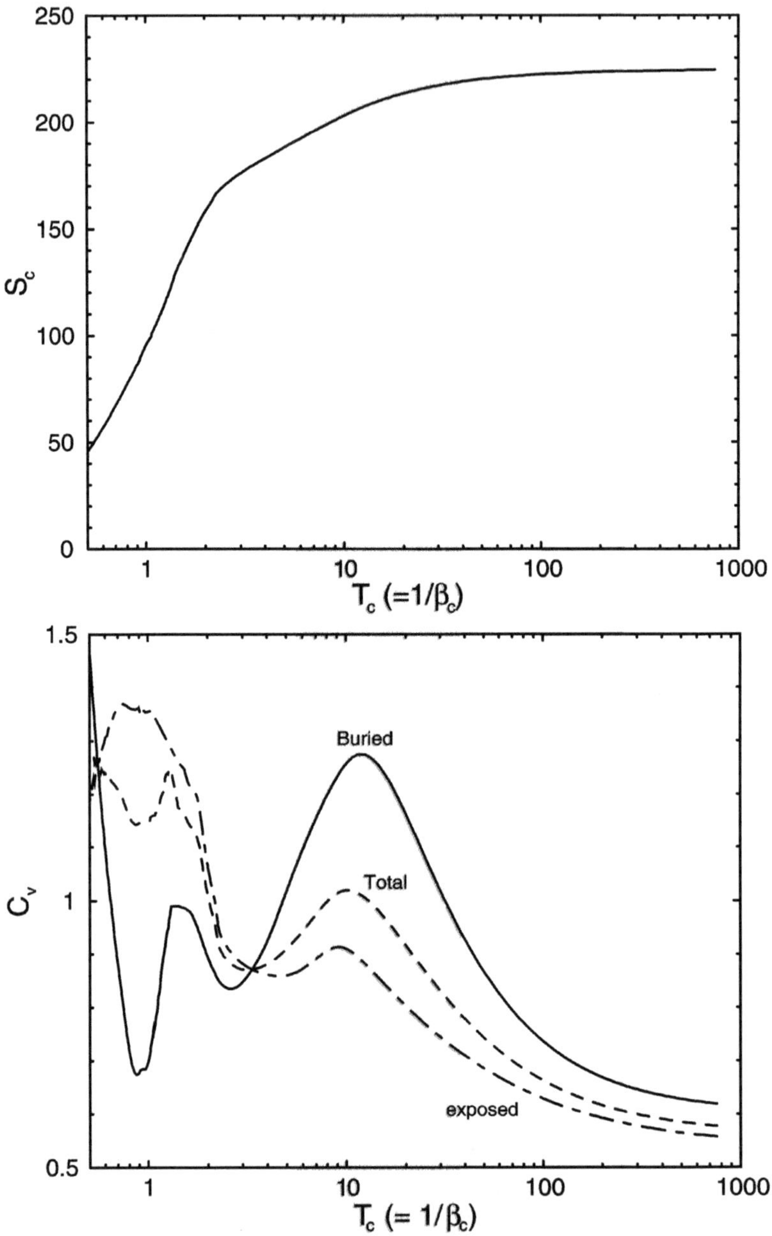

Fig. 1. Sequence–conformation entropy, S_c, of the SH3 domain is plotted against effective temperature, T_c (upper panel). Effective heat capacities per residue C_v for all buried and exposed residues are plotted against effective temperature (lower panel). Temperatures are given in arbitrary units determined by the molecular potential used, here moles per kilocalorie.

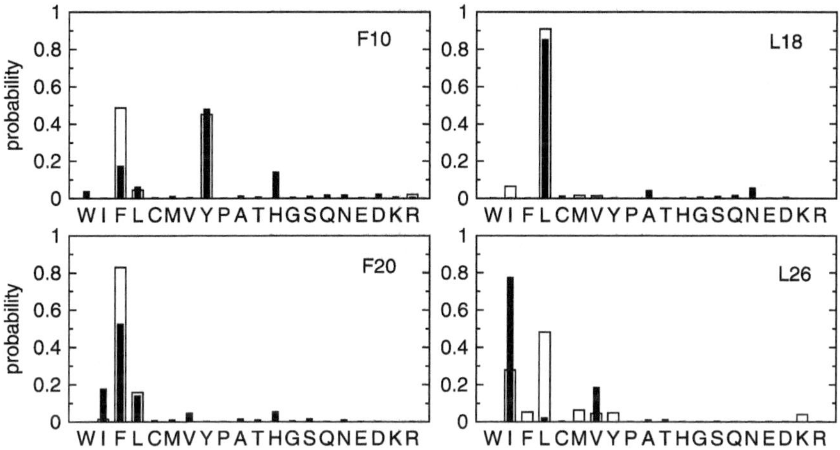

Fig. 2. Amino acid probabilities of the SH3 domain (PDB:1CKA). Calculated profile (▮); sequence alignment-based profile (▯). F10, L18, F20, and L26 are located at hydrophobic core sites with fractional solvent accessible surface areas less than 20%.

amino acids and rotamer states per site decreases with decreasing energy. As shown in **Fig. 1**, C_v reaches a valley at approximately T_c (= $1/\beta_c$) = 2 mol/kcal, after passing a peak at approx 10 mol/kcal, at which point, the identities and conformations of residues in the interior become relatively well defined ***(59)***; whereas surface residues, although predominantly hydrophilic, have a large number of rotamer states of comparable probability. This is consistent with the conformational variability of surface-exposed residues among proteins in the structural database. Thus, the heat capacity is helpful in determining at which "effective temperature" the probabilities should be examined. In addition, direct comparisons of the calculated profile with those obtained by sequence alignments (e.g., the homology-derived secondary structure of proteins database; **ref. *66***) yields good agreement, especially for buried regions (**Fig. 2**).

2.4. Gene Libraries for Protein Profiles

In **Subheading 1.3.**, we discussed how to make use of a particular structure's sequence profile. If a specific sequence is identified, either as a consensus sequence or a directed search, this may be straightforwardly realized in an experiment using peptide synthesis or by expressing a synthetic gene that codes for the sequence. Larger proteins are most often realized using gene expression. If the probabilistic sequence information is to be used to construct a combinatorial library (*see* **Note 4**), methods are needed that transcribe the protein profiles into libraries of partially random gene sequences. Nonuniform distributions of nucleotides are necessary to encode peptide sequences that have such a bias toward specific amino acids. Pseudo-independent nucleotide probabilities at each position of a set of partially random genes can be determined computationally, such that the protein library encoded by the gene library best reproduces the desired amino acid profile. The calculated gene library can then be used in context of standard DNA synthesis.

Let $P_1(n_1)$, $P_2(n_2)$, $P_3(n_3)$ be the probabilities of each of the four possible nucleotides (n_i = A, T, G, C) in the first, second, and third position of a codon, respectively. If these are treated as independent, the probability that amino acid α will appear as encoded by the codon $n_1n_2n_3$ is $P(\alpha|n_1,n_2,n_3) = P_1(n_1)P_2(n_2)P_3(n_3)\delta(\alpha|n_1,n_2,n_3)$, where $\delta(\alpha|n_1,n_2,n_3) = 1$ only if n_1,n_2,n_3 is a codon for amino acid α, and is zero otherwise. If the codons of amino acid α are equally likely (no codon bias), the probability of an amino acid α is the sum of codon probabilities corresponding to this amino acid:

$$P_{calc}(\alpha) = \sum_{n_1,n_2,n_3} P_1(n_1)P_2(n_2)P_3(n_3)\delta(\alpha \mid n_1,n_2,n_3)$$

Objective functions quantify the difference between a desired amino acid probability distribution and the amino acid probability distribution encoded from a given set of nucleotide probabilities ***(67,68)***. To find the nucleotide probabilities that not only best reproduce the desired amino acid frequencies but also prevent the occurrence of stop codons, a new objective function has been presented ***(69)***. The objective function comprises both a χ^2 function, which quantifies the absolute difference between the desired and calculated amino acid probabilities, and a relative entropy term. Such relative entropies are commonly used to quantify the "distance" between two probability distributions, and are strong indicators of cases in which information in one distribution is not contained in the other ***(50)***:

$$H = \sum_{\alpha=1}^{21}\left\{P_{calc}(\alpha)\ln\frac{P_{calc}(\alpha)+\varepsilon}{P_{des}(\alpha)+\varepsilon} + 0.5[P_{des}(\alpha) - P_{calc}(\alpha)]^2\right\}$$

Here, ε is introduced as an arbitrary small constant ($\varepsilon = 10^{-6}$), to avoid numerical instability if $P_{des}(\alpha)$ vanishes. Stop codons are treated as an "effective amino acid." The objective function is optimized (minimized), subject to the usual constraints on the nucleotide probabilities: $0 \le P_i(n_i) \le 1$, and $\sum_{n_i} P_i(n_i) = 1$. This may be done using a Lagrange multiplier method or computational packages available for constrained minimization ***(69)***. Codons optimized for a particular organism or expression system may also be included in an objective function of this type ***(69)***.

Illustrated in **Fig. 3** is a nucleotide design for a particular amino acid position in a protein, here site 54 of the SH3 domain. Shown are the desired frequencies of the amino acids (open bar in the upper panel of **Fig. 3**) and the amino acid frequencies as encoded (filled bar in the upper panel of **Fig. 3**) by

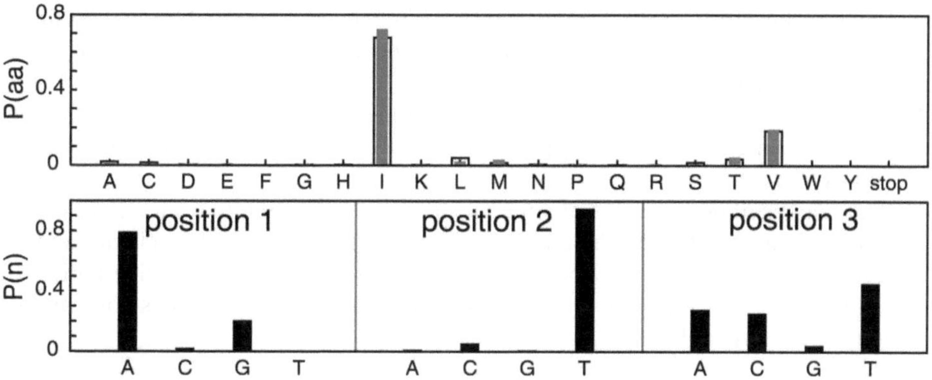

Fig. 3. Probability distributions of amino acids (upper) and nucleotides (lower) for site 54 of a SH3 domain (PDB:1CKA). For the amino acid probabilities, the desired probability distribution is shown by open bars, and that encoded by calculated gene library is shown by filled bars. An oligonucleotide library with the frequencies of the nucleotides specified in the lower panel encodes for the site-specific amino acid probabilities in the upper panel.

the computed nucleotide probabilities (shown in the lower panel of **Fig. 3**). The agreement between the two is excellent in this case. In general, the calculated probabilities agree well with the desired probabilities. In many cases, an exact match between the desired and calculated probability distribution cannot be achieved because of the partial degeneracy of codons to amino acids. This computational method provides excellent yields of complete sequences (those not containing stop codons): for test proteins of 50 to 60 residues, in which all sites are subject to selective randomization, the yield of complete sequences is 96% or more. High yields are particularly important when a large fraction (or all) of a gene is subject to combinatorial mutation.

3. Notes

1. Self-consistent methods are not the best for finding global optima ***(16)***. In protein design, this is generally because of the approximate manner for treating correlations between residue identities and conformational states in mean-field-like theories. Stochastic sampling and elimination methods generally provide better optimization results.
2. In designing sequences, backbone flexibility should be considered. In natural proteins, small backbone adjustments can accommodate mutations ***(70,71)***. However, most computational methods treat the backbone as rigid because including the flexibility is computationally demanding. The probabilistic nature of the statistical theory suggests that its predicted profiles are less sensitive to backbone choice. Amino acid profiles from 21 slightly different backbone structures of protein L are similar to one another for energies (or effective temperatures) above the peak of

the heat capacity, C_v *(59)*. Sequence properties that are robust with respect to small structural fluctuations (1 Å root mean square deviation) occur at values of E_c at or above this heat capacity peak.

3. In performing the constrained optimization of the entropy, we find that root-finding methods using these equations perform as well as constrained minimization algorithms *(62)*.
4. The calculated probabilities of amino acids can also be used to evaluate structural tolerance to mutation. Such mutations are most likely to accumulate in an in vitro directed protein evolution experiment *(72)* at sites that have a broad distribution of amino acids, i.e., sites at which many amino acids have probabilities comparable to the most probable amino acid.

References

1. Go, N. (1983) Theoretical studies of protein folding. *Annu. Rev. Biophys. Bioeng.* **12,** 183–210.
2. Shea, J. E. and Brooks, C. L. 3rd. (2001) From folding theories to folding proteins: a review and assessment of simulation studies of protein folding and unfolding. *Annu. Rev. Phys. Chem.* **52,** 499–535.
3. Brooks, C. L. 3rd. (2002) Protein and peptide folding explored with molecular simulations. *Acc. Chem. Res.* **35,** 447–454.
4. Kraemer-Pecore, C. M., Wollacott, A. M., and Desjarlais, J. R. (2001) Computational protein design. *Curr. Opin. Chem. Biol.* **5,** 690–695.
5. Bryson, J. W., Betz, S. F., Lu, H. S., et al. (1995) Protein design: a hierarchic approach. *Science* **270,** 935–941.
6. Dunbrack, R. (2002) Rotamer libraries. *Curr. Opin. Struct. Biol.* **12,** 431–440.
7. Shakhnovich, E. I. and Gutin, A. M. (1993) A new approach to the design of stable proteins. *Protein Eng.* **6,** 793–800.
8. Jones, D. T. (1994) De novo protein design using pairwise potentials and a genetic algorithm. *Protein Sci.* **3,** 567–574.
9. Hellinga, H. W. and Richards, F. M. (1994) Optimal sequence selection in proteins of known structure by simulated evolution. *Proc. Natl. Acad. Sci. USA* **91,** 5803–5807.
10. Desjarlais, J. R. and Handel, T. M. (1995) De-novo design of the hydrophobic cores of proteins. *Protein Sci.* **4,** 2006–2018.
11. Johnson, E. C., Lazar, G. A., Desjarlais, J. R., and Handel, T. M. (1999) Solution structure and dynamics of a designed hydrophobic core variant of ubiquitin. *Struct. Fold. Des.* **7,** 967–976.
12. Jiang, X., Farid, H., Pistor, E., and Farid, R. S. (2000) A new approach to the design of uniquely folded thermally stable proteins. *Protein Sci.* **9,** 403–416.
13. Jiang, X., Bishop, E. J., and Farid, R. S. (1997) A de novo designed protein with properties that characterize natural hyperthermophilic proteins. *J. Am. Chem. Soc.* **119,** 838, 839.
14. Bryson, J. W., Desjarlais, J. R., Handel, T. M., and DeGrado, W. F. (1998) From coiled coils to small globular proteins: design of a native-like three-helix bundle. *Protein Sci.* **7,** 1404–1414.

15. Walsh, S. T. R., Cheng, H., Bryson, J. W., Roder, H., and DeGrado, W. F. (1999) Solution structure and dynamics of a denovo designed three-helix bundle protein. *Proc. Natl. Acad. Sci. USA* **96,** 5486–5491.
16. Voigt, C. A., Gordon, D. B., and Mayo, S. L. (2000) Trading accuracy for speed: a quantitative comparison of search algorithms in protein sequence design. *J. Mol. Biol.* **299,** 789–803.
17. Gordon, D. B. and Mayo, S. L. (1998) Radical performance enhancements for combinatorial optimization algorithms based on the dead-end elimination theorem. *J. Comput. Chem.* **19,** 1505–1514.
18. Gordon, D. B. and Mayo, S. L. (1999) Branch-and terminate: a combinatorial optimization algorithm for protein design. *Struct. Fold. Des.* **7,** 1089–1098.
19. Pierce, N. A., Spriet, J. A., Desmet, J., and Mayo, S. L. (2000) Conformational splitting: a more powerful criterion for dead-end elimination. *J. Comput. Chem.* **21,** 999–1009.
20. Looger, L. L. and Hellinga, H. W. (2001) Generalized dead-end elimination algorithms make large-scale protein side-chain structure prediction tractable: implications for protein design and structural genomics. *J. Mol. Biol.* **307,** 429–445.
21. Dahiyat, B. I. and Mayo, S. L. (1997) De novo protein design: fully automated sequence selection. *Science* **278,** 82–87.
22. Marshall, S. A. and Mayo, S. L. (2001) Achieving stability and conformational specificity in designed proteins via binary patterning. *J. Mol. Biol.* **305,** 619–631.
23. Malakauskas, S. M. and Mayo, S. L. (1998) Design, structure, and stability of a hyperthermophilic protein variant. *Nat. Struct. Biol.* **5,** 470–475.
24. Strop, P. and Mayo, S. L. (1999) Rubredoxin variant folds without iron. *J. Am. Chem. Soc.* **121,** 2341–2345.
25. Shimaoka, M., Shifman, J. M., Jing, H., Takagi, L., Mayo, S. L., and Springer, T. A. (2000) Computational design of an integrin i domain stabilized in the open high affinity conformation. *Nat. Struct. Biol.* **7,** 674–678.
26. Benson, D. E., Wisz, M. S., Liu, W., and Hellinga, H. W. (1998) Construction of a novel redox protein by rational design: conversion of a disulfide bridge into a mononuclear iron-sulfur center. *Biochemistry* **37,** 7070–7076.
27. DeGrado, W. F., Summa, C. M., Pavone, V., Nastri, F., and Lombardi, A. (1999) De novo design and structural characterization of proteins and metalloproteins. *Annu. Rev. Biochem.* **68,** 779–819.
28. Bolon, D. N. and Mayo, S. L. (2001) Enzyme-like proteins by computational design. *Proc. Natl. Acad. Sci. USA* **98,** 14,274–14,279.
29. Street, A. G. and Mayo, S. L. (1999) Computational protein design. *Struct. Fold. Des.* **7,** R105–R109.
30. Saven, J. G. (2001) Designing protein energy landscapes. *Chem. Rev.* **101,** 3113–3130.
31. Gromiha, M. M., Uedaira, H., An, J., Selvaraj, S., Prabakaran, P., and Sarai, A. (2002) Protherm, thermodynamic database for proteins and mutants: developments in version 3.0. *Nucleic Acids Res.* **30,** 301, 302.

32. Calhoun, J. R., Kono, H., Lahr, S., Wang, W., DeGrado, W. F., and Saven, J. G. (2003) Computational design and characterization of a monomeric helical dinuclear metalloprotein. *J. Mol. Biol.* **334,** 1101–1115.
33. Slovic, A. M., Kono, H., Lear, J. D., Saven, J. G., and DeGrado, W. F. (2004) Computational design of water-soluble analogues of the potassium channel kcsa. *Proc. Natl. Acad. Sci. USA* **101,** 1828–1833.
34. Zou, J. and Saven, J. G. (2003) Using self-consistent fields to bias monte carlo methods with applications to designing and sampling protein sequences. *J. Chem. Phys.* **118,** 3843–3854.
35. Park, S., Kono, H., Wang, W., Boder, E. T., and Saven, J. G. (2005) Progress in the development and application of computational methods for probabilistic protein design. *Comp. Chem. Eng.* **24,** 407–421.
36. Keefe, A. D. and Szostak, J. W. (2001) Functional proteins from a random-sequence library. *Nature* **410,** 715–718.
37. Rojas, N. R. L., Kamtekar, S., Simons, C. T., et al. (1997) De novo heme proteins from designed combinatorial libraries. *Protein Sci.* **6,** 2512–2524.
38. Roy, S., Ratnaswamy, G., Boice, J. A., Fairman, R., McLendon, G., and Hecht, M. H. (1997) A protein designed by binary patterning of polar and nonpolar amino acids displays native-like properties. *J. Am. Chem. Soc.* **119,** 5302–5306.
39. Roy, S., Helmer, K. J., and Hecht, M. H. (1997) Detecting native-like properties in combinatorial libraries of de novo proteins. *Fold. Des.* **2,** 89–92.
40. Finucane, M. D., Tuna, M., Lees, J. H., and Woolfson, D. N. (1999) Core-directed protein design. I. An experimental method for selecting stable proteins from combinatorial libraries. *Biochemistry* **38,** 11,604, 11,612.
41. Xu, G. F., Wang, W. X., Groves, J. T., and Hecht, M. H. (2001) Self-assembled monolayers from a designed combinatorial library of de novo beta-sheet proteins. *Proc. Natl. Acad. Sci. USA* **98,** 3652–3657.
42. Case, M. A. and McLendon, G. L. (2000) A virtual library approach to investigate protein folding and internal packing. *J. Am. Chem. Soc.* **122,** 8089, 8090.
43. Arndt, K. M., Pelletier, J. N., Müller, K. M., Alber, T., Michnick, S. W., and Plückthun, A. (2000) A heterodimeric coiled-coil peptide pair selected in vivo from a designed library-versus-library ensemble. *J. Mol. Biol.* **295,** 627–639.
44. Zhao, H. M. and Arnold, F. H. (1997) Combinatorial protein design: strategies for screening protein libraries. *Curr. Opin. Struct. Biol.* **7,** 480–485.
45. Giver, L. and Arnold, F. H. (1998) Combinatorial protein design by in vitro recombination. *Curr. Opin. Chem. Biol.* **2,** 335–338.
46. Hoess, R. H. (2001) Protein design and phage display. *Chem. Rev.* **101,** 3205–3218.
47. Moffet, D. A. and Hecht, M. H. (2001) De novo proteins from combinatorial libraries. *Chem. Rev.* **101,** 3191–3203.
48. Holm, L. and Sander, C. (1998) Touring protein fold space with dali/fssp. *Nucleic Acids Res.* **26,** 316–319.
49. Luthy, R., Bowie, J. U., and Eisenberg, D. (1992) Assessment of protein models with 3-dimensional profiles. *Nature* **356,** 83–85.

50. Durbin, R., Eddy, S., Krogh, A., and Mitchison, G. (1998) *Biological Sequence Analysis: Probabilistic Models of Proteins and Nucleic Acids.* Cambridge University Press, Cambridge, UK.
51. Raha, K., Wollacott, A. M., Italia, M. J., and Desjarlais, J. R. (2000) Prediction of amino acid sequence from structure. *Protein Sci.* **9,** 1106–1119.
52. Kuhlman, B. and Baker, D. (2000) Native protein sequences are close to optimal for their structures. *Proc. Natl. Acad. Sci. USA* **97,** 10,383–10,388.
53. Koehl, P. L. M. (1999) De novo protein design. I. In search of stability and specificity. *J. Mol. Biol.* **239,** 1161–1181.
54. Koehl, P. L. M. (1999) De novo protein design. II. Plasticity in sequence space. *J. Mol. Biol.* **293,** 1183–1193.
55. Kraemer-Pecore, C. M., Lecomte, J. T., and Desjarlais, J. R. (2003) A de novo redesign of the ww domain. *Protein Sci.* **12,** 2194–2205.
56. Larson, S. M., England, J. L., Desjarlais, J. R., and Pande, V. S. (2002) Thoroughly sampling sequence space: large-scale protein design of structural ensembles. *Protein Sci.* **11,** 2804–2813.
57. Hayes, R. J., Bentzien, J., Ary, M. L., Hwang, M. Y., Jacinto, J. M., Vielmetter, J., Kundu, A., and Dahiyat, B. I. (2002) Combining computational and experimental screening for rapid optimization of protein properties. *Proc. Natl. Acad. Sci. USA* **99,** 15,926–15,931.
58. Zou, J. and Saven, J. G. (2000) Statistical theory of combinatorial libraries of folding proteins: energetic discrimination of a target structure. *J. Mol. Biol.* **296,** 281–294.
59. Kono, H. and Saven, J. G. (2001) Statistical theory for protein combinatorial libraries. Packing interactions, backbone flexibility, and the sequence variability of a main-chain structure. *J. Mol. Biol.* **306,** 607–628.
60. Dunbrack, R. and Cohen, F. E. (1997) Bayesian statistical analysis of protein side-chain retainer preferences. *Protein Sci.* **6,** 1661–1681.
61. McQuarrie, D. A. (1976) *Statistical mechanics*. Harper and Row, New York.
62. Press, W. H., Teukolsky, S. A., Vetterling, W. T., and Flannery, B. P. (1992) *Numerical recipes*. 2nd ed., Cambridge University Press, Cambridge, UK.
63. Weiner, S. J., Kollman, P. A., Case, D. A., et al. (1984) A new force field for molecular mechanical simulation of nucleic acids and proteins. *J. Am. Chem. Soc.* **106,** 765–784.
64. Kono, H. and Doi, J. (1996) A new method for side-chain conformation prediction using a hopfield network and reproduced rotamers. *J. Comput. Chem.* **17,** 1667–1683.
65. Wernisch, L., Hery, S., and Wodak, S. J. (2000) Automatic protein design with all atom force-fields by exact and heuristic optimization. *J. Mol. Biol.* **301,** 713–736.
66. Sander, C. and Schneider, R. (1991) Database of homology-derived protein structures and the structural meaning of sequence alignment. *Proteins* **9,** 56–68.
67. Jensen, L. J., Andersen, K. V., Svendsen, A., and Kretzschmar, T. (1998) Scoring functions for computational algorithms applicable to the design of spiked oligonucleotides. *Nucleic Acids Res.* **26,** 697–702.

68. Wolf, E. and Kim, P. S. (1999) Combinatorial codons: a computer program to approximate amino acid probabilities with biased nucleotide usage. *Protein Sci.* **8,** 680–688.
69. Wang, W. and Saven, J. G. (2002) Designing gene libraries from protein profiles for combinatorial protein experiments. *Nucleic Acids Res.* **30,** e120.
70. Eriksson, A. E., Baase, W. A., Zhang, X. J., Heinz, D. W., Blaber, M., Baldwin, E. P., and Matthews, B. W. (1992) Response of a protein structure to cavity-creating mutations and its relation to the hydrophobic effect. *Science* **255,** 178–183.
71. Axe, D. D., Foster, N. W., and Fersht, A. R. (1996) Active barnase variants with completely random hydrophobic cores. *Proc. Natl. Acad. Sci. USA* **93,** 5590–5594.
72. Voigt, C. A., Mayo, S. L., Arnold, F. H., and Wang, Z. G. (2001) Computational method to reduce the search space for directed protein evolution. *Proc. Natl. Acad. Sci. USA* **98,** 3778–3783.

2

Global Incorporation of Unnatural Amino Acids in *Escherichia coli*

Jamie M. Bacher and Andrew D. Ellington

Summary

The incorporation of amino acid analogs is becoming increasingly useful. Site-specific incorporation of unnatural amino acids allows the application of chemical biology to protein-specific investigations and applications. However, the global incorporation of unnatural amino acids allows for tests of proteomic and genetic code hypotheses. For example, the adaptation of organisms to unnatural amino acids may lead to new genetic codes. To understand and quantify changes from such perturbations, an understanding is required of the microbiological and proteomic responses to the incorporation of unnatural amino acids. Here we describe protocols to characterize the effects of such proteome-wide perturbations.

Key Words: Unnatural amino acids; genetic code ambiguity; amino acid misincorporation; amino acylation errors; amino acid analogs; genetic code evolution.

1. Introduction

The overexpression of proteins that contain unnatural amino acids is becoming increasingly interesting for a number of reasons. As one example, the site-specific incorporation of unnatural amino acids allows specific tests of chemical hypotheses regarding protein structure and function. Similarly, the site-specific incorporation of unnatural amino acids with novel chemistries may foment new protein functions, such as crosslinking with keto-substituted amino acids *(1)*. Global incorporation of unnatural amino acids throughout a protein can also lead to novel physical or functional properties. For example, the global replacement of methionine with heavy-atom analogs, such as selenomethionine, has become a useful tool for obtaining difference maps in X-ray crystallography *(2)*. Global perturbation of organismal proteomes with unnatural amino acids has also been used as a means of experimentally probing the evolution of

From: *Methods in Molecular Biology, vol. 352: Protein Engineering Protocols*
Edited by: K. M. Arndt and K. M. Müller © Humana Press Inc., Totowa, NJ

the genetic code. Both bacteria and phage have been adapted to incorporate unnatural amino acids, and the number and type of mutations that were required for these evolutionary transitions have been examined ***(3–6)***.

Although there is a body of literature regarding the forced growth of bacteria on unnatural amino acids and the subsequent isolation of proteins containing unnatural amino acids (for examples using tryptophan analogs, *see* **refs. *7–11***), for the most part, these are just technical descriptions; there is no consideration of how changes in protocol will affect the outcome of these experiments. We, therefore, present a more-detailed account of methods for whole-protein or whole-cell incorporation of unnatural amino acids, with special emphasis on the incorporation of tryptophan analogs.

2. Materials

1. *E. coli* strains C600 Δ *trp*E (*thi*-1 *thr*-1 *leu*B6 *lac*Y1 *ton*A21 *sup*E44 *mcr*A Δ *trp*E) and derivative strains, C600p (C600 Δ *trp*E + pUC18), C600pGSR (C600 Δ *trp*E + pGSR), C600F (C600 Δ *trp*E F′KanR), and C600F(DE3) (C600F λDE3 lysogen). Strains used for transformation include DH5αF′ and TOP10 (Invitrogen, Carlsbad, CA).
2. Luria-Bertani media (per liter: 10 g tryptone, 5 g yeast extract, 10 g NaCl, and 1.5% bacto-agar for plates) and minimal media M9 (5X stock solution, per liter: 30 g Na_2HPO_4, 15 g KH_2PO_4, 5 g NH_4Cl, 2.5 g NaCl, and 1.5% bacto-agar for plates), supplemented with 20 μg/mL amino acids (*see* **Note 1**) and 0.0005% thiamine. Rich and minimal media are supplemented with antibiotics, 50 μg/mL ampicillin (Ap) or kanamycin (Kn), as indicated.
3. Tryptophan analogs: 4-, 5-, and 6-fluorotryptophan (fW), Sigma (St. Louis, MO).
4. Plasmids: pGEX-KG ***(12)***, pET100/D-topo (Invitrogen), for high-level expression of proteins. For polymerase chain reaction (PCR) amplification of genes, pGFPuv (Clontech, La Jolla, CA) or another source of the gene encoding GFPuv and a plasmid source of the Kn kinase gene (such as p182Sfi-, Kan, K. A. Marshall and A. D. Ellington, unpublished).
5. Vent and Taq DNA polymerases, restriction endonucleases, DNAse, T4 kinase, and T4 DNA ligase.
6. Oligonucleotide primers and dNTPs (Invitrogen).
7. Bacterial protein extraction reagent (B-PER) and B-PER II (Pierce, Beverly, MA).
8. 100 m*M* stock solution of isopropyl-β-D-thiogalactopyranoside (IPTG) in water.
9. Microcon concentrators, 10,000 molecular weight cutoff (Microcon, Rockford, IL).
10. Glutathione-Sepharose beads (Amersham Biosciences, Piscataway, NJ).
11. Ni-nitrilotriacetic acid (NTA) resin (Novagen), and protein purification columns (Bio-Rad, Hercules, CA).
12. Centri-Sep size-exclusion columns (Princeton Separations, Adelphia, NJ).
13. L-tosylamido-2-phenylethyl chloromethyl ketone-treated trypsin (Pierce, Beverly, MA).

14. Phosphate-buffered saline (PBS): 10X stock solution, per liter: 80 g NaCl, 2 g KCl, 11.5 g $Na_2HPO_4 \cdot 7H_2O$, and 2 g KH_2PO_4.
15. 1 *M* $MgCl_2$.
16. 50 m*M* Tris-HCl, pH 8.0, and 5 m*M* reduced glutathione.
17. Buffers for Ni-NTA purification:
    a. Binding buffer, 8X stock: 160 m*M* Tris-HCl, pH 7.9, 4 *M* NaCl, and 40 m*M* imidazole.
    b. Wash buffer, 8X stock: 160 m*M* Tris-HCl, pH 7.9, 4 *M* NaCl, and 480 m*M* imidazole.
    c. Elution buffer, 8X stock: 160 m*M* Tris-HCl, pH 7.9, 4 *M* NaCl, and 2 *M* imidazole.
18. Buffers for high-performance liquid chromatography (HPLC)–HPLC analysis:
    a. Buffer A: 50 m*M* NH_4OAc, pH 5.0.
    b. Buffer B: 50 m*M* NH_4OAc, pH 5.0 and 50% MeOH.
    c. Buffer C: 0.1 *M* NaH_2PO_4, pH 2.5.
    d. Buffer D: 0.1 *M* NaH_2PO_4, pH 2.5 and 50% MeOH.
19. Agarose DNA gel equipment and sodium dodecylsulfate (SDS) polyacrylamide gel electrophoresis (PAGE) equipment.
20. HPLC, HPLC-electrospray ionization (ESI), and mass spectrometry equipment.
21. Microplate reader.

3. Methods

The methods described below outline:

1. The growth of E. coli on tryptophan analogs.
2. The expression and purification of proteins with fW-substituted amino acids.
3. Methods for the analysis of levels of incorporation of unnatural amino acids.

3.1. Growth of E. coli *on Unnatural Amino Acids*

Incorporation of tryptophan analogs into *E. coli* is straightforward. Because tryptophanyl-transfer RNA synthetase has no editing domain, discrimination between the natural amino acid and analogs is based solely on structural determinants. Fluorinated tryptophan analogs are charged relatively effectively by the *Bacillus subtilis* tryptophanyl-transfer RNA synthetase; 4fW is charged six-fold less efficiently than W and 5fW is charged 74-fold less efficiently than W ***(13)***. Similarly, tryptophan analogs are known to enter the cell and support growth, for at least a few generations, before toxicity of the analog takes effect. To achieve efficient incorporation of tryptophan analogs, bacterial strains with mutations preventing tryptophan biosynthesis are used (*see* **Note 2**; **refs. *4–6*, *9***, and ***11***). In the examples presented below, the strain used was *E. coli* C600 Δ *trp*E, and derivatives. Minimal media supplemented with threonine, leucine, and thiamine, as well as tryptophan, analog, or some ratio of unnatural to natural amino acid. By convention, media is named by its supplements.

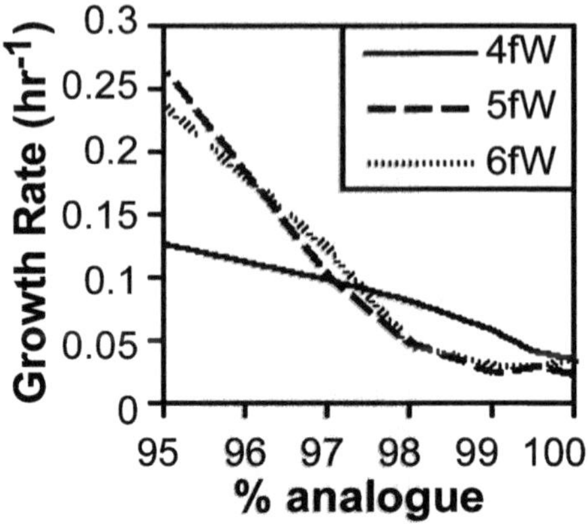

Fig. 1. Effect of various ratios of analog to natural tryptophan on the growth of *E. coli*. Strain C600p was grown on various ratios of 4fW (———), 5fW (– – –), and 6fW (''''''''''''''). Growth was followed by spectrophotometry in microplates, and rates were determined using **Eq. 1.** (Data is reprinted with permission from **ref. *5***, but reanalyzed according to this equation.)

M9B1TL95% 4fW+Ap, for example, indicates minimal M9 media supplemented with vitamin B1 (thiamine), threonine, leucine, 19:1 4fW:W, and Ap.

The effect of unnatural amino acid incorporation on the growth capabilities of bacteria can be determined by analyzing growth curves. Instruments such as the ELX808 Microplate Reader ***(14)*** or the Bioscreen C ***(5)*** can take multiple growth curves in parallel in a microplate format. For testing in the Bioscreen C, an overnight culture of C600p in minimal permissive medium was diluted 1:100 to inoculate 1.5 mL of each medium to be tested for growth inhibition. Of these 1.5 mL media, 350 µL was aliquoted into triplicate wells. Growth was tested with constant shaking at 37°C until all cultures reached the stationary phase. The growth rate was then calculated by fitting the logarithmic portion of the growth curve to the equation:

$$N_{(t)} = N_{(0)} \times e^{(-rt)} \tag{1}$$

where t = time, $N_{(t)}$ = the population (or optical density) at time t, and $N_{(0)}$ is the initial population density, and solving for r, which is the intrinsic growth rate. For example, fitting the exponential portion of growth to **Eq. 1** showed that 4fW causes little change in growth rate at ratios less than 97% relative to W, whereas the effects of 5fW and 6fW are more dramatic (**Fig. 1**). As such, for routine growth of bacteria, a ratio of 19:1 4fW:W can be used in the media. Growth of *E. coli* can be continued for several generations if only 4fW is supplied in the media. In addition, a more complete understanding of the growth curve can be achieved by fitting the entire curve (as opposed to only the logarithmic phase) to the logistic growth equation, which takes into account the carrying capacity:

$$N_{(t)} = \frac{K}{1+\{[(K/N_{(0)})-1]\times e^{(-rt)}\}} \qquad (2)$$

where variables are the same as in **Eq. 1**, with the addition of *K*, which is the carrying capacity of the culture. This equation has been used to quantify the effect of amino acid analog incorporation in *E. coli* under conditions of amino acylation errors ***(15)***.

3.2. Expression and Purification of Proteins With 4fW Incorporation

This section describes the steps taken to construct two expression plasmids (**Subheadings 3.2.1.1.** and **3.2.1.2.**), and to express and purify the proteins under conditions of high substitution of W by 4fW (**Subheadings 3.2.2.1.** and **3.2.2.2.**). In terms of expression and purification, the protocol adaptations required for growth on amino acid analogs are slight relative to growth only on natural amino acids. Finally, this section will outline a method to isolate total cellular protein from bacteria that have incorporated unnatural amino acids (**Subheading 3.2.3.**).

3.2.1. Expression Vector Construction

3.2.1.1. pGSR

Because of the pUC18 plasmid in C600p, pGEX-KG, a glutathione-*S*-transferase (GST) expression vector, required a form of selection other than Ap. Therefore, the plasmid was digested with *Sma*I and *Eco*RI. The Kn kinase gene of p182Sfi-Kan was amplified via PCR using the primers Kan1.39 (5′-CGCGGATCCGGCCACCATGGCCAAGCGAACCGGAAT) and Kan2.39 (5′-CCGGAATTCTGAGGCCTGACAGGCCTTAGAAGAACTCGT). The PCR product was digested with *Bsa*BI and *Eco*RI and ligated into *Sma*I- and *Eco*RI-digested pGEX-KG ***(16)***. The ligation was transformed into DH5αF′, and isolated by miniprep (QIAgen). The resulting plasmid, pGSR, is a GST expression vector that confers Kn resistance.

3.2.1.2. pET100GFPuv

The gene for the highly fluorescent GFPuv was amplified via PCR from the plasmid pGFPuv (Clontech) using Vent DNA polymerase (NEB) with the primers CFPA (5′-CACCACGGCCACTGTGGCCATGAGTAAAGGAGAAGAACTT-3′) and CFPB (5′-GGCCATCGGGGCCCTATTTATAGTTCATCCATGCC-3′); topoisomerase-mediated directional cloning requires a 5′-CACC on the forward primer. Overhanging adenosine residues were added to this product by incubation of 7.5 µL of the PCR product with 1 µL of 10X buffer, 1 µL of 4 m*M* dNTP, and 0.5 µL of Taq DNA polymerase at 72°C for 20 min. This reaction was used to clone the *GFPuv* gene into pET100/D-topo, as directed by the manufacturer. The topoisomerase reaction was used to transform chemically competent TOP10 cells, as directed. The resulting plasmid was isolated by miniprep (QIAgen). pET100GFPuv confers Ap resistance and expresses *GFPuv* under the control of a T7 ribonucleic acid (RNA) polymerase promoter, requiring that a host strain carry a λDE3 lysogen.

3.2.2. Protein Expression and Purification

To achieve high levels of incorporation of unnatural amino acids, initial growth is carried out in permissive conditions, followed by a switch to conditions that are more restrictive and include inducer ***(5)***. This can also be achieved by providing a limited amount of permissive amino acid, and an excess of unnatural amino acid that will be used once the supply of natural amino acid has been exhausted (*see* **Note 3**; **ref. *17)***.

3.2.2.1. Expression and Purification of GST With High Levels of Incorporation of 4fW

1. Grow a single colony of C600p transformed with pGSR and selected on Kn plates in M9B1TL95%4fW+Kn overnight.
2. Dilute the starter culture of **step 1** by 1:100 to inoculate a 100-mL culture of the same media. Grow bacteria to mid-log phase (optical density at 600 nm ≈ 0.5).
3. Centrifuge the culture at 5400*g* for 20 min, and resuspend in 100 mL M9B1TL3×100%4fW+Kn supplemented with 0.3 m*M* IPTG. Grow this culture for a further 16 h (*see* **Note 4**).
4. Centrifuge cells again, as indicated in **step 3** and lyse with 5 mL of B-PER (Novagen) lysing reagent.
5. After centrifugation to remove the insoluble fraction, add 10 m*M* $MgCl_2$ (from a 1 *M* stock solution) and 5 U DNAse to the lysate and incubate for 15 min at room temperature.
6. Add 500 µL of a 50% slurry of glutathione-Sepharose beads to the lysate and mix on a rotator for longer than 2 min at room temperature ***(18)***.
7. Spin the beads down briefly at maximum speed. Add 5 mL of PBS to wash the beads. Repeat this process twice more.
8. Perform a final wash of 1 mL of ice-cold PBS and transfer to microcentrifuge tubes.
9. Elute purified GST from the beads in three 0.5-mL fractions of 50 m*M* Tris-HCl, pH 8.0, plus 5 m*M* reduced glutathione by rotating for approx 2 min at room temperature.
10. Concentrate each fraction by centrifugation using Microcon concentrators.
11. Determine the purity of the proteins by SDS-PAGE; expect greater than 95% purity.

3.2.2.2. Expression and Purification of *GFPuv* Incorporating 6fW

1. Grow C600F(DE3) plus pET100GFPuv in 200 mL of M9B1TLW+Kn+Ap.
2. At mid-log phase, centrifuge cultures and resuspend pellets in 100 mL M9B1TL95%6fW+Kn+Ap with 1 m*M* IPTG and continue growth for an additional 3 h (*see* **Note 4**).
3. Spin down cultures and lyse pellets in 3 mL B-PER. Centrifuge the lysate for 30 min at 15,000*g* to clear insoluble material. To the soluble fraction, add 10 m*M* $MgCl_2$ (from a 1 *M* stock solution) plus 3 U DNAse, and allow the lysate to incubate at room temperature for 10 min.
4. Prepare a 3 mL Ni-NTA column (Novagen) by washing with 15 mL water and 9 mL binding buffer.
5. Pass the soluble fraction over the Ni-NTA column. Wash the column with 30 mL binding buffer and 18 mL wash buffer, and elute with 18 mL elution buffer. Collect fractions of 0.5 mL from the column, and analyze by SDS-PAGE and exposure to ultraviolet light (fractions with protein are visibly fluorescent). Pool fractions containing purified protein, and concentrate as described in **Subheading 3.2.2.1**.

3.2.3. Purification of Whole-Cell Protein Extracts

For incorporation of 4fW:

1. Grow the cultures of C600p to saturation in 25 mL of the appropriate media.
2. Pellet the cultures by centrifugation, and lyse in 200 µL B-PER.
3. Pass 50 µL of this lysate through a Centri-Sep size-exclusion column to remove unincorporated amino acids.

Similarly, to analyze the level of discrimination of bacteria against 6fW:

1. Grow C600F to saturation in 100 mL of culture on either M9B1TLW+Kn or M9B1TL95%6fW+Kn.
2. Pellet and lyse the bacteria in 200 µL B-PER II.
3. Pass half of this volume through a Centri-Sep column (*see* **Subheading 3.3.1.1.**).

3.3. Analysis of Incorporation Levels of Unnatural Amino Acids

In this section, two methods are discussed for determining the extent of unnatural amino acid incorporation. Complete hydrolysis followed by HPLC and mass spectrometry (**Subheading 3.3.1.1.**) or HPLC followed by HPLC under different conditions (**Subheading 3.3.1.2.**) can be used for purified protein and for whole-cell protein extracts. However, protease cleavage and fragment analysis (**Subheading 3.3.2.**) can only be used on purified proteins.

3.3.1. Analysis of Purified Proteins

Sample results pertaining to **Subheadings 3.3.1.1.** and **3.3.1.2.** are presented in **Table 1**.

Table 1
Incorporation of 6fW Into Protein

Method of analysis	Subheading	Incorporation (%)
Whole-cell protein extract		
HPLC-ESI	**3.3.1.1**	56.5
HPLC-HPLC	**3.3.1.2**	66.7
Purified GFPuv		
HPLC-ESI	**3.3.1.1**	68.6
Average incorporation		64.0

From **ref. *4***.

3.3.1.1. Hydrolysis and HPLC-ESI Analysis of Proteins Containing Unnatural Amino Acids

Acid hydrolysis of protein samples allows for the determination of the makeup of the protein. A global average is achieved, whether for whole-cell protein extracts or for purified proteins.

1. Lyophilize protein samples (e.g., half of the eluant from the Centri-Sep columns in **Subheading 3.2.3.**, or several micrograms of purified protein from **Subheading 3.2.2.**) to dryness.
2. Resuspend pellets in 1 mL of 5.4 *M* HCl with 10% thioglycolic acid to preserve tryptophan during hydrolysis.
3. Perform hydrolysis overnight, under a vacuum, at 110°C.
4. Lyophilize the hydrolysates again, and resuspend in 50 µL of water.

Hydrolysates can be analyzed by HPLC-ESI. In this case, the specific mass of the natural and unnatural amino acid can be detected and followed as they elute from the HPLC column. The relative ratios of the eluted masses can be determined as areas under the curves, and represent the relative amount of natural and unnatural amino acids. The actual molar amounts of amino acids can also be determined by generating a standard curve.

3.3.1.2. HPLC–HPLC Analysis of Hydrolyzed Protein Samples

An alternative approach is to run serial HPLC samples.

1. Inject hydrolysates onto a C-18 column and elute with the following program:
   a. 94% Buffer A and 6% Buffer B for 20 min.
   b. Switch by gradient to 98% Buffer A and 2% Buffer B during 10 min.
   c. Elute for an additional 15 min.
   d. Re-equilibrate the column to 94% Buffer A and 6% Buffer B in the last 5 min of the program.

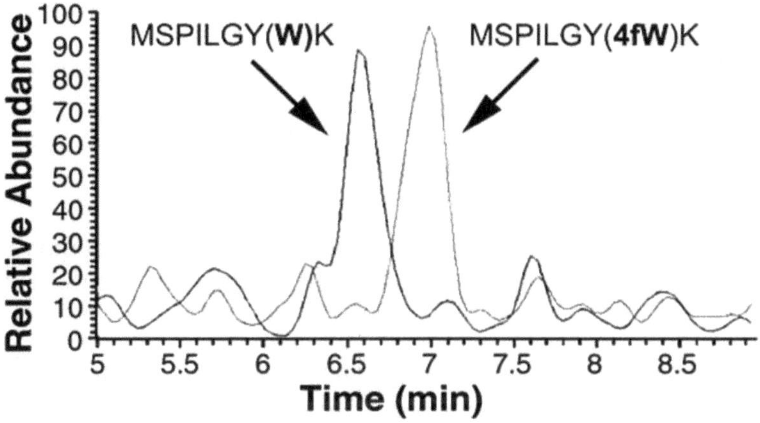

Fig. 2. Relative incorporation of 4fW and W. GST purified from C600pGSR grown in 95% 4fW was digested with trypsin and analyzed by HPLC-ESI. The peptide MSPILGY(W/4fW)K produces two peaks at masses 1094 and 1112 daltons, of roughly equivalent proportion (Data is reproduced with permission from **ref. 5**.)

2. Peaks eluting at times corresponding to standards (e.g., W and 6fW) are collected and lyophilized, and then resuspended in water.
3. Reinject samples on the same column, but under the buffer system of 80% buffer C and 20% Buffer D. Compare the elution times to standards.

The use of the reinjection strategy allows a confirmation of identity; coelution of different molecules may occur under one set of buffer conditions, but is unlikely under two. The area under the curve of the second buffer condition may be considered as a pure sample, and comparisons of these areas may be made between peaks corresponding to natural and unnatural amino acids.

In the case of tryptophan and tryptophan analogs, detection is made at 280 nm, without the need for derivatization (*see* **Note 5**). Phenylalanine and tyrosine give minor peaks from purified protein, whereas a large number of minor peaks are present in hydrolysates from whole-cell protein extracts. Nevertheless, quantitative data can be obtained.

3.3.2. Protease Digestion to Determine Unnatural Amino Acid Incorporation Levels

Because the masses of specific peptide fragments can be predicted based on the sequence of the protein, mass shifts in the predicted fragments are likely caused by the incorporation of the specific unnatural amino acid in question. As such, analysis of protease digestion products is a sensitive method for the determination of amino acid analog incorporation ***(5,19,20)***.

1. Purified GST (for example, 5 μg, e.g., from **Subheading 3.2.2.1.**) is lyophilized and resuspended in 0.1 *M* NH_4HCO_3.
2. Digest with immobilized L-tosylamido-2-phenylethyl chloromethyl ketone-treated trypsin (Pierce) at 37°C for 10 h.
3. Remove trypsin by centrifugation (*see* **Note 6**).
4. Lyophilize the digest, and resuspended in water to 210 μ*M*.
5. Analyze digestion products by HPLC-ESI.

Elution profiles of specified masses can be followed, and demonstrate the level of incorporation of unnatural amino acids into specific fragments (**Fig. 2**).

4. Notes

1. Unnatural amino acids are not often commercially available in an enantiomerically pure form. In the case of a racemic mixture, 20 μg/mL of the L-enantiomer should be used. If an unnatural amino acid is being mixed with a natural amino acid, double the amount of unnatural amino acid to be used, if it is racemic. For example, 95% 4fW requires 38 μg/mL DL-4fW plus 1 μg/mL L-W ***(5)***.
2. Auxotrophy of the bacterial strain to be used for incorporation is preferable, but may not be necessary for all applications. To select a strain that exhibited genetic code ambiguity, Döring et al. used a strain that was prototrophic for valine and cysteine, and selected variants that could substitute cysteine for valine at a valine codon; the result was an editing-deficient valyl-transfer RNA synthetase ***(19)***.
3. The vectors discussed here are typical of expression plasmids. One uses the T7 RNA polymerase system and is under the control of the lac operator, whereas the other directly uses the lac promoter for expression of targeted genes. Many other expression vectors have been used for the incorporation of unnatural amino acids into proteins of interest. Furthermore, these have been conducted for full, partial, and site-specific incorporation of the analog ***(1,5,19)***.
4. Optimization may be required to achieve maximal expression and incorporation of a gene of interest. Critical factors for optimization include the timing between inoculation of the culture and the switch to media with the inducer and unnatural amino acid, and the length of time of expression. For example, it was found that the expression of GST was maximal from a mid-log culture with overnight expression, whereas GFPuv was best expressed from a mid-log culture for only several more hours. More strikingly, the expression of enhanced cyan fluorescent protein for fluorescent assays, including fluorescence-activated cell sorting, was best achieved by inducing an overnight culture to express protein for 6 h (J. M. Bacher and A. D. Ellington, unpublished observations).
5. Analysis of hydrolyzed proteins by HPLC is simplified in these examples by examining the products for tryptophan, a naturally absorbing amino acid. If performing similar experiments for other amino acid analogs, the amino acids may require derivatization before HPLC analysis.
6. Trypsin treatment can effectively be achieved by performing the reaction in a Millipore spin column with a 0.45-μm pore-size filter (M. P. Robertson, personal communication, May 12, 2003). At the end of the reaction time, the reaction can be

purified of trypsin by centrifugation. The immobilized trypsin is trapped on the membrane, whereas the eluate carries the reaction products.

Acknowledgments

This work was supported by grants to A. D. E. from the National Aeronautics and Space Administration Astrobiology Institute, grant NCC2-1055; and the Army Research Office, MURI, DAAD, grant 19-9-1-0207. J. M. B. was partially supported as a Harrington Dissertation Fellow.

References

1. Wang, L., Zhang, Z., Brock, A., and Schultz, P. G. (2003) Addition of the keto functional group to the genetic code of Escherichia coli. *Proc. Natl. Acad. Sci. USA* **100,** 56–61.
2. Budisa, N., Steipe, B., Demange, P., Eckerskorn, C., Kellermann, J., and Huber, R. (1995) High-level biosynthetic substitution of methionine in proteins by its analogs 2-aminohexanoic acid, selenomethionine, telluromethionine and ethionine in Escherichia coli. Eur. J. Biochem. 230, 788–796.
3. Bacher, J. M., Hughes, R. A., Tze-Fei Wong, J., and Ellington, A. D. (2004) Evolving new genetic codes. *Trends Ecol. Evol.* **19,** 69–75.
4. Bacher, J. M., Bull, J. J., and Ellington, A. D. (2003) Evolution of phage with chemically ambiguous proteomes. *BMC Evolutionary Biol.* **3,** 24.
5. Bacher, J. M. and Ellington, A. D. (2001) Selection and characterization of escherichia coli variants capable of growth on an otherwise toxic tryptophan analogue. *J. Bacteriol.* **183,** 5414–5425.
6. Wong, J. T. (1983) Membership mutation of the genetic code: loss of fitness by tryptophan. *Proc. Natl. Acad. Sci. USA* **80,** 6303–6306.
7. Bronskill, P. M. and Wong, J. T. (1988) Suppression of fluorescence of tryptophan residues in proteins by replacement with 4-fluorotryptophan. *Biochem. J.* **249,** 305–308.
8. Parsons, J. F., Xiao, G., Gilliland, G. L., and Armstrong, R. N. (1998) Enzymes harboring unnatural amino acids: mechanistic and structural analysis of the enhanced catalytic activity of a glutathione transferase containing 5-fluorotryptophan. *Biochemistry* **37,** 6286–6294.
9. Pratt, E. A. and Ho, C. (1975) Incorporation of fluorotryptophans into proteins of escherichia coli. *Biochemistry* **14,** 3035–3040.
10. Zhang, Q. S., Shen, L., Wang, E. D., and Wang, Y. L. (1999) Biosynthesis and characterization of 4-fluorotryptophan-labeled Escherichia coli arginyl-tRNA synthetase. *J. Protein Chem.* **18,** 187–192.
11. Browne, D. R., Kenyon, G. L., and Hegeman, G. D. (1970) Incorporation of monoflurotryptophans into protein during the growth of Escherichia coli. *Biochem. Biophys. Res. Commun.* **39,** 13–19.
12. Guan, K. L. and Dixon, J. E. (1991) Eukaryotic proteins expressed in Escherichia coli: an improved thrombin cleavage and purification procedure of fusion proteins with glutathione S-transferase. *Anal. Biochem.* **192,** 262–267.

13. Xu, Z. J., Love, M. L., Ma, L. Y., et al. (1989) Tryptophanyl-tRNA synthetase from Bacillus subtilis. Characterization and role of hydrophobicity in substrate recognition. *J. Biol. Chem.* **264,** 4304–4311.
14. Hendrickson, T. L., Nomanbhoy, T. K., de Crécy-Lagard, V., et al. (2002) Mutational separation of two pathways for editing by a class I tRNA synthetase. *Mol. Cell.* **9,** 353–362.
15. Bacher, J. M., de Crécy-Lagard, V., and Schimmel, P. (2005) Inhibited cell growth and protein functional changes from an editing-defective tRNA synthetase. *Proc. Natl. Acad. Sci. USA* **102,** 1697–1701.
16. King, P. V. and Blakesly, R. W. (1986) Optimizing DNA ligations for transformations. *Focus* **8,** 30–32.
17. Minks, C., Alefelder, S., Moroder, L., Huber, R., and Budisa, N. (2000) Towards new protein engineering: in vivo building and folding of protein shuttles for drug delivery and targeting by the selective pressure incorporation (SPI) method. *Tetrahedron* **56,** 9431–9442.
18. Ausubel, F. M. (1987) Current Protocols in molecular Biology. Greene Pub. Associates and Wiley-Interscience, New York.
19. Döring, V., Mootz, H. D., Nangle, L. A., et al. (2001) Enlarging the amino acid set of Escherichia coli by infiltration of the valine coding pathway. *Science* **292,** 501–504.
20. Wang, L., Brock, A., Herberich, B., and Schultz, P. G. (2001) Expanding the genetic code of Escherichia coli. *Science* **292,** 498–500.

3

Considerations in the Design and Optimization of Coiled Coil Structures

Jody M. Mason, Kristian M. Müller, and Katja M. Arndt

Summary

Coiled coil motifs are, despite their apparent simplicity, highly specific, and play a significant role in the understanding of tertiary structure and its formation. The most commonly observed of the coiled coils, the parallel dimeric, is yet to be fully characterized for this structural class in general. Nonetheless, strict rules have emerged for the necessity of specific types of amino acids at specific positions. In this chapter, we discuss this system in light of existing coiled coil structures and in applying rules to coiled coils that are to be designed or optimized. Understanding and expanding on these rules is crucial in using these motifs, which play key roles in virtually every cellular process, to act as drug-delivery agents by sequestering other proteins that are not behaving natively or that have been upregulated (for example, by binding to coiled coil domains implicated in oncogenesis). The roles of the **a** and **d** "hydrophobic" core positions and the **e** and **g** "electrostatic" edge positions in directing oligomerization and pairing specificity are discussed. Also discussed is the role of these positions in concert with the **b**, **c**, and **f** positions in maintaining α-helical propensity, helix solubility, and dimer stability.

Key Words: Coiled coil; helix; heptad repeat; in vivo selection; leucine zipper; library design; protein design; protein engineering; protein fragment complementation assay; protein stability; rational design.

1. Introduction

The coiled coil is a common structural motif estimated to constitute 3 to 5% of the encoded residues in most genomes *(1)*. It consists of two to five α-helices that habitually twist around each other, typically left-handedly, to form a supercoil. Whereas regular α-helices go through 3.6 residues for each complete turn of the helix, the distortion imposed on each helix within a left-handed coiled coil lowers this value to 3.5. This equates to a seven amino acid repeat for every two turns of the helix *(2,3)*. The most frequently occurring type of coiled coil

From: *Methods in Molecular Biology, vol. 352: Protein Engineering Protocols*
Edited by: K. M. Arndt and K. M. Müller © Humana Press Inc., Totowa, NJ

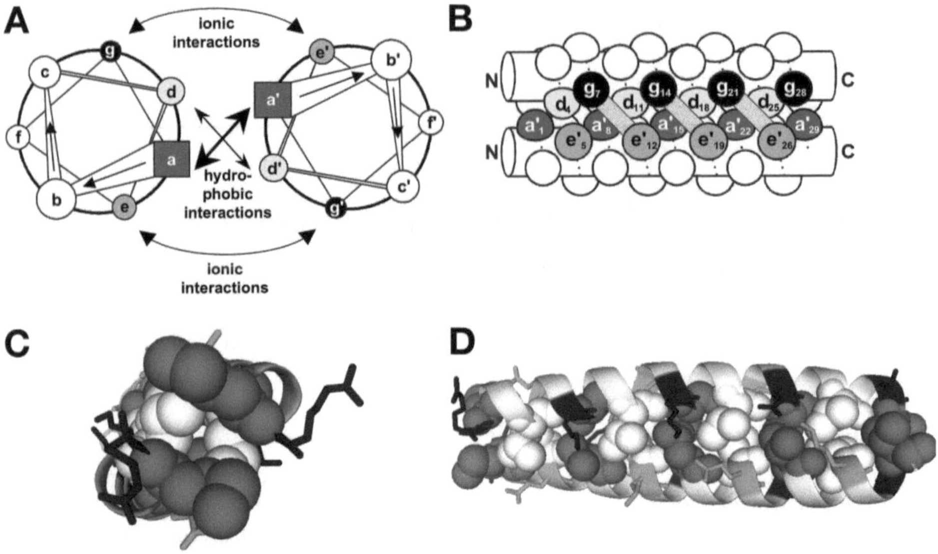

Fig. 1. Dimeric parallel coiled coil. **(A)** Helical wheel diagram looking down the helix axis from the N-terminus to the C-terminus. Heptad positions are labeled **a** to **g** and **a′** to **g′**, respectively. Positions **a**, **d**, **e**, and **g** are in different shades of gray. **(B)** Side view. The helical backbones are represented by cylinders, the side chains by knobs. The path of the polypeptide chain is indicated by a line wrapped around the cylinders. For simplicity, the supercoiling of the helices is not shown. While residues at positions **a** (dark gray) and **d** (light gray) make up the hydrophobic interface, residues at positions **e** (medium gray) and **g** (black) pack against the hydrophobic core. They can participate in interhelical electrostatic interactions between residue i (**g** position) of one helix and residue $i' + 5$ of the other helix (**e′** position, belonging to the next heptad), as indicated by the hatched bars. **(C,D)** Coiled-coil domain of the yeast transcription factor GCN4 (*see* **Note 1**) as ribbon plot (Protein Data Bank code: 2ETA; **ref.** *6*) to indicate supercoiling and **g/e′** interactions. The plot was made using Pymol *(7)*.

is the parallel (i.e., both helices run N to C alongside each other) dimeric, left-handed variety. In this class, the periodicity of each helix is seven, with anywhere from 2 (in designed coiled coils; **ref.** *4*) to 200 of these repeats in a protein *(5)*. In this repeat, the residues are designated **(a-b-c-d-e-f-g)**$_n$ in one helix, and **(a′-b′-c′-d′-e′-f′-g′)**$_n$ in the other (**Fig. 1**). In this model, **a** and **d** are usually nonpolar core residues found at the interface of the two helices, conversely, **e** and **g** are partially solvent exposed polar "edge" residues that give specificity between the two helices through electrostatic interactions. Finally, the remaining three residues (**b**, **c**, and **f**) are typically hydrophilic and exposed to the solvent. The apparent simplicity of the coiled coil structure, with its heptad periodicity, has led to extensive study. Remarkably, interaction between the helices remains a highly specific process. It is this interplay of seemingly simplistic structural periodicity, combined with both high specificities and impressive affinities, which make this ubiquitous structural class of proteins so fascinating.

In both natural and designed two- and three-stranded coiled coils, complementary charge pairs on the edge of the interface that relieve repulsive pairs in alternate oligomers are sufficient to promote formation of hetero-oligomers. We call this idea the peptide velcro (PV) hypothesis ***(8)***, in reference to the design by Kim and co-workers of an obligate heterodimeric coiled coil termed "Peptide Velcro" ***(9)***. This pair of peptides is identical except at the **e** and **g** positions, where one sequence contains Lys and the other sequence contains Glu (*see* **Note 2**). This peptide pair (like other similar pairs) forms a stable heterodimer in vitro. A recent study tested the PV hypothesis by directly comparing rational design and genetic selection strategies ***(8)***. Contrary to the PV hypothesis (but in agreement with the sequence properties of many natural coiled coils), the selected pairs neither maximized predicted attractive **g/e′** charge pairs nor eliminated predicted repulsive **g/e′** charge pairs (*see* **Subheading 2.2.3.**). A variety of factors can influence the contributions of **g/e′** ionic residues. Overall electrostatic potential, including intermolecular and intramolecular interactions, plays a major role; and interactions with the core residues, such as favorable packing or steric clashes, have also been proposed to modulate **g/e′** interactions. Other effects of the sequence context may arise from local helix flexibility or from interactions with **b**, **c**, or **f** residues. Examination of coiled coil structures also suggests that the **e** and **g** positions are structurally different, and these differences may accommodate specific charge pairs in different ways.

Here, we outline the importance of individual amino acids in maintaining α-helical structure and promoting the formation of a specific coiled coil structure of a desired oligomeric state and orientation required for a left-handed coiled coil. The aim of this chapter is to highlight the considerations required in the design or optimization of a coiled coil in this category. This chapter should, therefore, serve as a "protocol" to facilitate coiled coil design by explaining the most important aspects to obtain the desired oligomerization state (**Subheading 2.1.**), specificity (**Subheading 2.2.**), helix orientation (**Subheading 2.3.**), and stability (**Subheading 2.4.**). We discuss the influence of amino acids at the **a** and **d** hydrophobic core positions together with the **e** and **g** electrostatic edge positions, and the role of these together with the **b**, **c**, and **f** positions in maintaining α-helical propensity, helix solubility, and overall dimer stability. Additionally, N-capping and C-capping preferences are discussed. Unless specifically stated in the text, we use the dimeric parallel coiled coil motif as reference state. We also discussed stability and specificity of coiled coils in more general terms in a recent review ***(10)***.

Understanding of the rules governing association of helices has already permitted these proteins to be exploited in novel ways ***(11)***; for example, by fusing antibody Fv fragments to the helices to create a helix stabilized antibody ***(12)***, by fusing antibody scFv fragments to the helices to create miniantibodies ***(13)***, or as a thermo-sensor (in the case of TlpA, which was tagged to green fluorescent protein, with fluorescence changes monitored as a result of structural changes in TlpA during temperature changes). This could permit measurements of signal transduction processes involving coiled coil dimer formation ***(14)***.

2. Materials and Methods

The several different aspects comprising the specific design of coiled coils are discussed in this section. We aim to facilitate the choice of amino acids at the core and the edge positions to achieve the desired oligomerization state (**Subheading 2.1.**), specificity (**Subheading 2.2.**), and helix orientation (**Subheading 2.3.**). Here, we also relate different design solutions to the respective stabilities. The fourth section (**Subheading 2.4.**) relates to the overall stability, focusing on the outer (**b**, **c**, and **f**) positions. In this chapter, we attempt to break down all of the aspects to be considered into the relevant subheadings. Nonetheless, by the very nature of the interplaying factors leading to the formation of the coiled coil, discussions in these sections will invariably crossover in places. Subsections dealing with different effects exerted by the same residue positions should, therefore, be regarded as companion pieces.

2.1. Oligomeric State

Any protein must achieve the desired three-dimensional structure to function properly. Likewise, a coiled coil structure must adopt the correct oligomeric (quaternary) structure. In the following section, the main factors contributing to the adoption of this correct state are discussed.

2.1.1. The Core Residues

The **a** and **d** residues are called core residues because they form a hydrophobic strip that winds around each helix. The nonpolar nature of the **a** and **d** repeat facilitates oligomerization along one face of each helix. This is analogous to a hydrophobic core, which collapses during the folding of globular proteins, and represents a dominating contribution to the overall stability of the coiled coil. Consequently, the core residues exert a major influence on defining the oligomerization state.

The hydrophobic side chains in positions **a** and **d** bury into the neighboring helix in a "knobs-into-holes" manner, first described by Crick in 1953 ***(15)***. In this model, a side chain from one α-helix (the knob) packs into a space surrounded by four side chains of the opposite α-helix (the hole), and vice versa. These

Table 1
Influence of GCN4-p1 Core Mutations on Oligomerization State

Position **a–d**	Geometry of side chains at **a–d**	Oligomerization state	T_m[a]
GCN4-p1[b]	β–γ	Dimer	53°C
I–L	β–γ	Dimer	>100°C (77°C)
I–I	β–β	Trimer	>100°C (70°C)
L–I	γ–β	Tetramer	>100°C (94°C)
V–I	β–β	—[c]	73°C
L–V	γ–β	Trimers	81°C
V–L	β–γ	Mixture of dimers and trimers	95°C
L–L	γ–γ	Trimers	>100°C (76°C)

Adapted from **ref. *16*.**

[a]T_m measurements were performed in 50 m*M* phosphate buffer, pH 7.0; 150 m*M* NaCl; and 10 μ*M* peptide. In parentheses are T_m values measured in 3 *M* GuHCl.

[b]The wild-type GCN4-p1 differs from the V–L mutant by an Asn pair at the central core **a** position, which ensures dimer formation (*see* **Subheading 2.1.2.**).

[c]Species that could not be assigned.

packing geometries are defined by the angle that the Cα–Cβ bond of the knob forms with the Cα–Cα vector at the base of the hole on a projection from the end of the coiled helices. For a parallel dimer, a knob in position **a** (in heptad *i*) fits into the hole in the other helix that is lined up in a clockwise sequence (if looking into the hole) by residues in positions $\mathbf{a'}_i$, $\mathbf{d'}_i$, $\mathbf{g'}_{i-1}$, and $\mathbf{d'}_{i-1}$. Accordingly, a knob at $\mathbf{d}_i$ is in the hole lined by $\mathbf{d'}_i$, $\mathbf{a'}_i$, $\mathbf{e'}_i$, and $\mathbf{a'}_{i+1}$ (**Fig. 1**).

Exhaustive analysis of different core mutants of the coiled coil of GCN4 (a yeast homolog to the Jun transcription factor, sometimes referred to as GCN4-p1, *see* **Note 1**) revealed different packing geometries for different oligomerization states (**Table 1**; **ref.** ***16***). Comparison of the side-chain packing in the X-ray structure of the GCN4-p1 dimer and a designed tetrameric GCN4 mutant showed that the local geometries of the **a** and **d** layers are reversed in the two structures. Parallel knobs-into-holes packing was found at the **a** layer of the dimer and at the **d** layer of the tetramer. In contrast, perpendicular knobs-into-holes packing was observed at the **d** layer of the dimer and the **a** layer of the tetramer. A third class of knobs-into-holes interaction appeared at the **a** and **d** positions of the parallel trimeric variant ***(17)***. In both layers, the Cα–Cβ bond of each knob makes an approx 60° angle with the Cα–Cβ vector at the base of the corresponding hole. This arrangement was termed "acute" knobs-into-holes packing.

These different geometries account for a distinct preference of amino acids for a certain oligomerization state. The following list describes the outcome of several experiments in which various amino acids were tested in the context of stability and oligomerization specificity.

1. Specific hydrophobic residues are crucial in ordaining the oligomeric state of the coiled coil. Harbury et al. systematically changed (with the exception of the **a**1 Met) all **a** and **d** residues of GCN4 to Leu, Val, or Ile (*see* **Note 2**; **ref.** ***16***). As shown in **Table 1**, this led to coiled coils of different oligomeric states. The GCN4-IL (IL referring to Ile at positions **a** and Leu at positions **d**), II, and LI mutants were dimeric, trimeric, and tetrameric, respectively, and were independent of concentration over the entire concentration range investigated. The VI, VL, LV, and LL mutants gave rise to multiple oligomeric states. Each of the L, V, and I combinations gave rise to distinct packing preferences and, thus, to distinct geometries.
   a. β-branched residues (Val and Ile) are favored in the parallel knobs-into-holes packing (the **a** layer of the dimer and the **d** layer of the tetramer), whereas the γ-branched residue Leu is favored in the perpendicular geometry (the **d** layer of the dimer and the **a** layer of the tetramer). Conversely, the insertion of a β-branched amino acid into the perpendicular position would require adoption of a thermodynamically unfavorable rotamer ***(18)***.
   b. It also seems that Ile and Val, despite similar stereochemistry, are not equivalent in establishing the oligomerization states. A Val at the **a** positions is less able to specify dimers than Ile, which gave a higher dimer specific interaction, as opposed to a dimer–trimer mixture for Val ***(17)***.
   c. In contrast to the dimer and tetramer structures, the interior packing of the trimer can accommodate β-branched residues in the most preferred rotamer at both hydrophobic positions.
2. In a study by Woolfson and Alber, the role of the core residues were also studied and used to assign dimer and trimer propensities to unambiguous heptad registers ***(19)***. Dimers and trimers were analyzed for distinguishing features, and these features were used to identify new sequences. The frequency of buried Leu, Ile, Asn, Lys, and Gln were key in this prediction algorithm, called COILER. In total, 21 different proteins known to form parallel dimeric and trimeric proteins were used, equating to some 721 heptads in the database.
   a. Of the initially considered seven amino acids (Ala, Phe, Ile, Len, Met, Val), only Ile and Leu at the **a** site and Leu at the **d** site were observed with the required statistical significance. Dimers were found to be favored by enrichments of Ile at the **a** and Leu at the **d** positions, whereas Ile was strongly depleted at the **d** positions of dimers. Clearly, the selection of the said residues at these positions gives the best packing geometries for dimer formation (*see* **item 1** and **Table 1**).
   b. Val is distributed more evenly than Ile at the core positions of coiled coil sequences. It even occurs with frequencies less than those expected by chance at the **a** and **d** positions of dimers and trimers. These results are consistent with the observation that Val at the **a** and **d** sites discriminates little between dimeric and trimeric coiled coils (*see* **item 1** and **Table 1**).
   c. Packing at the **a** and **d** sites in a trimer are comparable, and particular residue selections for these positions are consequently less specific, leading to more evenly distributed hydrophobic amino acids ***(16,19)***.

3. In studying the positional effect of alanine substitutions in the core of a designed antiparallel coiled coil (*see* **Note 3**), Monera et al. found that when the alanine residues are in register (i.e., on the same rung), a dimer forms ***(20)***. If the alanines are out of register, the helices form a tetramer. The most likely explanation for this is that the cavity formed with alanines in register in the tetramer would be highly destabilizing, and, therefore, the dimer is favored, whereas the Leu–Ala repeats are able to distribute the cavity over a larger region and minimize loss of hydrophobic burial and van der Waals interactions. This demonstrates the oligomerization specificity that is generated by core residue packing.
4. The cartilage oligomeric matrix protein (COMP), which belongs to the thrombospondin family, contains an extremely stable five-stranded parallel α-helical coiled coil. The 46-amino acid-long coiled coil region (*see* **Note 4**) includes a ring of intermolecular (i.e., helix to helix) disulfide-bonded cysteines ***(21)***. The pentameric interface displays knobs into holes packing, with the knobs formed by the **a**, **d**, **e**, and **g** positions, which pack into holes created between side chains at positions **a′–g′**, **d′–e′**, **c′–d′**, and **a′–b′** of the adjacent subunit. Only the residue at position **f** remains completely exposed; the other six positions are significantly buried. The structures of the tetrameric GCN4-pLI mutant (*see* **item 1**) and the pentameric COMP both contain a large axial cavity. The channel in the tetramer varies from 1.0 to 1.3 Å ***(16)*** and, therefore, excludes water molecules (radius 1.4 Å). In contrast, several water molecules were found along the pore of the pentamer, which is consistent with the larger diameter of the channel (2–6 Å).
5. Studies in which the **d** residues have been changed to nonnatural amino acids that are even more hydrophobic in nature (trifluoroleucine and hexafluoroleucine) revealed an increase in stability (*see* **Note 5**; **refs.** *22* and *23*).
6. Hydrophobic burial at the **a/d** interface has been investigated using monomethylated, dimethylated, and trimethylated analogs of diaminopropionic acid (dap), which display increasing degrees of hydrophobicity ***(22)***. Addition of one methyl group to position 16 of one of the monomers (with aspartic acid at position 16 in the analogous peptide), stabilizes the subsequently heterodimeric fold of GCN4-p1 (*see* **Note 6**), possibly because of increased van der Waals interactions in the folded state and a lower desolvation penalty on folding. However, addition of three methyl groups results in destabilization, probably because the increased steric bulk is poorly accommodated. Curiously, addition of two methyl groups to the synthetic dap causes homotrimerization. This demonstrates how small changes in size and hydrophobicity can alter the stability and folding preferences.

In short, the best choices for amino acid at the core in dimeric coiled coils seem to be Leu at **d** positions, and β-branched Ile (or Val) at **a** positions. If Val is used, an Asn (as commonly found in natural coiled coils) should be incorporated at a central **a** position to add specificity to the interaction (*see* **Subheading 2.1.2.**). Trimers are best designed with an all Ile core, whereas tetramers favor Leu at the **a** and Ile at the **d** positions. Deviations from this β/γ side-chain branching arrangement will lead to unfavorable rotamer energies and to a lower stability of the desired structure, and could generate coiled coils of mixed oligomeric states and antiparallel alignment and, thus, of reduced specificity.

2.1.2. Polar Core Residues

Despite core residues being, on the whole, nonpolar, some 20% of core residues are charged *(25)*. These charged residues often serve to specify correct oligomeric states, presumably by ensuring a required helix alignment in structures that would otherwise exist as a mixture of dimers, trimers, and/or tetramers, as well as parallel or antiparallel arrangements. However, this gain in specificity is usually accompanied by a decrease in stability. The following list documents the role of these residues in terms of specificity and stability.

1. In their statistical analysis of dimers and trimers (*see also* **Subheading 2.1.1., item 2**) Woolfson and Alber *(19)* observed that:
   a. Lys and Asn are favored at **a** positions in dimers but depleted from trimers. Asn is three times more likely to be found at an **a** position within a dimer than at the same position in a trimer.
   b. Trimers are enriched in Gln residues at **a** sites, and Ser and Thr residues are enriched at either **a** or **d** sites.
   c. Buried lysines at **a** positions are often found in conjunction with glutamates at the flanking **e** or **g** positions. This is, for example, seen in the X-ray structures of the Jun/Fos heterodimer *(26)* as well as in the GCN4 Asn16Lys mutant (*see* **Subheading 2.1.2., item 5**; **ref.** *27*).
2. The buried Asn pair confers dimer specificity (and in-register alignment), possibly through interhelical hydrogen bond formation between Asn side chains, and this is indeed observed in X-ray *(6)* and nuclear magnetic resonance studies *(28)*. Other conformations that do not satisfy the hydrogen-bonding potential of the Asn side chains are, therefore, energetically disfavored.
3. If the core **a** Asn of GCN4-p1 (*see* **Note 2**) is mutated to Val, the coiled coil experiences a huge increase in stability at the expense of dimerization specificity. Harbury et al. reported a mixture of dimers and trimers because of the lack of specificity conferred by a Val at the **a** position as compared with an Ile (*see* **Subheading 2.1.1., item 1** and **Table 1**; **ref.** *16*), whereas Potekhin et al. reported trimer formation *(29)*.
4. In another case, the core Asn pair of the parallel heterodimer Peptide Velcro (*see* **Note 1**) was mutated to Leu (yielding the peptides, Acid-pLL and Base-pLL), which resulted in a mixture of parallel and antiparallel tetramers *(30)*.
5. In GCN4 (*see* **Note 2**), Alber's group exchanged the Asn pair at the core **a** residue to Gln and Lys, to investigate whether these too were able to confer oligomeric specificity. Lys formed dimers similar to the wild-type, whereas Gln formed a mixture of dimers and trimers *(27)*. They reasoned that the structural uniqueness dictated by the polar group is not merely caused by polar burial, but is also dependent on correct interaction of the side chain with the surroundings. These context effects are much more difficult to predict than mere residue frequencies within a given heptad position.

6. During selection for heterodimeric coiled coils with a protein fragment complementation assay (*see* **Subheading 2.5.3.**) using dihydrofolate reductase, Arndt et al. found a core Asn pair to be favored over Asn–Val or Val–Val combinations in an otherwise Val–Leu core (*see* **Note 7**; **ref. *31***). This is in good agreement with many naturally occurring coiled coils.
7. In another study, an **a** or **d** position of the dimeric GCN4-pVL (*see* **Subheading 2.1.1., item 1**) was mutated to a single polar residue Asn, Gln, Ser, or Thr, respectively (*see* **Note 8**; **ref. *25***). Only Asn pairs at an **a** position and Thr pairs at a **d** position were capable of conferring the correct state. It is likely that the desolvation penalty on burying the residues in the core is vanquished by their interaction energy, which may also serve to ensure correct alignment.
8. Differences in packing environments yield different preferences for hydrophobic residues, even within the **a** and **d** positions. In two exhaustive studies using a model peptide with Val at **a** and Leu at **d** positions, the central **a** and **d** positions were systematically changed to every amino acid to assess their effects on stability and oligomerization state (*see* **Note 9**; **refs. *32*** and ***33***). These changes were the first comprehensive quantitative assessment of the effect on the stability of two-stranded coiled coils of side-chain substitution within the hydrophobic core, and permitted a relative thermodynamic stability scale to be constructed for the 19 naturally occurring amino acids in the **a** and **d** positions. **Table 2** lists those amino acids that gave rise to a well-defined oligomerization state if placed either at the central **a** or **d** position ***(32)***. Leu-, Tyr-, Gln-, and His-substituted **a** site analogs were found to be exclusively three stranded, whereas the Asn-, Lys, Orn-, Arg-, and Trp-substituted analogs formed exclusively two-stranded monomers. When substituting for the central **d** position, Ile and Val (the β-branched residues) induced the three-stranded oligomerization state (as detailed in **Subheading 2.1.1., item 1**), whereas Tyr, Lys, Arg, Orn, Glu, and Asp induced the two-stranded state.
9. Ji et al. mutated gp41, a six-helix bundle envelope protein from simian immunodeficiency virus, that is, along with gp120, responsible for viral fusion with $CD4^+$ cells ***(34)***. Structurally, it consists of a trimer formed by antiparallel heterodimers. In this study, each of the four buried polar residues responsible for core hydrogen bonds and salt bridges (two Gln residues and two Thr residues) were individually mutated to Ile. Of these, three formed more-stable six-helix bundles, whereas one formed insoluble aggregates (*see* **Note 10**). These results demonstrate the role that such residues have in governing a structural stability balance and specificity. This is important because the protein undergoes a structural change before fusion and must have the correct stability balance between the two structures to render this permissible. These polar core residues aid in regulating this conformational stability and, hence, in membrane fusion itself.

In general, Asn pairs at the core **a** position clearly dominate in dimers, especially if the cores deviate from the optimal Ile–Leu **a–d** residues, because this combination appears to result in parallel dimers without the need for Asn pairs within the core. A core **a** position Gln may be a good choice for trimers, although, to confer exclusive specificity for trimers, additional factors may be required.

Table 2
Systematic Change of the Central a and d Positions to Every Amino Acid Using a Model Peptide That Otherwise Has a Val at the a and Leu at the d Positions[a]

	Position **a**		Position **d**	
	Oligomerization[b]	Normalized stability[c]	Oligomerization	Normalized stability
Leu	Trimer	100	(69% Dimer)	100
Ile	(61% Dimer)	105	Trimer	89
Val	(57% Trimer)	108	Trimer	63
Tyr	Trimer	74	Dimer	67
Trp	Dimer	55	(78% Dimer)	47
Gln	Trimer	41	(61% Dimer)	56
Asn	Dimer	56	(72% Trimer)	41
Lys	Dimer	37	Dimer	25
Orn	Dimer	10	Dimer	7
Arg	Dimer	31	Dimer	9
His	Trimer	28	(55% Dimer)	37
Glu	(54% Dimer)	10	Dimer	12
Asp	Dimer[d]	—	Dimer	24

[a]Amino acids that lead to a defined oligomerization state are shown. (From **refs.** ***32*** and ***33***.)

[b]Helices were disulfide bridged via an N-terminal Cys–Gly–Gly linker when assessing the oligomerization state (*see* **Note 9**). However, the reported oligomerization state relates to the number of helices.

[c]Normalized stability represents the stability of each substituted analog relative to Gly = 0 and Leu = 100.

[d]The Asp analog was 100% folded at 5°C. At room temperature, the analog was only ~20% folded.

2.1.3. Edge Residues

The **e** and **g** (edge) positions of the heptad repeat flank the **a** and **d** residues in coiled coil interfaces (**Fig. 1**). Burial of these positions highly depends on the oligomerization state. Consequently, the choice of amino acids at the **e** and **g** sites may be influenced by the oligomerization state. **Table 3** shows the calculated percent buried surface area expressed as the fraction of accessible side-chain surface area in the isolated helix that becomes buried in the oligomer ***(16)***.

1. Compared with the corresponding sites of dimers, the **e** and **g** positions of trimers are enriched for hydrophobic residues (Ile, Leu, Val, Phe, Tyr, and Trp) and depleted of specific hydrophilic residues (Glu, Gln, Ser, and Arg; **ref.** ***19***). These patterns are consistent with the extension of the hydrophobic interface of trimers, relative to that in dimers. This increase in percentage of hydrophobic residues causes the width of the narrow hydrophobic face to increase, and, with it, the likelihood of higher oligomerization states, in which more nonpolar burial can occur than in a two helix coiled coil. This can be seen in **Table 3**, in which the percentage of hydrophobic burial at the **e** and **g** positions is increased by approx 40%.

Table 3
Percentage of Buried Surface Areas for GCN4-p1 Dimer and p-LI Tetramer

Position	GCN4-p1 dimer	GCN4-pLI tetramer
a	87	92
b	0	10
c	1	19
d	87	99
e	26	72
f	0	0
g	27	66

From **ref.** ***14***.

The decrease in oppositely charged $\mathbf{g}_i$ to $\mathbf{e}'_{i+1}$ pairs in trimers (12%) compared with dimers (23%) is consistent with this ***(19)***.

2. Fairman et al. mutated the C-terminal homotetrameric coiled coil domain of the lac repressor to generate a heterotetramer ***(35)***. Peptides containing either all Lys or all Glu at the **b** and **c** positions, which flank the **e** and **g** positions weakly associate, but, if these are mixed, a highly stable tetramer is formed (*see* **Note 11**). This demonstrates that, at least for tetramers, the **b** and **c** residues also play a significant role in the stability of the coiled coil. This is a role akin to the **g/e′** ionic interactions found in dimeric coiled coils, but with the widened hydrophobic interface of the tetramer extending to these residues, the ionic role falls further outwards from the core to the **b** and **c** residues. By changing pH and salt levels, these ion pair interactions between Glu and Lys were shown to be responsible for the increased stability. Additionally, these charges direct against homo-oligomers, and it may be that this unfavorable charge repulsion in potential homodimeric interactions drives the heterodimer formation ***(9,35)***.

2.2. Pairing Specificity

The following section discusses the importance of pairing specificity, and some of the ways in which the coiled coil ensures that no other energetically favorable structures can be accessed. Remarkably, despite their similarity in sequence and structure, coiled coils interact preferentially with functional partners. This section analyzes the factors mediating such high selectivity.

2.2.1. Core Residues

The patterning of hydrophobic residues (mostly Leu, Ile, and Val), as outlined in **Subheading 2.1.1.**, is a dominant driving force behind the association of the helices. However, for this pattern to be observed so frequently, how can the coils, at the same time, use the core to direct against alternative structures forming? The answer is a complicated picture involving subtle changes, such as insertions of nonstabilizing, nonhydrophobic core residues, which will select against alternative structures.

1. Sharma et al. designed a peptide (anti-APCp1) that is targeted to bind a coiled coil sequence from the adenomatous polyposis coli (APC) tumor suppressor protein, which is implicated in colorectal cancers (*see* **Note 12**; **ref.** ***36***). In this, they used core changes together with **g/e′** interactions, rather than Asn pairings, to ascribe specificity to the interaction. They reasoned that the low requirement discovered for core residues in driving specificity is surprising considering their dominant influence on association, and that core mutations can have large effects on stability and specificity. To address this issue, they designed a peptide to bind to the first 55 amino acids of APC (APC55) and mutated this anti-APCp1 to generate the more-frequently observed **a–a′** and **d–d′** pairings based on covariation patterns at the **a** and **d** positions of keratin type I and type II heterodimers. They made three mutations (A41I at layer **a** and A2M and M44A at layer **d**) to change the wild-type Ala–Ala and Met–Met interactions to the more-frequently found Ala–Ile, Ala–Met, and Met–Ala interactions, respectively. Two further mutations (T6G in layer **a** and N30H in layer **d**) served to destabilize the respective homodimers with Gly–Gly and His–His pairs. Additional **e–g′** pairings optimized ionic interactions while directing against anti-APCp1 homodimerization. The resulting heterodimer, APCp1/APC55, was both stable and specific.
2. Schnarr and Kennan formed heterotrimeric proteins by steric matching of core hydrophobic residues ***(37)***. In their study, unnatural residues of various side-chain lengths were used to promote specific heterotrimer formation. The authors replaced a core **a** position of GCN4 with Ala or cyclohexylalanine (*see* **Note 13**). The result was a sterically mismatched core layer in the trimer, with either three Ala or three cyclohexylalanines, generating steric void or repulsion, respectively. Two Ala and a cyclohexylalanine, however, generated a heterotrimer with good steric matching. The use of nonnatural side chains can be used in this way to generate coiled coils. The additional bulk of the cyclohexylalanine complements the Ala core layers to provide a steric match, whereas bulkier side chains only serve to destabilize the molecule.

2.2.2. Polar Core Residues

The roles of polar core residues in directing specific oligomerization states of coiled coils were discussed in **Subheading 2.1.2**. Heterotypic core contacts that permit generation of heterospecificity in coiled coil pairings were mentioned in **Subheading 2.1.1**. Further specificity can be obtained by interaction of polar core interactions with the outer residues.

1. Next to Asn, Lys at position **a** is the most common buried polar residue in natural dimeric coiled coils ***(19)***. Lys at position **a** can form an *intra*helical electrostatic interaction with an **e** position residue of the preceding heptad ***(27)*** as well as an *inter*helical **g′–a** polar interaction with a **g′** position polar residue of the preceding heptad of the opposing helix in a parallel dimer ***(26)***.

2. Campbell and Lumb placed two position **a** Lys residues into the context of the Base-pLL peptide of Peptide Velcro to enable interhelical polar interactions between these Lys residues and the **e′** or **g′** Glu residues of Acid-pLL of Peptide Velcro (*see* **Note 14**; **ref.** ***38***). As expected, the dimeric state was favored, most likely because the desolvation penalty would be higher in a higher oligomeric state. In addition, such an interaction is less destabilizing than an Asn–Asn (**a–a′** contact), presumably because there is a greater desolvation penalty to pay for burying the latter. However, no discrimination between parallel and antiparallel arrangement occurred, presumably because the **a–g′** interactions in the parallel orientation and the **a–e′** interactions in the antiparallel orientation were energetically similar.
3. Harbury's group used a computational design approach (*see* **Subheading 2.5.4., item 4**) to generate specificity by considering not only the desired structure but also alternate undesired structures ***(39)***. The **a**, **d**, **e**, and **g** positions of the central heptad of GCN4 (*see* **Note 15**) were varied; all nonproline residues and specific homotypic and heterotypic sequences were selected *in silico* and experimentally verified. Next to volume complementarity and charge complementation at **g/e′** pairs, it was observed that Glu at **d** paired preferentially with an Arg residue at position **e′**.

2.2.3. Edge Residues

Pairing specificity is greatly influenced by the nature of the electrostatic **e** and **g** residues (in parallel dimeric helices between **g** of one heptad and **e′** of the following heptad on the other helix; this is termed $i \rightarrow i' + 5$). These residues are commonly found to be Glu and Lys, respectively. Such complementary polar interactions add specificity and consolidate the stability provided by the core hydrophobic interactions. The charge pattern on the outer contacting edges of a coiled coil will dictate whether the protein will form a homomeric or heteromeric protein, and whether the orientation of the coiled coil is to be parallel or antiparallel. However, the PV hypothesis (*see* **Heading 1**; **ref.** ***8***) is an oversimplification; residues that serve little or no role in the stability of the complex serve their purpose by directing the helices away from homologs or undesirable interactions that would otherwise compromise the specificity of the molecule (negative design). Alternatively, some coiled coils are likely to have no evolutionary pressure because they are already specific and need not be any more stable than they already are.

1. As expected, replacing favorable **g/e′** Gln–Gln pairings with repulsive Glu–Glu pairs has been shown to destabilize the coiled coil conformation ***(40)***.
2. Hodges and co-workers estimated the salt bridges between **g/e′** pairs to contribute 0.37 kcal/mol to the stability of the coiled coil (*see* **Note 16**; **ref.** ***41***).
3. Careful placing of charges within the **e** and **g** positions can permit heterodimer formation, and additionally ensure that formation of the homodimer is unfavorable ***(9,42,43)***, as implicated in the PV hypothesis.

Table 4
Coupling Energies ($\Delta\Delta\Delta G_{int}$) for g-e′ Pairings Calculated Using a Double Mutant Alanine Thermodynamic Cycle

g–e	Glu	Gln	Arg	Lys
Glu	+0.7 ± 0.2	+0.2 ± 0.1	–0.5 ± 0.1	–0.3 ± 0.15
Gln	+0.2 ± 0.1	0.0 ± 0.1	+0.3 ± 0.1	+0.3 ± 0.1
Arg	–1.1 ± 0.1	+0.4 ± 0.1	+0.8 ± 0.1	+0.8 ± 0.1
Lys	–0.9 ± 0.1	+0.3 ± 0.1	+0.6 ± 0.1	+0.6 ± 0.1

From **ref. *44***.
Values in kcal/mol.

4. The four most common amino acids found at the **e** and **g** positions are Glu, Gln, Arg, and Lys ***(44)***. These residues contain long hydrophobic side chains that are able to interact with **a** and **d** core residues, and terminate with a charged (Glu, Arg, or Lys) or polar (Gln) group. By mutating first **e**, then **g**, then both residues of two interacting heptad pairs to Ala, Krylov et al. were able to establish the coupling energies ($\Delta\Delta\Delta G_{int}$) of those contacting residues for chicken vitellogenin-binding protein ***(44,45)*** (*see* **Note 17**). This double mutant thermodynamic cycle was used to permit a thermodynamic scale to be generated for outer residue contact preferences (**Table 4**). At 150 m*M* KCl and pH 7.4, Glu–Arg attractions are found to be slightly more stable than Glu–Lys attractions, presumably because Arg side chains are longer and interact better with the glutamate, and the respective methylene groups shielding the core more effectively from the solvent. This may, in turn, increase the effect of the charged end groups, which give a greater contribution in less aqueous surroundings. As expected, high salt weakens these interactions, as does low pH, in which polar interactions are weakened and the hydrophobic effect is increased. The Glu–Arg interaction, followed by Glu–Lys and Gln–Gln, are the most stabilizing (regardless of orientation), with the Glu–Glu and Arg–Arg, Arg–Lys, Lys–Arg, and Lys–Lys same-charge interactions being considerably less favored.
5. Arndt et al. designed a peptide library based on the Jun-Fos heterodimer, in which the **b**, **c**, and **f** residues are from their respective wild-type proteins, the **a** and **d** positions are Val and Leu (with the exception of **a**3 Asn inserts in the core to direct desired helix orientation and oligomerization state), and the **e** and **g** residues are varied using trinucleotides to yield equimolar mixtures of Arg, Lys, Gln, and Glu (*see* **Note 7**; **ref.** ***8***). Unexpectedly, even the best-selected winner, the Winzip-A2B1 heterodimer (*see* **Note 18**), lacked fully complementary charged residues at **g/e′** pairs, despite an exhaustive selection process. Rather, two of the six **g/e′** pairs are predicted to be repulsive, suggesting that sequence solutions deviate from simple charge complementarity rules (PV hypothesis). Presumably, the overall electrostatic potential (including intramolecular and intermolecular interactions) plays a major role, and interactions with core residues, such as favorable packing or steric clashes could also modulate these **g/e′** interactions (*see* **refs.** ***8*** and ***31*** and references therein). Such observations are in agreement with naturally occurring coiled coils, which

usually have a complicated interaction pattern. These coiled coils have to fulfill a number of criteria, such as biostability and extremely high specificity within a family and almost no crossreactivity with coiled coils of other families *(46)*.

2.3. Helix Orientation

The majority of coiled coils fold into a parallel alignment, however, a growing number of structurally characterized proteins contain antiparallel coiled coil domains *(47)*. Despite the growing recognition of the biological importance of antiparallel coiled coils, the study of this class of molecules has been hampered by the lack of well-behaved model systems. None of the short antiparallel coiled coil domains found in proteins such as seryl–transfer RNA (tRNA) synthetase or hepatitis delta-virus antigen have been shown to be sufficient for dimerization without undergoing further self-association.

Hodges and co-workers were the first to report the characterization of a *de novo* designed coiled coil that was constrained in an antiparallel orientation by an interior disulfide bond *(48)*. This and other designed antiparallel coiled coils were more stable than their respective parallel counterparts, with nearly equivalent interhelical interactions *(49,50)*. These data suggest that, assuming everything is equal, the helix–dipole interactions (*see* **Subheading 2.4.3.**) favor the antiparallel orientation.

2.3.1. Core Residues

The core residues that pack against each other are **a**–**a′** and **d–d′** in parallel coiled coils. Antiparallel coiled coils have **a–d′** and **d–a′** central packing, yielding identical packing layers *(51)*.

1. It has been shown that the relative position of Ala residues in the core of a *de novo* designed coiled coil can control the parallel or antiparallel orientation (*see* **Subheading 2.1.1., item 3**; **ref.** *52*). By careful placement of the Ala in the middle heptad, either all-parallel or all-antiparallel tetramers are formed. This was achieved by an alternating pair of Ala and Leu residues (Ala–Leu–Ala–Leu) in each of the two planes at the central heptad core positions of the molecule. Such an alternating core of small and large residues (Ala–Leu–Ala–Leu) is the best way to accommodate these small side chains. In the parallel arrangement, the packing would be all Ala in one plane and all Leu in the other plane and would, thus, result in a large cavity that would solvate the core and destabilize the molecule (*see* **Note 19**).
2. In a similar experiment, the core Asn of the dimeric GCN4-p1 (*see* **Note 2**) was exchanged to Ala. The result was an antiparallel trimer to avoid a core cavity *(53)*. However, parallel trimers were obtained in the presence of benzene, which bound to the core cavity *(54)*.
3. In a recent design of an antiparallel homodimeric coiled coil, termed APH, steric matching of β-branched (Ile at position **d**) and truncated (Ala at the opposing **a′** position) side chains in the hydrophobic core were used, along with other features

(*see also* **Subheading 2.3.3., item 3**), to promote antiparallel orientation *(55)*. The parallel arrangement should be energetically disfavored by an Ile **d** layer, which is poorly accommodated in dimeric coiled coils (*see* **Subheading 2.1.1.**) and by an Ala–Ala "hole" in the hydrophobic core (*see* **Note 20**).

2.3.2. Polar Core Residues

Similar to the parallel coiled coils, buried polar residues also play a key role in dictating the helix orientation.

1. In parallel coiled coils, **a–a′** Asn pairing is commonly observed (*see* **Subheading 2.1.2.**), however, this has not been seen for the corresponding **a–d′** interaction in an antiparallel coiled coil (for a review, *see* **ref.** *47*). Indeed, Leu–Leu packing interactions found in the antiparallel coiled coils are more stable than in parallel core-packing interactions *(49)*, and it is likely to be the Asn–Asn interactions and electrostatic interactions (*see* **Subheadings 2.2.3.** and **2.3.3.**) that drive the specificity of the parallel conformation over the antiparallel.
2. Despite no native identifications of Asn–Asn **a–d′** pairs, Oakley and Kim modified the parallel heterodimeric coiled coil Peptide Velcro such that the buried polar interaction was only expected to occur if the helices are in an antiparallel orientation (i.e., **a–d′** pairings; **ref.** *56*). It was estimated that this single buried polar interaction conferred a modest antiparallel preference of approx 2.3 kcal/mol (*see* **Note 21**). However, an exclusive formation of antiparallel coiled coils was obtained only by combining the buried polar interaction with the introduction of charge repulsions in the parallel orientation (*see* **Subheading 2.3.3., item 2**).
3. Comparable to the **a–g′** or **d–e′** interactions found in parallel dimeric coiled coils (*see* **Subheading 2.2.2.**), **a–e′** or **d–g′** interactions can occur in antiparallel coiled coils, as, for example, observed in the seryl tRNA synthetase coiled coil between Arg-54 at a **d** position on one strand and Glu-74 at a **g′** position in the other *(57)*, or in an **a–e′** interaction in the GreA coiled coil *(58)*. It was assumed that this type of buried polar interaction might also play a role in orientation specificity. Oakley's group replaced the core Asn pair in Peptide Velcro, such that an **a** position Arg in Acid-p1 could interact with a **g′** position Glu in Base-p1 in the antiparallel orientation (*see* **Note 22**; **ref.** *59*). The parallel conformation should be destabilized by a potentially repulsive interaction between an **e′** position Lys. However, while the introduction of Arg in the core was able to promote the dimeric state, in accordance with the study of Campbell and Lumb (*see* **Subheading 2.2.2., item 2** and **Note 14**; **ref.** *38*), no clear preference for the antiparallel or parallel orientation was found. The free energy difference between both states was estimated to be only 0.1 ± 0.1 kcal/mol. One possible explanation could be the formation of an interhelical interaction with a neighboring **g** position Glu.

2.3.3. Edge Residues

In parallel coiled coils, the polar **e** and **g′** positions interact favorably. In the antiparallel coiled coil, **e** interacts with **e′**, and **g** with **g′**, the result being that one helix is effectively rotated by 180°. These changes in preference must play an important role in orientation selection.

1. Monera et al. designed parallel and antiparallel coiled coils predicted to have either interchain attractions or repulsions (*see* **Note 23**; **ref. *49***). It was indeed found that the major orientation found was the one resulting in electrostatic interactions between oppositely charged amino acids.
2. Oakley's group refined a previously designed antiparallel heterodimeric coiled coil (*see* **Subheading 2.3.2., item 2**) further. In their first design ***(56)***, one peptide contained only Glu residues, the other only Lys at both the **e** and **g** positions (*see* **Note 21**). In their new design, they substituted a single residue at a **g** position in each peptide such that all potentially attractive interactions are expected in the antiparallel orientation (*see* **Note 24**; **ref. *60***). In contrast, two potentially repulsive Coulombic interactions are expected in the parallel orientation, and indeed, a strong preference for the antiparallel arrangement was found.
3. In the recently designed antiparallel homodimeric coiled coil, APH (*see also* **Subheading 2.3.1., item 3**), Glu residues at the N-terminal **e** and **g** positions and Lys at the C-terminal **e** and **g** positions have been used to direct antiparallel orientation, resulting in eight potential Coulombic interactions in the antiparallel arrangement and eight potential repulsions in the parallel arrangement (*see* **Note 20**; **ref. *55***).

2.4. Stability

Achieving a favorable stability is the net result of large and opposing enthalpic and entropic forces. The result is a protein of modest stability that has evolved to interact with its partner protein under physiological conditions, but not to be so stable as to never dissociate. Achieving this balance is, once again, the result of core and edge residues and their interactions, helix length, helical propensity, solubility, and helical capping. These stabilizing factors are discussed in this section.

One important and delicate consideration is solubility. The participating helices must have a nonpolar core to permit a favorable interaction, but the overall helices must not be so nonpolar as to aggregate under working conditions. Residues **a** and **d** must form a hydrophobic strip that connects the two helices, and **e** and **g** should typically be polar residues involved in ensuring that only the true binding partners interact with the helix (i.e., to be destabilizing for noninteracting partners), and consolidating the stability introduced from the core. This leaves the remaining residues in the heptad, the **b**, **c**, and **f** positions, to address the charge balance and to ensure that the helix is both stable and soluble. Glu and Lys, also of reasonable helical propensity, are well suited. Charged residues may also interact favorably with the helix dipole (*see* **Subheading 2.4.3.**) and form favorable interactions with charged residues one turn away in the helix. This gives an additional advantage in the selection of these residues at solvent exposed sites away from the dimer interface. The insertion of a Tyr at a solvent-exposed position is advantageous for concentration determination of the peptides. Crudely speaking, for the archetypal dimeric coiled coils, such as GCN4 and Jun proteins, there is a preponderance of Lys, Asp, Arg, Glu, and Asn at these positions. Also present, albeit infrequently, is Ala, which presumably adds extra stability to the helix, perhaps in cases in which residues in close spatial proximity have caused the overall α-helical propensity to lower.

2.4.1. Helical Length

Generally speaking, as the length of the coiled coil chain increases, a (nonlinear) increase in stability is observed ***(61)***. This is because the sequence of the coiled coil will play an additional major role. For example, Lau and Hodges constructed a 29-mer with greater stability than tropomyosin, a 284-residue coiled coil (*see* **Note 25**; **ref.** ***62***). One would expect the coiled coil structure to become more stable with length on the grounds of increased hydrophobic burial, hydrogen bonds, and polar interactions, and, again, this was proven to be the case in a length study of a designed homodimeric peptide (*see* **Note 26**; **ref.** ***63***), although a minimum length of two heptads (*see* **Note 27**; **ref.** ***64***) is also required to permit association in the case of a homotrimer. The coiled coil domain of the Lac repressor has been used as the basis to assess the effects of chain length on stability and folding ***(65)***. Unsurprisingly, the dissociation constant of the tetramers decreased as the number of heptads in each helix increased (*see* **Note 28**). Somewhat more surprising was the fact that a tetramer with as few as four heptads in each helix folded cooperatively, with no evidence for a dimeric intermediate.

Long coiled coils, such as tropomyosin and myosin heavy-chain domain are, in contrast to short coiled coils, not enriched in the core exclusively with bulky nonpolar amino acids. Rather, because of the stability afforded from the length of the protein, such proteins contain clusters of small nonpolar or charged residues ***(66)***. These residues account for approx 40% of the core. Such destabilizing clusters consist mainly of Ala, because this residue is less destabilizing to the core than a polar amino acid (being easier to pack than a polar side chain), and is able to contribute favorably to the overall helical propensity. This means that although a stability increase proportional to the gain in heptads is not observed in these long coiled coils, from an evolutionary point of view, such a vast gain in stability is not required. Instead, hydrophobic stabilizing clusters afford the coiled coil the necessary stability, whereas destabilizing clusters do not, but do maintain the helical structure. These clusters, predominant in Ala, increase the flexibility and local unfolding in such regions without affecting the overall stability of the coiled coil, presumably allowing the protein to exercise its specific biological function.

2.4.2. Helix Propensity

Litowski and Hodges have reported that increasing the α-helical propensity of noncore residues by exchanging Ser for Ala (the amino acid of highest helical propensity), can stabilize the whole coiled coil *(67)*. In their model, the Glu/Lys coiled coil (*see* **Note 29**), this led to stabilizations of approx 0.4 to 0.5 kcal/mol per substitution. This is in good agreement with an earlier study by O'Neil and DeGrado *(68)*. This number is less than for single α-helices, presumably because of the additional stabilizing interactions involved in maintaining the coiled coil.

The outer residues, **b**, **c**, and **f** should, generally speaking, be able to form hydrogen bonds with the solvent, and, in doing so, compensate for those buried potential hydrogen bonding partners that are unable to do so *(69)*. These residues are also responsible for helping to maintain the α-helical propensity, with each residue having a distinct conformational preference that will either stabilize or destabilize the helix *(68)*. The α-helical propensity of Ala is known to be the highest of all amino acids, and, although all common core hydrophobic residues (with the exception of Asn) have good α-helical propensities, the solvent-exposed Arg and Lys also play a considerable role. This is because both Lys and Arg make good hydrogen-bonding partners and are well-suited to improving the solubility of the molecule. Indeed we found helix propensity to be an important factor in coiled coil design (*70*, http://www.molbiotech.uni-freiburgode/bCIPA).

2.4.3. Interactions With the Helix Macrodipole

Statistical analysis of the composition of α-helices in protein structures revealed that different amino acids prefer different regions of the helix. In particular, potentially negatively charged side chains (Asp or Glu) strongly prefer positions near the N-terminal end of helices, whereas potentially positively charged side chains (His, Arg, or Lys) had a less pronounced preference for the C-terminal end *(71,72)*. Explanations for the preferences of polar side chains for the ends of helices fall into two principal models. First, because the first four backbone NH groups and final four backbone CO groups of the α-helix are not able to form the $i \rightarrow i + 4$ hydrogen bond to other backbone groups, polar side chains at the end of helices can substitute as hydrogen bonding partner. This is termed helix capping and is described in **Subheading 2.4.4**. Second, electrostatic "charge–helix dipole" interactions between the charged side chains and the net dipole moment of the α-helix formed by the alignment of individual peptide backbone dipoles may also stabilize or destabilize the protein.

1. Hodges' group investigated the positional dependence of negatively charged Glu side chains on the stability of a designed homodimeric coiled coil with no intrahelical or interhelical interactions (*see* **Note 30**; **ref.** *73*). A Glu substituted for Gln near the N-terminus in each chain of the coiled coil stabilized the coiled coil at pH 7.0, consistent with the charge–helix dipole interaction model. In contrast, Glu substitution in the middle of the helix destabilized the coiled coil because of the lower helical propensity and hydrophobicity of Glu compared with Gln at pH 7.0. A Glu substitution at the C-terminus destabilized the coiled coil even more, because of

the combined effects of intrinsic destabilization and unfavorable charge–helix dipole interaction with the negative pole of the helix dipole.

2. During selection for heterodimeric coiled coils from two designed libraries with an equimolar mixture of Glu, Gln, Arg, and Lys residues at four **e** and four **g** positions (*see* **Note 7**), Arndt et al. observed an enrichment of negatively charged Glu and neutral Gln at the N-terminal part and positively charged Lys and Arg at the C-terminal part of selected coiled coil sequences ***(31)***. This bias is in good agreement with the proposed charge–helix dipole interactions.

2.4.4. Helix Capping

A helix can be labeled as N′-Ncap-N1-N2-N3-N4-mid-C4-C3-C2-C1-Ccap-C′. Of these positions, N′, Ncap, Ccap, and C′ have nonhelical ψ and ϕ angles, and only N1-N2-N3...C3-C2-C1 participate in the $i \rightarrow i + 4$ hydrogen bonding that is characteristic of the α-helix. The N1, N2, N3, C1, C2, and C3 residues are unique because their amide groups participate in $i \rightarrow i + 4$ backbone–backbone hydrogen bonds using either only their CO (at the N terminus) or their NH (at the C terminus) groups (*see also* **Subheading 2.4.3.**). The need for these groups to form hydrogen bonds has powerful effects on helix structure and stability ***(74)***. From N4 (or C4) upward, the residues can satisfy both NH and OH backbone hydrogen bonds.

1. In helix design, the most selective position for the stability at the N-terminus is the Ncap position. The six best amino acids for this position are Ser, Asp, Thr, Asn, Gly, and Pro, with another 11 (Val, Ile, Phe, Ala, Lys, Leu, Tyr, Arg, Glu, Met, and Gln) being strongly avoided ***(75)***. Of the six preferred residues for Ncap, Ser, Asp, and Thr are the best.
2. A good example of an Ncap motif is Ser-Xaa-Xaa-Glu (Ncap-N1-N2-N3), in which a reciprocal side chain/main chain interaction pattern (OH of the Ncap Ser to NH of Glu, and the carboxyl group of Glu to NH of Ser) stabilizes the helix. Further stabilization can be achieved by hydrophobic residues before and after the Ser-Xaa-Xaa-Glu motif ***(76)***. Lu et al. introduced this capping motif in the GCN4 sequence and were thereby able to stabilize the coiled coil by 1.2 kcal/mol (*see* **Note 31**; **ref.** *77*).
3. The N1 and C′ positions strongly favor Pro (it is sterically compatible because the proceeding residues have nonhelical dihedral backbone angles), which is indeed a common helix termination motif, but should be avoided both in the main body of the helix and in the C3, C2, and Ccap positions. Pro, being the most water soluble of all amino acids ***(78)***, is compatible with solvent-exposed positions at the helix ends, and also requires no hydrogen bonding acceptor because it lacks a backbone NH group ***(76)***.
4. C-terminal capping motifs involve backbone–backbone hydrogen bonds, rather than the side chain to backbone hydrogen bonding that is observed between Ncap and N3. At the C-terminus backbone, hydrogen bonds are satisfied by posthelical backbone groups (e.g., C′ and the following C″ in the Schellman motif ***(76)***. This means that the C-terminus need only select for C′ residues that can adopt positive ϕ angles, for example, Gly.

2.5. Rational Design vs Selection Strategies

The complex nature of protein design means that confidence in designing or preselecting a sequence that is to be of greatest stability is a daunting task. Rather, if procedures permit, it can be more fruitful to generate a library of interactions; a collection of changes to the molecule that can then be assayed for the most successful interacting sequences. This can be regarded as a semi-rational approach, whereby balanced changes at specific positions are introduced into the library (and often also keeping wild-type residues) and the resulting large combination of molecules can be screened. Normally, in this type of design, to change all heptad positions to all amino acids would generate a large and unrealistic library size (e.g., a four heptad coiled coil with only two amino acid options at each position would generate $2^{28} = 2.68 \times 10^8$ library members). However, to generate only one sequence that is to be synthesized and tested for binding strength would be time consuming, cumbersome, and, often, disappointing. Using a semi-rational design, one is afforded a reasonable compromise, whereby a molecule can be included within a library to test whether it is indeed a good candidate for binding, and, at the same time, affording the versatility to generate new, unintuitive, and, often, novel binding partners. Selection in a cell-based system has the added bonus of concomitantly screening out sequences that are susceptible to proteases found within that host organism.

2.5.1. Degenerate Codons

By using degenerate codons, one is able to encode a mixture of amino acids at positions at which a change is desired. Again, by carefully choosing the corresponding nucleotide positions that are to be randomized, one is able to generate a codon that is degenerate, but only for the desired bases and, hence, desired amino acids. As has been discussed, the coiled coil has a preference for "types" of amino acids at various positions. For example, the **e** and **g** residues are commonly found to be polar but complementary (*see* **Table 4**), therefore, an Glu–Lys interaction could be exchanged for an Glu–Arg interaction with little perturbation to the structure of the molecule (because Lys and Arg are rather similar in terms of both bulk and charge). In this situation, Arg could be introduced and Lys could be retained in the library to observe which interaction is more favorable in context of the whole molecule. Likewise, the hydrophobic core's preference for β-branched amino acids at the **a** position means that although a Val may be found in the wild-type molecule, an Ile might be preferred. It should be noted, however, that adhering to these preferences is an oversimplification, and apparent non-favored residues are required to design against nonspecific structures or even overtly stable structures (*see* **ref. *8***). Generally speaking, however, keeping the overall "binary patterning" will ensure that all resulting library members have a reasonable chance of forming a coiled coil structure rather than alternate structures being energetically favored. The coiled coil library members will be able to bury their hydrophobic residues and surround them with the polar **e** and **g** residues, in the same way that the wild-type structure was able to. For a coiled coil, the maintenance of the hydrophobic core, termed the "3-4 repeat" (**a**-to-**d** and **d**-to-**a** is a spacing of three and four residues, respectively), is essential. Finally, it is possible to polymerize specific synthetic trinucleotides such that the final gene sequence codes for those, and only those, amino acid sequences that are desired in the final peptide *(79)*. This will also facilitate in keeping the library size low, allowing more desired changes to be screened. This is not true of degenerate codon usage using regular degenerate oligonucleotides, which, depending on the codon usage table, may imply that to have two desired amino acids, another undesirable amino acid must be included at that position (*see* **Subheading 2.5.2.** and **Fig. 2**). It is also possible that two amino acids from opposite ends of the codon table are desired, but realistically can only be included by using such presynthesized trinucleotide codons. Such trinucleotide building blocks have recently become commercially available (Glen Research, Sterling, VA; Metkinen Oy, Finland).

2.5.2. Codon Bias

Codon bias is the probability that one codon rather than another will code for a particular amino acid (**Fig. 2A**). For example, in the case of *Escherichia coli,* CGT is found to code for Arg approximately five times more frequently than CGA. Obviously, when transforming cells (e.g., *E. coli*) with plasmids containing a gene coding for library peptides, one must select triplets that are frequently observed in the host organism over infrequently used codons, which may lead to frame shifts *(80–82)* or poorly represented sequences in the library. In addition, when designing a library that is to be screened, one may, for example, wish to introduce all combinations of β-branched amino acids into the **a** position, but depending on the codon usage, this may or may not be possible. **Fig. 2** shows some of our most favored codon combinations for different positions in coiled coils. Most of the combinations include only a few options to keep the libraries to a reasonable size. By examining **Fig. 2** (the codon usage table for *E. coli*; *see* www.kazusa.or.jp/codon for this and other organisms), we see that to include Val, Ile, and Thr at an **a** position of the coiled coil would require the codons GTN, ATH (if Met should be avoided), or ACN. This would mean requiring G or A at position one of the codon, T or C at position two, and any nucleotide other than G at the third position, because ATG would code for Met, resulting in the degenerate codon RYH (all resulting codons in this case are used significantly by *E. coli*). For such a Val, Ile, and Thr combination, however, it is not possible to rule out amino acids that are not required in the library,

	T			C			A			G		
T	UUU	F	0.59	UCU	S	0.17	UAU	Y	0.60	UGU	C	0.47
	UUC	F	0.41	UCC	S	0.14	UAC	Y	0.40	UGC	C	0.53
	UUA	L	0.15	UCA	S	0.15	UAA	*	0.60	UGA	*	0.31
	UUG	L	0.13	UCG	S	0.13	UAG	*	0.09	UGG	W	1.00
C	CUU	L	0.12	CCU	P	0.19	CAU	H	0.59	CGU	R	0.34
	CUC	L	0.10	CCC	P	0.14	CAC	H	0.41	CGC	R	0.34
	CUA	L	0.04	CCA	P	0.21	CAA	Q	0.34	CGA	R	0.07
	CUG	L	0.46	CCG	P	0.47	CAG	Q	0.66	CGG	R	0.12
A	AUU	I	0.49	ACU	T	0.19	AAU	N	0.52	AGU	S	0.17
	AUC	I	0.37	ACC	T	0.38	AAC	N	0.48	AGC	S	0.23
	AUA	I	0.14	ACA	T	0.19	AAA	K	0.73	AGA	R	0.08
	AUG	M	1.00	ACG	T	0.24	AAG	K	0.27	AGG	R	0.05
G	GUU	V	0.29	GCU	A	0.19	GAU	D	0.64	GGU	G	0.34
	GUC	V	0.20	GCC	A	0.26	GAC	D	0.36	GGC	G	0.35
	GUA	V	0.17	GCA	A	0.24	GAA	E	0.67	GGA	G	0.15
	GUG	V	0.34	GCG	A	0.31	GAG	E	0.33	GGG	G	0.16

A/G = R C/G = S C/T = Y A/T = W A/C = M G/T = K

A/C/T = H C/G/T = B A/C/G = V A/G/T = D A/C/G/T = N

Fig. 2. Codons and possible combinations for the design of coiled coil libraries. Codons and their bias as represented in the *E. coli.* Shown is the standard textbook triplet code, followed by the amino acid represented and the fraction of this codon for this organism (http://www.kazusa.or.jp/codon; **ref.** ***83***). Also shown are the letters that represent a mixture of oligonucleotides, and will subsequently lead to the degenerate base positions required in the generation of the library. The frames give examples of possible combinations for coiled-coil library design.

because a mixture of above mentioned oligonucleotides would also include GCH, which codes for Ala. Another example is that if one wanted to include the amino acids Gln, Asn, and Lys, one must also include His, by default. This is a problem that can be overcome by polymerizing synthetic trinucleotides that code only for those amino acids desired ***(31,79)***. In addition, one must be careful to rule out codons that are poorly represented, e.g., when representing Arg, because amino-acyl tRNA synthetases for the AGG and AGA codons are in short supply for *E. coli*. Finally, it should be noted that if a position is to be represented without bias in the library, then a 1:1 ratio between all amino acids at that position (and likewise for the respective codons) should be selected. Representing an amino acid more than once over another at a degenerate position will naturally put a bias toward that change into the system, because it becomes overrepresented.

codon	NNK	NTN	NTK	DTK	VTY	RSY	VAN	VAR	GAN	MRG	
Gly	2	–	–	–	–	2	–	–	–	–	aliphatic side chains
Ala	2	–	–	–	–	2	–	–	–	–	
Val	2	4	2	2	2	–	–	–	–	–	
Leu	3	6	3	1	2	–	–	–	–	–	
Ile	1	3	1	1	2	–	–	–	–	–	
Met	1	1	1	1	–	–	–	–	–	–	
Pro	2	–	–	–	–	–	–	–	–	–	
Phe	1	2	1	1	–	–	–	–	–	–	
Trp	1	–	–	–	–	–	–	–	–	–	
Ser	3	–	–	–	–	2	–	–	–	–	polar side chains
Thr	2	–	–	–	–	2	–	–	–	–	
Asn	1	–	–	–	–	–	2	–	–	–	
Gln	1	–	–	–	–	–	2	2	–	1	
Tyr	1	–	–	–	–	–	–	–	–	–	
Cys	1	–	–	–	–	–	–	–	–	–	
Lys	1	–	–	–	–	–	2	2	–	1	charged side chains
Arg	3	–	–	–	–	–	–	–	–	2	
His	1	–	–	–	–	–	2	–	–	–	
Asp	1	–	–	–	–	–	2	–	2	–	
Glu	1	–	–	–	–	–	2	2	2	–	
stop (TAG)	1	–	–	–	–	–	–	–	–	–	
total	32	16	8	6	3	4	6	3	2	4	

Fig. 3. Listed are amino acid frequencies corresponding to the codon randomization scheme shown in **Fig. 2**. Additionally, the distribution for NNK is given. This combination is widely used when all 20 amino acids should be included and stop codons should be minimized. "Total" gives the number of amino acids to be used when calculating the library size. Hydrophobic amino acid combinations encoded by NTN (black line in **Fig. 2**) or VTY (gray long dashes), respectively. To minimize overrepresentation of Ile and Leu, NTN can be reduced to NTK or DTK. Different polar and charged combinations are encoded by VAN (gray line), VAR (black long dashes), GAN (black dots), or MRG (gray short dashes), respectively. Possible loop regions, if desired, can be obtained by the codon RSY (black short dashes). Please note that, in our examples, the third base is also mixed to increase variety in codon usage, which might be important for repetitive sequences. However, if desired, the last base can be kept constant in most instances. This is especially important in the case of the Gln, Arg, and Lys mixture (short gray dashes), in which the codon MRG yields only one rare Arg codon (AGG), whereas the combination MRR yields three rare Arg codons (CGG, AGA, AGG). Combinations NTN and VAN have been applied previously in the binary design pattern developed in the Hecht group (*see* **refs.** ***84*** and ***85*** and Chapter 9).

2.5.3. Selection Systems

The most common selection system open to the coiled coil is the protein-fragment complementation assay selection. In this assay, interacting proteins, e.g., two coiled coil fragments, are tethered to two halves of a reporter protein that only becomes active after association of the two fused proteins or peptides. This has been used for dihydrofolate reductase ***(8,31,70,86–88)***, ubiquitin ***(89)***, β-galactosidase ***(90)***, β-lactamase ***(91,92)***, and green fluorescent protein ***(93)***. Such intracellular assays have the additional benefit of concomitantly selecting against protease susceptible or toxic peptides.

Another selection system for coiled coil interactions is the yeast two-hybrid system. In this system, one helix is fused to the binding domain (which binds to a promoter) and the other to the activating domain (which interacts with the polymerase) of the Gal4p transcription factor. Only interaction between the two helices will bring the two chimeric proteins into close proximity and permit the transcriptional activator to function, thus, switching on a reporter gene (e.g., β-galactosidase) by which the experiment can then be assayed for activity. Furthermore, novel interacting partners can be found by screening a single protein or domain against a library of other proteins using this system (for a recent review, *see* **ref. *94***).

The λ repressor system is based on reconstituting the activity of the *E. coli* λ repressor by replacing the C-terminal domain with a heterologous oligomerization domain. The interaction is detected when the C-terminal domain forms a dimer (or higher-order oligomer) with itself (homotypic interaction) or with a different domain from another fusion (heterotypic interaction). Functional repressors can be detected either by monitoring immunity to phage infection or by assaying repression of reporters, e.g., β-galactosidase ***(95,96)***.

Finally, using phage display, proteins can be displayed on the surface coat proteins of filamentous bacteriophage. Peptide libraries displayed on the phage can be selected for binding and enriched by several rounds of selection. This technique also permits selection for binding to nonnatural compounds ***(97)***.

2.5.4. Calculations Concerning Designs

Instead of using in vivo selection systems, some groups also used an *in silico* approach to generate promising sequences for coiled coil design. Some aspects will be discussed in this section.

1. Keating et al. developed a computational method to predict hydrophobic core mutation effects on interaction specificity ***(98)***. This is achieved using an algorithm that considers van der Waals packing interaction energies, a solvation term, and α-helical propensities. From this, partnering preferences arising from core packing was predicted. Coiled coils were designed with a core **a** and **d** residue mutated to Leu, Ile, and Val to yield six different heterodimers with a range of stabilities (*see* **Note 32**). The algorithm was able to predict stability with good agreement to the experimental data.
2. Mayo's group used a design algorithm to assess the surface position interactions. They tested three scoring functions: a hydrogen-bond potential, a hydrogen-bond potential in conjunction with a penalty for uncompensated burial of polar hydrogens, and a hydrogen-bond potential in combination with helix propensity ***(69)***.

The algorithm was used to find the optimal amino acid sequence for each of the three scoring functions, using GCN4-p1 (*see* **Note 2**), and the corresponding peptides were consequently synthesized. All resultant peptides were dimeric, close to 100% helical at 1°C and had melting temperatures of 69°C to 72°C compared with 15°C for a GCN4 peptide with random hydrophilic surface residues. The data suggest that helix propensity is the key factor in sequence design for the surface residues of the coiled coil peptides.

3. For the generation of coiled coils with a right-handed superhelical twist, an algorithm was developed that incorporated main-chain flexibility *(99)*. For main-chain fixing to be used, a naturally existing example is needed to supply the coordinates, and even then, these values may not be valid for close homology-adjusted sequences. By allowing backbone flexibility, a small subset of main-chain conformations could be sampled. These samplings were then coupled with side-chain packing sampling. Consequently, right-handed dimeric, trimeric, and tetrameric-coiled coils were computationally designed (*see* **Note 33**). The overall protein fold was specified by hydrophobic polar residue patterning. The oligomerization state, main-chain conformation, and side-chain rotamers were computationally selected by best packing in alternate backbone structures. The resulting designed peptides formed the correct oligomeric state ensembles in accord with the design goals, and the X-ray structure of the tetramer matched the design structure in atomic detail.
4. A paper by Harbury used a combination of positive (toward the desired structure) and negative design (away from undesired alternate structures) to optimize interaction specificity (*see* **Subheading 2.2.2., item 3**; **ref.** ***39***). The **a**, **d**, **e**, and **g** positions of the central heptad of GCN4 were varied with all nonproline residues to generate approx 8×10^9 possible sequences (*see* **Note 15**). The algorithm used was akin to a computational double mutant cycle in which peptides are sequence optimized rather than structurally optimized. Competing energetic states considered were:

   a. The folded homodimeric desired structure.
   b. Energetically favored heterodimers.
   c. Unfolded energetically destabilized states.
   d. Aggregated states of poor solubility.

   The free energies were evaluated in each of these four competing states for candidate sequences. This so-called "multistate" design has an advantage over single state design systems, which only look for the lowest free energy states of the target. Unlike these designs, the multistate design structures often deviate from the PV hypothesis, and, at the same time, take into consideration factors such as too much hydrophobic exposure causing aggregation and too much polar burial being destabilizing. Rather, all of these factors are considered in the design, which only requires change when these competing forces dominate over the selection of the target state. Selection is, therefore, the result of a balance between stability and specificity, and not of target stability alone.

3. Notes

1. Sequence of GCN4-p1: Ac-R **M**KQ**L**EDK **V**EE**L**LSK **N**YH**L**ENE **V**AR**L**KKL **V**GER-COOH. In boldface are the **a** and **d** residues that have been mutated in the studies by Harbury et al. ***(16,17)***. Underlined is the core Asn16, which has been subjected to many mutations.
2. The designed heterodimeric Peptide Velcro ***(9)*** consists of the synthetic peptides Acid-p1 (Ac-AQLEKE LQALEKE NAQLEWE LQALEKE LAQ-NH_2) and Base-p1 (Ac-AQLKKK LQALKKK NAQLKWK LQALKKK LAQ-NH_2). In this and the following notes, the individual heptads of the respective sequence are separated by a space. The core Asn residue, which has been mutated as described, e.g., in **Subheading 2.1.2.**, **item 4**, is underlined.
3. The sequence of the two antiparallel cysteine disulfide-bridged peptides 2H (Ac-K **C**EALEGK LEALEGK LEAAEGK LEALEGK LEALEG-NH_2 and Ac-E LAELKGE LAELKGE AAELKGE LAELKGE LAE**C**KG-NH_2) and 4H (Ac-K **C**EALEGK LEALEGK LEAAEGK LEALEGK LEALEG-NH_2 and Ac-E LAELKGE LAELKGE LAEAKGE LAELKGE LAE**C**KG-NH_2). Cysteines are shown in bold to indicate the point of covalent bonding between the two chains in 2H and 4H, respectively, and the Ala residue that specifies the oligomerization state is underlined ***(20)***.
4. Sequence of the coiled coil domain (amino acids 27–72) of the rat COMP protein ***(21)***: GDL APQMLRE LQETNAA LQDVREL LRQQVKE ITFLKNT VMEC-DAC G. In the expressed fragment of rat COMP, Gly 27 was replaced by Met.
5. Studies were based on the peptide A1 (MRGSHHHHHHGSMA SGD**L**ENE YAQ**L**ERE VRS**L**EDE AAE**L**EQK VSR**L**KNE IED**L**AEI GD**L**NNTSGIRRPAA KLN). Incorporation of trifluoroleucine and hexafluoroleucine for Leucine (in boldface) was achieved by induction of gene expression in leucine-free culture media supplemented either with trifluoroleucine or with hexafluoroleucine ***(22,23)***.
6. Kretsinger et al. based their study on the GCN4-p1 peptide as given in **Note 2** with the differences that the C-terminus was amidated, and that they reported a sequence with an additional Ser between the Asp and Lys in the first full heptad ***(24)***. Because this addition would shift the heptad repeat, we presume that it is an error in the figure. The underlined Asn (*see* **Note 2**) was subjected to exchange to Asp dap as well as monomethylated, dimethylated, and trimethylated analogs of dap.
7. In this selection, heterodimers were selected from two designed coiled coil libraries: LibA: VAQL#E# VKTL#A# §YEL#S# VQRL#E# VAQL and LibB: VDEL#A# VDQL#D# §YAL#T# VAQL#K# VEKL, where # denotes an equimolar mixture of E, Q, K, and R, and § denotes an equimolar mixture of V and N ***(31)***. The sequences for the core **a** and **d** positions were taken from GCN4 (*see* **Note 2**) and for the **b**, **c**, and **f** positions (underlined) were taken from the coiled coil domains of c-Jun (IARLEEK VKTLKAQ NYELAST ANMLREQ VAQL) and c-Fos (TDTLQAE TDQLEDE KYALQTE IANLLKE KEKL), respectively.
8. The sequence of the GCN4-pVL variant is Ac-R **MKQLEDK VEE#LSK** **§**YH**L**ENE **V**AR**L**KKL **V**GER, where the **a** and **d** positions are in boldface.

Position 12(**d**), indicated by #, is either a Leu or a polar residue (N, Q, S, or T), and position 16(**a**), indicated by §, is either a Val or a polar residue *(25)*.

9. Both studies used a disulfide-bridged coiled coil based on the model sequence VGALKKE, with some modification to avoid intrachain and interchain charge–charge interactions with the site of substitution (**X**) and to adjust the overall charge. The sequences were Ac-CGGE VGALKAQ VGALQAQ **X**GALQKE VGALKKE VGALKK-NH_2 ***(33)*** and Ac-CGGE VGALKAE VGALKAQ IGA**X**-QKQ IGALQKE VGALKK-NH_2 ***(32)***, respectively.
10. Ji et al. used a recombinant model of the simian immunodeficiency virus gp41 core, designated N36(L6)C34, where the amino-terminal helices (N36) form a central, trimeric coiled coil, while the carboxyl-terminal helices (C34) pack in an antiparallel orientation into hydrophobic grooves on the surface of this coiled coil trimer. N36 and C34 are separated by a short linker (L6; **ref. *34***). Mutations of polar core residues (in boldface) to Ile were made in the N36 domain: AGIVQQ **Q**QQLLDV VKR**Q**QEL LRL**T**VWG **T**KNLQTR VT. The Q → I mutation that formed insoluble aggregates is underlined.
11. The sequence of the parent peptide, Lac21 is: Ac-MKQLADS LMQLARQ VSR-LESA-NH_2 (*see also* **Note 28**). Underlined residues were mutated to E or K to result in the peptides Lac21E and Lac21K, respectively, which formed a heterotetramer ***(35)***.
12. Sequence of APC-55: AAAS YDQLLKQ VEALKME NSNLRQE LEDNSNH LTKLETE ASNMKEV LKQLQGS I and of anti-APCp1: **M**AAK **G**DQLKKE VEALE*Y*E NSNLR*K*K LED**H**K*K*K LTKLKTE **I**SNA**K***KM* LKQL*YAS* I ***(36)***. Core changes in anti-APCp1 compared with APC-55 are marked in boldface; changes at the **e** and **g** positions are underlined; and changes to increase stability, to increase the net charge to facilitate purification, and to add chromophores are marked in italics.
13. The three peptides were T_9: Ac-R MKQLEKK **X**EELLSK **A**QQLEKE **A**AQLKKL VG-NH_2, T_{16}: Ac-R MKQLEKK **A**EELLSK **X**QQLEKE **A**AQLKKL VG-NH_2, T_{23}: Ac-R MKQLEKK **A**EELLSK **A**QQLEKE **X**AQLKKL VG-NH_2 ***(37)***. Residues that were different in all three peptides are marked in boldface, **X** denotes the cyclohexylalanine residue.
14. Sequences were based on Acid-pLL and Base-pLL, which are identical to Acid-p1 and Base-p1 (*see* **Note 1**) but with the core Asn mutated to Leu (*see also* **Subheading 2.1.2., item 4**). Two L → K mutations (in boldface) were made to Base-pLL to yield Base-pK: Ac-AQLKKK LQALKKK **K**AQLKWK **K**QALKKK LAQ-NH_2 ***(38)***.
15. Studies were based on an N-terminally capped variant of GCN4 (*see* **Note 31**; **ref.** *7*) with the Asn16 shifted by one heptad level to position 9, yielding the peptide p-CAP: **S** VKELEDK **N**EELLS**X** **X**YH**XX**NE VARLKKL VGER. Changes compared with GCN4-p1 (*see* **Note 2**) are marked in boldface. **X** denotes positions allowed to vary in the design calculations ***(39)***.
16. Studies were based on the designed homodimeric coiled coil EK: Ac-K CGALEKK LGALEKK AGALEKK LGALEKK LGALEK-NH_2. Three mutant coiled coils were made in which:

    a. Five Glu residues at **e** positions in EK (underlined) were mutated to Gln residues (peptide QK).

 b. Five Lys residues at **g** positions (underlined) were mutated to Gln residues (peptide EQ).
 c. Both were combined (peptide QQ).

 Using a double-mutant cycle analysis, the energetic contribution of interhelical ionic attractions to coiled coil stability was calculated ***(41)***.

17. Sequence of parental vitellogenin-binding protein ER_{34}: ITIR AAFLEKE NTALRTE VAELRKE VGRCRNI VSKYETR YGPL. Underlined **e** and **g** residues are altered in subsequent peptides ***(45)***.
18. The best winner WinZip-A2B1 from the selection (*see* **Note 7**) consists of peptides WinZip-A2 (Ac-STT VAQLRER VKTLRAQ NYELESE VQRLREQ VAQL AS-NH_2) and WinZip-B1 (Ac-STS VDELQAE VDQLQDE NYALKTK VAQLRKK VEKL SE-NH_2) ***(8)***.
19. Studies were based on designed E- and K-peptides. E-Peptide: Ac-E LGALEKE LGALEKE **L**GA**L**EKE LGALEKE LGALEK-NH_2; K-Peptide: Ac-K LGALKEK LGALKEK **L**GA**L**KEK LGALKEK LGALKE-NH_2. Positions 16 or 19 (bold) were changed to Ala to create different Leu–Ala core combinations, and positions 2 or 33 (underlined) were exchanged to Cys to allow disulfide bridging in the parallel or antiparallel state ***(52)***.
20. Sequence of APH: MKQLEKE LKQLEKE LQA**I**EKQ LAQLQWK **A**QA*R*KKK LAQLKKK LQA ***(55)***. Sterically matched core residues (Ile and Ala) are shown in bold; designed Coulombic interactions between N-terminal glutamines and C-terminal lysines are underlined. In addition, a single Arg residue was incorporated at a **d** position (in italics) to promote dimer formation.
21. The resulting peptides were termed Acid-a1 (Ac-AQLEKE LQALEKE **L**AQLEWE **N**QALEKE LAQ-NH_2) and Base-a1 (Ac-AQLKKK LQA**N**KKK **L**AQLKWK LQALKKK LAQ-NH_2) ***(56)***. Changes compared with Acid-p1 and Base-p1 (*see* **Note 1**) are in boldface.
22. Peptides Acid-**R**d**L** (Ac-AQLEKE LQALEKE LAQ**R**EWE LQALEKE LAQ-NH_2) and Base-**E**g**L** (Ac-AQLKKK LQALKK**E** LAQLKWK LQALKKK LAQ-NH_2) were based on Acid-a1 and Base-a1 (*see* **Note 21**), with the designed polar interaction in boldface ***(59)***.
23. Five different peptides were synthesized with an N- or C-terminal Cys and a core Ala residue. Three peptides (C2A16, C33A16, and C33A19) were based on the heptad repeat, LEALEGK: Ac-K LEALEGK LEALEGK **L**EA**L**EGK LEALEGK LEALEG-NH_2, with the Cys either at position 2 or 33 (underlined) and the Ala either at position 16 or 19 (bold). Two peptides (C33A16 and C33A19) were based on the heptad repeat, LAELKGE: Ac-E LAELKGE LAELKGE **L**AE**L**KGE LAELKGE LAECKG-NH_2, with the Cys at position 33 and the Ala either at position 16 or 19 ***(49)***.
24. Peptide Acid-**K**g (Ac-AQLEKE LQALEK**K** LAQLEWE NQALEKE LAQ-NH_2) was based on Acid-a1 (*see* **Note 21**), and peptide Base-**E**g (Ac-AQLKKK LQANKK**E** LAQLKWK LQALKKK LAQ-NH_2) was based on Base-a1 with the changes marked in boldface.

25. Synthetic peptides with the following sequence were constructed: Ac-(K LEA-LEG)$_n$-K-NH_2, with n = 1 to 5 ***(62)*** and compared with carboxamidomethylated α-tropomyosin at cysteine 190 (CM-tropomyosin).
26. A series of polypeptides containing 9, 12, 16, 19, 23, 26, 30, 33, and 35 amino acid residues were designed with the sequence: Ac-E iealkae iealkae iealkae iealkae ieacka-NH_2 for the 35-mer peptide ***(63)***. Shorter peptides were comprised of the respective number of amino acids counted from the C-terminus.
27. The peptide Succ-DELERR IRELEAR IK-NH_2 was used in this study ***(64)***, Succ indicated the succinylated N-terminus.
28. Investigated peptides were Lac 21: Ac-MKQLADS LMQLARQ VSRLESA-NH_2, Lac 28: Ac-LMQLARQ MKQLADS LMQLARQ VSRLESA-NH_2, and Lac 35: Ac-LMQLARQ LMQLARQ MKQLADS LMQLARQ VSRLESA-NH_2 ***(65)***.
29. Studies were based on the mutants of the E/K heterodimer comprised of the E-peptide with the sequence Ac-(E #§ALEK)$_n$-NH_2 and the K-peptide with the sequence Ac-(K #§ALKE)$_n$-NH_2 with n = 3 or 4. # indicates I or V; § indicates A or S ***(67)***.
30. The peptide sequence was Ac-Q CGALQKQ VGALQKQ VGALQKQ VGALQKQ VGALQK-NH_2. Positions 1, 6, 15, 20, and 34 that were mutated to Gln are underlined ***(73)***.
31. This study worked with recombinant GCN4-pMSE peptide (**MS** VKELEDK VEELLSK NYHLENE VARLKKL VGER). The capping motif is in boldface and mutations compared with GCN4-p1 (*see* **Note 2**) are underlined. Other peptides from this study were GCN4-pSE, which lacked the initiator methionine, and GCN4-pAA, in which the Ser and Glu of GCN4-pSE were mutated to Ala ***(77)***. Stabilities of GCN4-pAA were comparable to GCN4-p1, whereas the GCN4-pSE and GCN4-pMSE variants were stabilized by 0.5 kcal/mol and 1.2 kcal/mol, respectively, relative to GCN4-pAA. Thus, the hydrophobic contribution of the terminal Met residue is 0.7 kcal/mol.
32. Heterodimeric coiled coils, denoted GABH, for GCN4 (*see* **Note 2**) Acid/Base Heterodimer were designed with the acidic sequence A (Ac-E VKQLEAE VEE#ESE #WHLENE VARLEKE NAECEA-NH_2) and the basic sequence B (Ac-K VKQLKAK VEE#KSK #WHLKNK VARLKKK NAECKA-NH_2) ***(98)***. Positions **d**12 and **a**16 (#) were mutated to Val, Ile, and Leu, respectively, to yield the peptides A_{LL}, A_{IV}, and A_{LI}, and B_{LL} and B_{LV}.
33. Designed sequences were dimeric RH2 (Ac-AE **I**EQ**L**KKE§AYL **I**KK**L**KAE**K**LAE **I**KK**L**KQE**K**A-NH_2), trimeric RH3 (Ac-AE #EQ#KKE**I**AYL #KK#KAE**I**LAE#K K#KQE**I**A-NH_2), and tetrameric RH4 (Ac-AE **L**EQ#KKE**I**AYL **L**KK#KAE**I**L AE **L**KK#KQE**I**A-NH_2) ***(99)***. The hydrophobic residues (**a**, **d**, and **h**) of the undecatad repeat (**a** to **k**) are marked in boldface; § indicates norvaline; and # indicates alloisoleucine residues.

Acknowledgment

This work was funded in the Emmy Noether program of the Deutsche Forschungsgemeinschaft (grant Ar 373).

References

1. Wolf, E., Kim, P. S., and Berger, B. (1997) MultiCoil: a program for predicting two- and three-stranded coiled coils. *Protein Sci.* **6**, 1179–1189.
2. Lupas, A. (1996) Coiled coils: new structures and new functions. *Trends Biochem. Sci.* **21**, 375–382.
3. Landschulz, W. H., Johnson, P. F., and McKnight, S. L. (1988) The leucine zipper: a hypothetical structure common to a new class of DNA binding proteins. *Science* **240**, 1759–1764.
4. Burkhard, P., Stetefeld, J., and Strelkov, S. V. (2001) Coiled coils: a highly versatile protein folding motif. *Trends Cell Biol.* **11**, 82–88.
5. Kohn, W. D., Mant, C. T., and Hodges, R. S. (1997) Alpha-helical protein assembly motifs. *J. Biol. Chem.* **272**, 2583–2586.
6. O'Shea, E. K., Klemm, J. D., Kim, P. S., and Alber, T. (1991) X-ray structure of the GCN4 leucine zipper, a two-stranded, parallel coiled coil. *Science* **254**, 539–544.
7. DeLano, W. L. (2002) The PyMOL molecular graphics system. DeLano Scientific, San Carlos, CA; http://www.pymol.org.
8. Arndt, K. M., Pelletier, J. N., Müller, K. M., Plückthun, A., and Alber, T. (2002) Comparison of in vivo selection and rational design of heterodimeric coiled coils. *Structure* **10**, 1235–1248.
9. O'Shea, E. K., Lumb, K. J., and Kim, P. S. (1993) Peptide 'Velcro': design of a heterodimeric coiled coil. *Curr. Biol.* **3**, 658–667.
10. Mason, J. M. and Arndt, K. M. (2004) Coiled coil domains: stability, specificity, and biological implications. *Chem. Biochem.* **5**, 170–176.
11. Müller, K. M., Arndt, K. M., and Alber, T. (2000) Protein fusions to coiled-coil domains. *Methods Enzymol.* **328**, 261–282.
12. Arndt, K. M., Müller, K. M., and Plückthun, A. (2001) Helix-stabilized Fv (hsFv) antibody fragments: substituting the constant domains of a Fab fragment for a heterodimeric coiled-coil domain. *J. Mol. Biol.* **312**, 221–228.
13. Pack, P., Müller, K. M., Zahn, R., and Plückthun, A. (1995) Tetravalent miniantibodies with high avidity assembling in Escherichia coli. *J. Mol. Biol.* **246**, 28–34.
14. Naik, R. R., Kirkpatrick, S. M., and Stone, M. O. (2001) The thermostability of an alpha-helical coiled-coil protein and its potential use in sensor applications. *Biosens. Bioelectron.* **16**, 1051–1057.
15. Crick, F. H. S. (1953) The packing of α-helices: simple Coiled Coils. *Acta Crystallogr.* **6**, 689–697.
16. Harbury, P. B., Zhang, T., Kim, P. S., and Alber, T. (1993) A switch between two-, three-, and four-stranded coiled coils in GCN4 leucine zipper mutants. *Science* **262**, 1401–1407.
17. Harbury, P. B., Kim, P. S., and Alber, T. (1994) Crystal structure of an isoleucine-zipper trimer. *Nature* **371**, 80–83.
18. Betz, S. F., Bryson, J. W., and DeGrado, W. F. (1995) Native-like and structurally characterized designed alpha-helical bundles. *Curr. Opin. Struct. Biol.* **5**, 457–463.
19. Woolfson, D. N. and Alber, T. (1995) Predicting oligomerization states of coiled coils. *Protein Sci.* **4**, 1596–1607.

20. Monera, O. D., Sonnichsen, F. D., Hicks, L., Kay, C. M., and Hodges, R. S. (1996) The relative positions of alanine residues in the hydrophobic core control the formation of two-stranded or four-stranded alpha-helical coiled-coils. *Protein Eng.* **9**, 353–363.
21. Malashkevich, V. N., Kammerer, R. A., Efimov, V. P., Schulthess, T., and Engel, J. (1996) The crystal structure of a five-stranded coiled coil in COMP: a prototype ion channel? *Science* **274**, 761–765.
22. Tang, Y. and Tirrell, D. A. (2001) Biosynthesis of a highly stable coiled-coil protein containing hexafluoroleucine in an engineered bacterial host. *J. Am. Chem. Soc.* **123**, 11,089–11,090.
23. Tang, Y., Ghirlanda, G., Petka, W. A., Nakajima, T., DeGrado, W. F., and Tirrell, D. A. (2001) Fluorinated coiled-coil proteins prepared in vivo display enhanced thermal and chemical stability. *Angew. Chem. Int. Ed.* **40**, 1494–1496.
24. Kretsinger, J. K. and Schneider, J. P. (2003) Design and application of basic amino acids displaying enhanced hydrophobicity. *J. Am. Chem. Soc.* **125**, 7907–7913.
25. Akey, D. L., Malashkevich, V. N., and Kim, P. S. (2001) Buried polar residues in coiled-coil interfaces. *Biochemistry* **40**, 6352–6360.
26. Glover, J. N. and Harrison, S. C. (1995) Crystal structure of the heterodimeric bZIP transcription factor c-Fos-c-Jun bound to DNA. *Nature* **373**, 257–261.
27. Gonzalez, L., Jr., Woolfson, D. N., and Alber, T. (1996) Buried polar residues and structural specificity in the GCN4 leucine zipper. *Nat. Struct. Biol.* **3**, 1011–1018.
28. Junius, F. K., Mackay, J. P., Bubb, W. A., Jensen, S. A., Weiss, A. S., and King, G. F. (1995) Nuclear magnetic resonance characterization of the Jun leucine zipper domain: unusual properties of coiled-coil interfacial polar residues. *Biochemistry* **34**, 6164–6174.
29. Potekhin, S. A., Medvedkin, V. N., Kashparov, I. A., and Venyaminov, S. (1994) Synthesis and properties of the peptide corresponding to the mutant form of the leucine zipper of the transcriptional activator GCN4 from yeast. *Protein Eng.* **7**, 1097–1101.
30. Lumb, K. J. and Kim, P. S. (1995) A buried polar interaction imparts structural uniqueness in a designed heterodimeric coiled coil. *Biochemistry* **34**, 8642–8648.
31. Arndt, K. M., Pelletier, J. N., Müller, K. M., Alber, T., Michnick, S. W., and Plückthun, A. (2000) A heterodimeric coiled-coil peptide pair selected in vivo from a designed library-versus-library ensemble. *J. Mol. Biol.* **295**, 627–639.
32. Tripet, B., Wagschal, K., Lavigne, P., Mant, C. T., and Hodges, R. S. (2000) Effects of side-chain characteristics on stability and oligomerization state of a de novo-designed model coiled-coil: 20 amino acid substitutions in position "d". *J. Mol. Biol.* **300**, 377–402.
33. Wagschal, K., Tripet, B., Lavigne, P., Mant, C., and Hodges, R. S. (1999) The role of position a in determining the stability and oligomerization state of alpha-helical coiled coils: 20 amino acid stability coefficients in the hydrophobic core of proteins. *Protein Sci.* **8**, 2312–2329.

34. Ji, H., Bracken, C., and Lu, M. (2000) Buried polar interactions and conformational stability in the simian immunodeficiency virus (SIV) gp41 core. *Biochemistry* **39**, 676–685.
35. Fairman, R., Chao, H. G., Lavoie, T. B., Villafranca, J. J., Matsueda, G. R., and Novotny, J. (1996) Design of heterotetrameric coiled coils: evidence for increased stabilization by Glu(–)-Lys(+) ion pair interactions. *Biochemistry* **35**, 2824–2829.
36. Sharma, V. A., Logan, J., King, D. S., White, R., and Alber, T. (1998) Sequence-based design of a peptide probe for the APC tumor suppressor protein. *Curr. Biol.* **8**, 823–830.
37. Schnarr, N. A. and Kennan, A. J. (2002) Peptide tic-tac-toe: heterotrimeric coiled-coil specificity from steric matching of multiple hydrophobic side chains. *J. Am. Chem. Soc.* **124**, 9779–9783.
38. Campbell, K. M. and Lumb, K. J. (2002) Complementation of buried lysine and surface polar residues in a designed heterodimeric coiled coil. *Biochemistry* **41**, 7169–7175.
39. Havranek, J. J. and Harbury, P. B. (2003) Automated design of specificity in molecular recognition. *Nat. Struct. Biol.* **10**, 45–52.
40. Kohn, W. D., Kay, C. M., and Hodges, R. S. (1995) Protein destabilization by electrostatic repulsions in the two-stranded alpha-helical coiled-coil/leucine zipper. *Protein Sci.* **4**, 237–250.
41. Zhou, N. E., Kay, C. M., and Hodges, R. S. (1994) The net energetic contribution of interhelical electrostatic attractions to coiled-coil stability. *Protein Eng.* **7**, 1365–1372.
42. Graddis, T. J., Myszka, D. G., and Chaiken, I. M. (1993) Controlled formation of model homo- and heterodimer coiled coil polypeptides. *Biochemistry* **32**, 12,664–12,671.
43. Moll, J. R., Olive, M., and Vinson, C. (2000) Attractive interhelical electrostatic interactions in the proline- and acidic-rich region (PAR) leucine zipper subfamily preclude heterodimerization with other basic leucine zipper subfamilies. *J. Biol. Chem.* **275**, 34,826–34,832.
44. Krylov, D., Mikhailenko, I., and Vinson, C. (1994) A thermodynamic scale for leucine zipper stability and dimerization specificity: e and g interhelical interactions. *EMBO J.* **13**, 2849–2861.
45. Krylov, D., Barchi, J., and Vinson, C. (1998) Inter-helical interactions in the leucine zipper coiled coil dimer: pH and salt dependence of coupling energy between charged amino acids. *J. Mol. Biol.* **279**, 959–972.
46. Newman, J. R. and Keating, A. E. (2003) Comprehensive identification of human bZIP interactions with coiled-coil arrays. *Science* **300**, 2097–2101.
47. Oakley, M. G. and Hollenbeck, J. J. (2001) The design of antiparallel coiled coils. *Curr. Opin. Struct. Biol.* **11**, 450–457.
48. Monera, O. D., Zhou, N. E., Kay, C. M., and Hodges, R. S. (1993) Comparison of antiparallel and parallel two-stranded alpha-helical coiled-coils. Design, synthesis, and characterization. *J. Biol. Chem.* **268**, 19,218–19,227.

49. Monera, O. D., Kay, C. M., and Hodges, R. S. (1994) Electrostatic interactions control the parallel and antiparallel orientation of alpha-helical chains in two-stranded alpha-helical coiled-coils. *Biochemistry* **33**, 3862–3871.
50. Oakley, M. G. and Kim, P. S. (1997) Protein dissection of the antiparallel coiled coil from Escherichia coli seryl tRNA synthetase. *Biochemistry* **36**, 2544–2549.
51. Kohn, W. D. and Hodges, R. S. (1998) De novo design of α-helical coiled coils and bundles: models for the development of protein-design principles. *Trends Biotechnol.* **16**, 379–389.
52. Monera, O. D., Zhou, N. E., Lavigne, P., Kay, C. M., and Hodges, R. S. (1996) Formation of parallel and antiparallel coiled-coils controlled by the relative positions of alanine residues in the hydrophobic core. *J. Biol. Chem.* **271**, 3995–4001.
53. Holton, J. and Alber, T. (2004) Automated protein crystal structure determination using ELVES. *Proc. Natl. Acad. Sci. USA* **101**, 1537–1542.
54. Gonzalez, L., Jr., Plecs, J. J., and Alber, T. (1996) An engineered allosteric switch in leucine-zipper oligomerization. *Nat. Struct. Biol.* **3**, 510–515.
55. Gurnon, D. G., Whitaker, J. A., and Oakley, M. G. (2003) Design and characterization of a homodimeric antiparallel coiled coil. *J. Am. Chem. Soc.* **125**, 7518–7519.
56. Oakley, M. G. and Kim, P. S. (1998) A buried polar interaction can direct the relative orientation of helices in a coiled coil. *Biochemistry* **37**, 12,603–12,610.
57. Cusack, S., Berthet-Colominas, C., Hartlein, M., Nassar, N., and Leberman, R. (1990) A second class of synthetase structure revealed by X-ray analysis of Escherichia coli seryl-tRNA synthetase at 2.5 A. *Nature* **347**, 249–255.
58. Stebbins, C. E., Borukhov, S., Orlova, M., Polyakov, A., Goldfarb, A., and Darst, S. A. (1995) Crystal structure of the GreA transcript cleavage factor from Escherichia coli. *Nature* **373**, 636–640.
59. McClain, D. L., Gurnon, D. G., and Oakley, M. G. (2002) Importance of potential interhelical salt-bridges involving interior residues for coiled-coil stability and quaternary structure. *J. Mol. Biol.* **324**, 257–270.
60. McClain, D. L., Woods, H. L., and Oakley, M. G. (2001) Design and characterization of a heterodimeric coiled coil that forms exclusively with an antiparallel relative helix orientation. *J. Am. Chem. Soc.* **123**, 3151–3152.
61. Litowski, J. R. and Hodges, R. S. (2001) Designing heterodimeric two-stranded alpha-helical coiled-coils: the effect of chain length on protein folding, stability and specificity. *J. Pept. Res.* **58**, 477–492.
62. Lau, S. Y., Taneja, A. K., and Hodges, R. S. (1984) Synthesis of a model protein of defined secondary and quaternary structure. Effect of chain length on the stabilization and formation of two-stranded alpha-helical coiled-coils. *J. Biol. Chem.* **259**, 13,253–13,261.
63. Su, J. Y., Hodges, R. S., and Kay, C. M. (1994) Effect of chain length on the formation and stability of synthetic alpha-helical coiled coils. *Biochemistry* **33**, 15,501–15,510.
64. Burkhard, P., Meier, M., and Lustig, A. (2000) Design of a minimal protein oligomerization domain by a structural approach. *Protein Sci.* **9**, 2294–2301.
65. Fairman, R., Chao, H. G., Mueller, L., Lavoie, T. B., Shen, L., Novotny, J., and Matsueda, G. R. (1995) Characterization of a new four-chain coiled-coil: influence of chain length on stability. *Protein Sci.* **4**, 1457–1469.

66. Kwok, S. C. and Hodges, R. S. (2004) Stabilizing and destabilizing clusters in the hydrophobic core of long two-stranded α-helical coiled-coils. *J. Biol. Chem.* **279**, 21,576–21,588.
67. Litowski, J. R. and Hodges, R. S. (2002) Designing heterodimeric two-stranded alpha-helical coiled-coils. Effects of hydrophobicity and alpha-helical propensity on protein folding, stability, and specificity. *J. Biol. Chem.* **277**, 37,272–37,279.
68. O'Neil, K. T. and DeGrado, W. F. (1990) A thermodynamic scale for the helix-forming tendencies of the commonly occurring amino acids. *Science* **250**, 646–651.
69. Dahiyat, B. I., Gordon, D. B., and Mayo, S. L. (1997) Automated design of the surface positions of protein helices. *Protein Sci.* **6**, 1333–1337.
70. Mason, J. M., Schmitz, M. A., Müller, K. M., and Arndt, K. M. (2006) Semi-rational design of Jon-Fos coiled coils with increased affinity: universal implications for leucine zipper prediction and design. *Proc. Natl. Acad. Sci. USA*, in press.
71. Richardson, J. S. and Richardson, D. C. (1988) Amino acid preferences for specific locations at the ends of alpha helices. *Science* **240**, 1648–1652.
72. Dasgupta, S. and Bell, J. A. (1993) Design of helix ends. Amino acid preferences, hydrogen bonding and electrostatic interactions. *Int. J. Pept. Protein Res.* **41**, 499–511.
73. Kohn, W. D., Kay, C. M., and Hodges, R. S. (1997) Positional dependence of the effects of negatively charged Glu side chains on the stability of two-stranded alpha-helical coiled-coils. *J. Pept. Sci.* **3**, 209–223.
74. Doig, A. J. (2002) Recent advances in helix-coil theory. *Biophys. Chem.* **101-102**, 281–293.
75. Kumar, S. and Bansal, M. (1998) Dissecting alpha-helices: position-specific analysis of alpha-helices in globular proteins. *Proteins* **31**, 460–476.
76. Aurora, R. and Rose, G. D. (1998) Helix capping. *Protein Sci.* **7**, 21–38.
77. Lu, M., Shu, W., Ji, H., Spek, E., Wang, L., and Kallenbach, N. R. (1999) Helix capping in the GCN4 leucine zipper. *J. Mol. Biol.* **288**, 743–752.
78. Sober, H. A. (1977) *CRC Handbook of Biochemistry and Molecular Biology*. 3rd ed., The Chemical Rubber Co, Cleveland, OH.
79. Virnekäs, B., Ge, L., Plückthun, A., Schneider, K. C., Wellnhofer, G., and Moroney, S. E. (1994) Trinucleotide phosphoramidites: ideal reagents for the synthesis of mixed oligonucleotides for random mutagenesis. *Nucleic Acids Res.* **22**, 5600–5607.
80. Spanjaard, R. A. and van Duin, J. (1988) Translation of the sequence AGG-AGG yields 50% ribosomal frameshift. *Proc. Natl. Acad. Sci. USA* **85**, 7967–7971.
81. Jung, S., Arndt, K. M., Muller, K. M., and Pluckthun, A. (1999) Selectively infective phage (SIP) technology: scope and limitations. *J. Immunol. Methods* **231**, 93–104.
82. Arndt, K. M., Jung, S., Krebber, C., and Pluckthun, A. (2000) Selectively infective phage technology. *Methods Enzymol.* **328**, 364–388.
83. Nakamura, Y., Gojobori, T., and Ikemura, T. (2000) Codon usage tabulated from international DNA sequence databases: status for the year 2000. *Nucleic Acids Res.* **28**, 292; http://www.kazusa.or.jp/codon.
84. Kamtekar, S., Schiffer, J. M., Xiong, H., Babik, J. M., and Hecht, M. H. (1993) Protein design by binary patterning of polar and nonpolar amino acids. *Science* **262**, 1680–1685.

85. West, M. W. and Hecht, M. H. (1995) Binary patterning of polar and nonpolar amino acids in the sequences and structures of native proteins. *Protein Sci.* **4**, 2032–2039.
86. Pelletier, J. N., Campbell-Valois, F. X., and Michnick, S. W. (1998) Oligomerization domain-directed reassembly of active dihydrofolate reductase from rationally designed fragments. *Proc. Natl. Acad. Sci. USA* **95**, 12,141–12,146.
87. Pelletier, J. N., Arndt, K. M., Plückthun, A., and Michnick, S. W. (1999) An in vivo library-versus-library selection of optimized protein-protein interactions. *Nat. Biotechnol.* **17**, 683–690.
88. Arndt, K. M., Jouaux, E. M., and Willemsen, T. (2004) Der richtige Dreh—Coiled Coils auf dem Weg zur Anwendung. *BioForum* **10**, 48–49.
89. Johnsson, N. and Varshavsky, A. (1994) Split ubiquitin as a sensor of protein interactions in vivo. *Proc. Natl. Acad. Sci. USA* **91**, 10,340–10,344.
90. Rossi, F., Charlton, C. A., and Blau, H. M. (1997) Monitoring protein-protein interactions in intact eukaryotic cells by beta-galactosidase complementation. *Proc. Natl. Acad. Sci. USA* **94**, 8405–8410.
91. Wehrman, T., Kleaveland, B., Her, J. H., Balint, R. F., and Blau, H. M. (2002) Protein-protein interactions monitored in mammalian cells via complementation of beta -lactamase enzyme fragments. *Proc. Natl. Acad. Sci. USA* **99**, 3469–3474.
92. Galarneau, A., Primeau, M., Trudeau, L. E., and Michnick, S. W. (2002) Beta-lactamase protein fragment complementation assays as in vivo and in vitro sensors of protein protein interactions. *Nat. Biotechnol.* **20**, 619–622.
93. Ghosh, I., Hamilton, A. D., and Regan, L. (2000) Antiparallel leucine zipper-directed protein reassembly: application to the green fluorescent protein. *J. Am. Chem. Soc.* **122**, 5658–5659.
94. Miller, J. and Stagljar, I. (2004) Using the yeast two-hybrid system to identify interacting proteins. *Methods Mol. Biol.* **261**, 247–262.
95. Marino-Ramirez, L., Campbell, L., and Hu, J. C. (2003) Screening peptide/protein libraries fused to the lambda repressor DNA-binding domain in E. coli cells. *Methods Mol. Biol.* **205**, 235–250.
96. Hu, J. C., O'Shea, E. K., Kim, P. S., and Sauer, R. T. (1990) Sequence requirements for coiled-coils: analysis with lambda repressor-GCN4 leucine zipper fusions. *Science* **250**, 1400–1403.
97. Willats, W. G. (2002) Phage display: practicalities and prospects. *Plant Mol. Biol.* **50**, 837–854.
98. Keating, A. E., Malashkevich, V. N., Tidor, B., and Kim, P. S. (2001) Side-chain repacking calculations for predicting structures and stabilities of heterodimeric coiled coils. *Proc. Natl. Acad. Sci. USA* **98**, 14,825–14,830.
99. Harbury, P. B., Plecs, J. J., Tidor, B., Alber, T., and Kim, P. S. (1998) High-resolution protein design with backbone freedom. *Science* **282**, 1462–1467.

4

Calcium Indicators Based on Calmodulin–Fluorescent Protein Fusions

Kevin Truong, Asako Sawano, Atsushi Miyawaki, and Mitsuhiko Ikura

Summary

Calmodulin (CaM) is an ubiquitous protein involved in Ca^{2+}-mediated signal transduction. On Ca^{2+} influx, CaM acquires a strong affinity to various cellular proteins with one or more CaM recognition sequences, resulting in the onset or termination of Ca^{2+}-regulated cascades. Through nuclear magnetic resonance and crystallographic structural studies of these Ca^{2+}–CaM complexes, we have gained a deep understanding of CaM target recognition mechanisms. One immediate application is the creation of protein-based Ca^{2+} sensors using CaM complexes and green fluorescent proteins, previously named "chameleon." The major advantage of chameleons is that they can be expressed in single cells and targeted to the specific organelles or tissues to measure localized Ca^{2+} changes. This chapter describes the methods involved in cloning chameleons, characterizing their biochemical and biophysical properties, and imaging them in single cells using a digital fluorescence microscope.

Key Words: Fluorescence resonance energy transfer; calmodulin; green fluorescent protein; chameleon; calcium signaling.

1. Introduction

Because Ca^{2+} concentration ($[Ca^{2+}]$)changes are involved in many cellular processes, such as cell development, differentiation, and apoptosis, the study of these changes within their spatial and temporal context can provide valuable insights into their biological significance. Synthetic dyes (such as Fura and Indo) and aequorin are common tools for studying $[Ca^{2+}]$ changes, however, most synthetic dyes leak rapidly from cells, and aequorin has a weak bioluminescence and is not ratiometric *(1)*. In contrast, protein-based Ca^{2+} sensors based on Ca^{2+}–calmodulin (CaM) complexes and green fluorescent protein (GFP), previously named chameleons, overcome both previous limitations (*see* **Note 1**). In this chapter, we describe the methods involved in cloning chameleons,

From: *Methods in Molecular Biology, vol. 352: Protein Engineering Protocols*
Edited by: K. M. Arndt and K. M. Müller © Humana Press Inc., Totowa, NJ

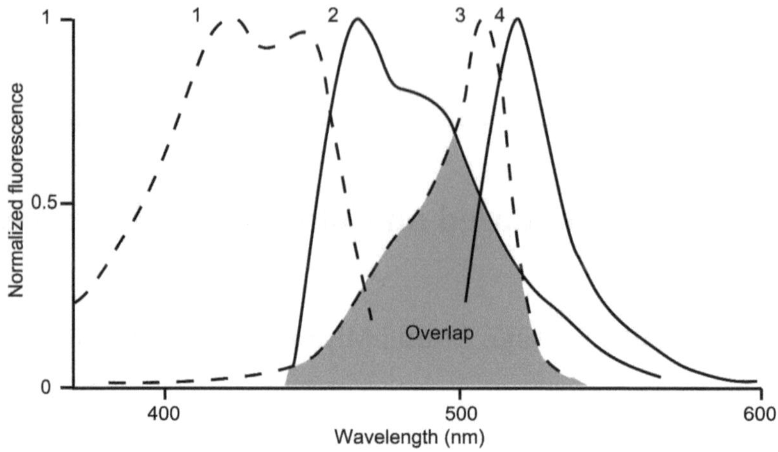

Fig. 1. Emission (solid) and excitation (dotted) spectrums of CFP (labeled 1 and 2) and YFP (labeled 3 and 4). The overlap between the CFP emission spectrum and YFP excitation spectrum allows FRET to occur between these fluorophores.

characterizing their biochemical properties, and imaging them in single cells using a digital fluorescence microscope. Before we can describe the methods, it is necessary to understand the theory of fluorescence resonance energy transfer (FRET; *see* **Subheading 1.1.**) and the general design of chameleons (*see* **Subheading 1.2.**).

1.1. Concept of FRET

FRET is the transfer of energy from the donor to acceptor fluorophore that can occur if the donor's emission spectrum strongly overlaps with the acceptor's excitation spectrum. This transfer of energy occurs at distances less than 80 Å and is most efficient when the fluorophores are closest and in a parallel orientation ***(2,3)***. In the design of chameleons, GFP mutants are used as fluorophore partners: cyan fluorescent protein (CFP) as donor and yellow fluorescent protein (YFP) as acceptor (*see* **Fig. 1**; **ref.** ***4***). Intermolecular FRET can be used to detect protein–protein interactions if both fluorophores are fused to different interacting partners, whereas intramolecular FRET can be used to detect protein cleavage and conformational changes if both fluorophores are fused to the same protein ***(3)***.

1.2. General Design of Chameleons

Using the concept of intramolecular FRET, a chameleon consists of a tandem fusion of N- and C-terminal domains of CaM (N-CaM and C-CaM, respectively) and CaM recognition sequences (CRS), sandwiched between CFP and YFP ***(5–7)***. Depending on the binding orientation of the CRS, it can

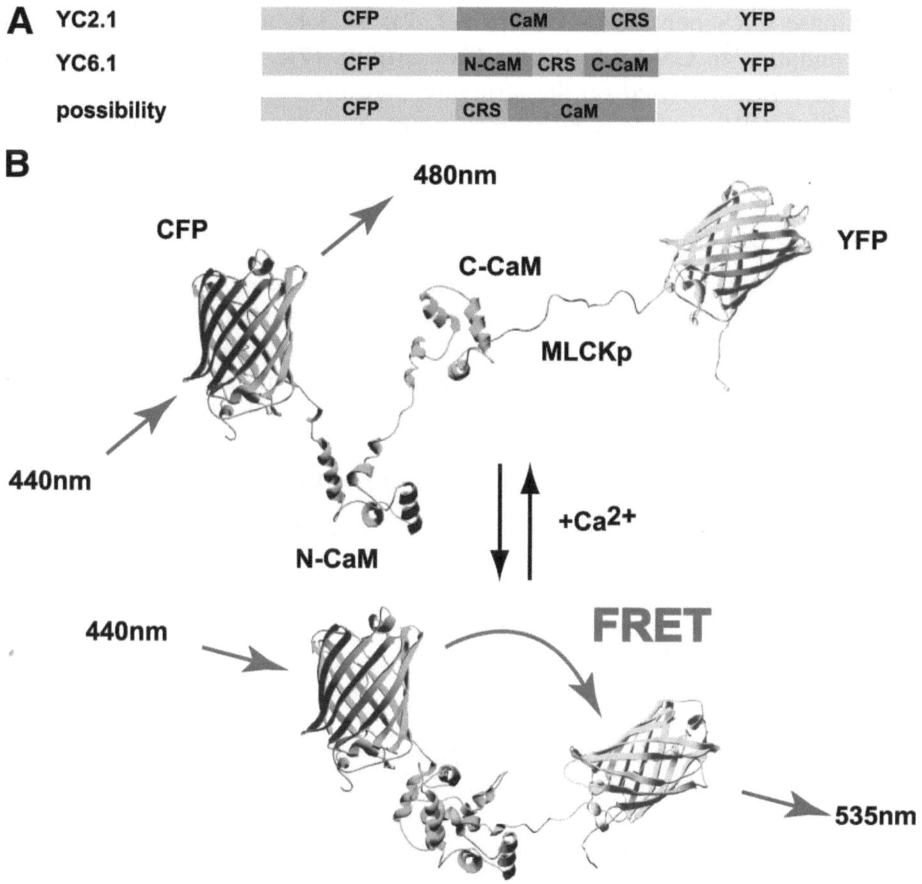

Fig. 2. **(A)** The possibilities for fusing the CRS to CaM. **(B)** A model of the possible the conformational change of YC2.1. Before the Ca^{2+} influx, the relative position of the fluorophores is far apart and, therefore, there is minimal FRET. After the Ca^{2+} influx, the fluorophores are brought closer by the Ca^{2+} CaM–CRS complex, resulting in an increased FRET.

either be placed N-terminal to CaM, C-terminal to CaM, or in between N-CaM and C-CaM (*see* **Fig. 2A** and **Note 2**). After a Ca^{2+} influx, CaM binds the CRS, causing a conformational change that brings CFP and YFP closer, for an increased FRET efficiency (*see* **Fig. 2B**). Therefore, a change in FRET efficiency is directly correlated to a change in [Ca^{2+}]. A number of yellow chameleons (YC), which use the CFP–YFP FRET partners, have been created that have varying Ca^{2+} binding affinities, subcellular localizations, and dynamic ranges. The methods presented in this chapter apply to all published YCs for Ca^{2+} imaging, including: YC2.1, based on the structure of CaM bound to the myosin light chain kinase CRS peptide (MLCKp; **ref.** *7*); YC3.1, a mutant with a single E104Q mutation to CaM that lowers Ca^{2+} affinity *(7)*; YC2nu, targeted to the nucleus *(6)*; YC6.1, based on the structure of CaM bound to CaM-dependent kinase kinase CRS peptide (CKKp) that improves dynamic range *(5)*.

2. Materials

2.1. Cloning of Chameleons

1. PRSETB bacterial expression plasmid (Invitrogen): His-tag for Ni-nitrilotriacetic acid (NTA) agarose protein purification; ampicillin resistance gene for selection; and isopropyl-β-D-thio-galactopyranoside (IPTG) for inducible protein expression.
2. pcDNA3 mammalian transient expression plasmid (Invitrogen): cytomegalovirus promoter and ampicillin resistance gene for bacterial selection.

2.2. Biochemical and Biophysical Characterization

1. Luria-Bertani (LB) medium: 10 g tryptone, 5 g yeast extract, and 5 g NaCl; add water to a final volume of 1 L.
2. IPTG.
3. Phenylmethylsulfonyl fluoride.
4. Ni-NTA agarose (Qiagen).
5. *Escherichia coli* strain BL21 (DE3).
6. Sonicator.
7. EGTA buffer: 100 m*M* KCl, 50 m*M* HEPES pH 7.4, and 10 m*M* EGTA.
8. $CaC1_2$ buffer: 100 m*M* KCl, 50 m*M* HEPES, pH 7.4, 10 m*M* EGTA; and 10 m*M* $CaCl_2$.
9. Spectrofluorophotometer (Shimadzu).

2.3. Imaging With Chameleons

1. Dulbecco's modified Eagle's medium (DMEM) with 10% fetal bovine serum (FBS).
2. Hanks' Balanced Salts Solution (HBSS) with Ca^{2+} (Gibco).
3. 37°C CO_2 incubator.
4. HeLa cell strain.
5. GeneJuice (Novagen).
6. Olympus IX70 inverted epifluorescence microscope.
7. Olympus Xenon lamp.
8. MicroMax 1300YHS charge-coupled device (CCD) camera and Sutter Lambda 10-2 filter changers controlled by Metafluor 4.5r2 software (Universal Imaging).
9. CFP–YFP FRET filter set (Omega Optical; *see* **Note 3**): 440AF21 excitation filter (CFP excitation), 455DRLP dichroic mirror, 480AF30 emission filter (CFP emission), and 535AF26 emission filter (YFP emission).
10. Neutral density (ND) filter set (Omega Optical).
11. UApo ×40 oil Iris/340 objective (Olympus).

12. U-MNIBA band pass mirror cube unit (Olympus).
13. Histamine, ionomycin, EGTA, and BAPTA-AM (Sigma).
14. 35-mm-diameter glass-bottom dishes (Maltek).

3. Methods

The methods described below outline the cloning of chameleons for expression in bacterial and mammalian cells (*see* **Subheading 3.1.**), the biochemical and biophysical characterization of chameleons (*see* **Subheading 3.2.**), and the use of chameleons in FRET in vivo Ca^{2+} imaging (*see* **Subheading 3.3.**).

3.1. Cloning of Chameleons

Using standard recombinant DNA methods *(8)*, chameleons are assembled modularly using five different components: CFP, YFP, N-CaM, C-CaM, and CRS. CFP and YFP plasmids are available from Clontech; CaM from Dr. Ikura; CRS from polymerase chain reaction oligonucleotide synthesis. The DNA manipulations to construct a chameleon are not described here in detail because there is no single strategy that applies to all possibilities (*see* **Note 4**). After the construction of a chameleon, the fusion fragment is cloned into the PRSETB plasmid for sufficient protein expression needed to perform biochemical and biophysical characterization. For transient mammalian expression of chameleons used in in vivo Ca^{2+}-imaging experiments, we cloned into the pcDNA3 plasmid (*see* **Note 5**).

3.2. Biochemical and Biophysical Characterization

Before the chameleon can be used in in vivo Ca^{2+} imaging, it is purified (*see* **Subheading 3.2.1.** and **Note 6**), and in vitro experiments with a fluorometer are performed to determine its dynamic range (*see* **Subheading 3.2.2.**) and Ca^{2+}-binding curve (*see* **Subheading 3.2.3.**).

3.2.1. Purification of Chameleons From Bacterial Cells

1. Transform BL21 (DE3) cells with the bacterial plasmid using standard molecular biology methods *(8)*.
2. Plate the cells on LB plates containing ampicillin and incubate overnight at 37°C.
3. Select a single colony and grow in 100 mL of LB medium containing 100 μ*M* ampicillin at 37°C.
4. At an optical density at 600 nm of 0.7, induce with 0.5 m*M* IPTG for 3 h.
5. Centrifuge the cells for 30 min at 3000*g*.
6. Resuspend the cell pellet in 10 mL of 50 m*M* HEPES, pH 7.4, 10% glycerol, 100 m*M* KCl, 1 m*M* $CaCl_2$, and 1 m*M* phenylmethylsulfonyl fluoride.
7. Sonicate with the 0.5-in. horn at maximum power, 10% duty cycle, four times, for 4 min each. Cool the solution for 5 min between sonications.
8. Centrifuge the cell debris for 20 min at 30,000*g*.

9. Gently mix the supernatant with 1 mL of slurry Ni-NTA agarose for 30 min at 4°C.
10. Wash the column with 10 mL of 50 m*M* HEPES, pH 7.4, 100 m*M* KCl, and 5 m*M* imidazole.
11. Elute with 1 mL of 50 m*M* HEPES, 100 m*M* KCl, and 100 m*M* imidazole. It is unnecessary to cleave the His-Tag because it does not interfere with the fluorescent properties.
12. Dialyze the sample with 2 L of 50 m*M* HEPES, pH 7.4, and 100 m*M* KCl at 4°C.

3.2.2. Fluorescence Spectral Analysis

A change in FRET efficiency (or change in [Ca^{2+}], in the case of chameleons) is often observed by a change in the emission ratio (R is the peak emission intensity of the acceptor divided by the peak emission intensity of the donor). The dynamic range of a Ca^{2+} indicator is defined as the division of the maximum ratio, R_{max}, by the minimum ratio, R_{min}. A chameleon with a larger dynamic range is a more effective in vivo Ca^{2+} indicator.

1. Dilute the sample in 50 m*M* HEPES, pH 7.4; 100 m*M* KCl; and 20 μ*M* EGTA into a 1-mL cuvet. Any dilution factor is acceptable as long as the emission signal remains detectable. Record the fluorescence emission spectrum from 450 to 570 nm at 433 nm excitation.
2. Record the fluorescence emission spectrum of the blank sample.
3. Subtract the spectrum in **step 2** (background) from **step 1** to find the spectra of the chameleon in the absence of Ca^{2+}. Determine R_{min} from this spectrum.
4. Repeat **steps 1** to **3** in the presence of 1 m*M* $CaCl_2$ to find the spectra of the chameleon in the presence of Ca^{2+}. Determine R_{max} from this spectrum.
5. **Figure 3** shows the fluorescence spectrum of YC6.1 *(5)* before and after Ca^{2+}, with an R_{min} and an R_{max} of 1.1 and 2.4, respectively.

3.2.3. Ca^{2+}-Binding Properties

The Ca^{2+}-binding curve is used to assess the effective range of [Ca^{2+}] that the Ca^{2+} indicator can measure. Ca^{2+}/EDTA and Ca^{2+}/EGTA buffer are used as standards because at low [Ca^{2+}], Ca^{2+} contaminants can significantly distort the free [Ca^{2+}] levels (*see* **Note 7**).

1. In a 1-mL cuvet, dilute the sample in the EGTA buffer. Record the fluorescence emission spectrum from 450 to 570 nm at 433 nm excitation. Determine the emission ratio.
2. To obtain the Ca^{2+}-binding curve, add successive fractions of the $CaC1_2$ solution to the sample and determine the emission ratio. Because the experiment is performed at 20°C with the EGTA and $CaCl_2$ buffers described in this chapter, the free calcium can be calculated by solving the quadratic equation (for different conditions, please consult **ref.** ***9***):

$$([Ca^{2+}]_{free})^2 + \{(10{,}000{,}060.5 - [Ca^{2+}]_{total}) \times [Ca^{2+}]_{free}\} - (60.5 \times [Ca^{2+}]_{total}) = 0$$

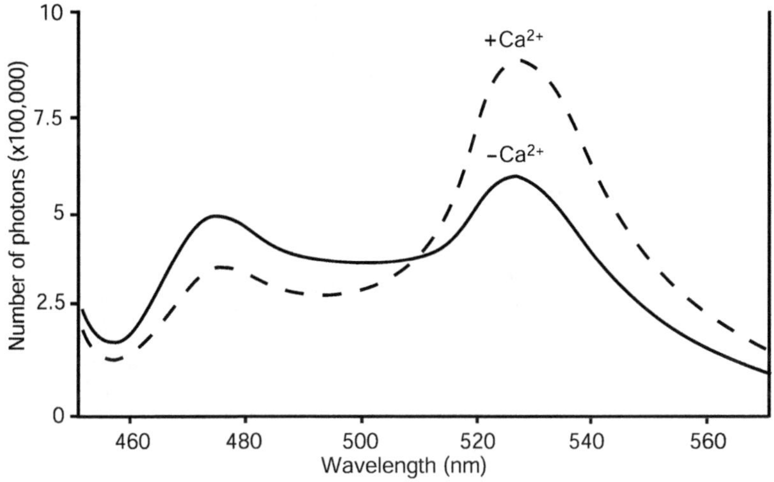

Fig. 3. Example fluorescence spectrum of YC6.1.

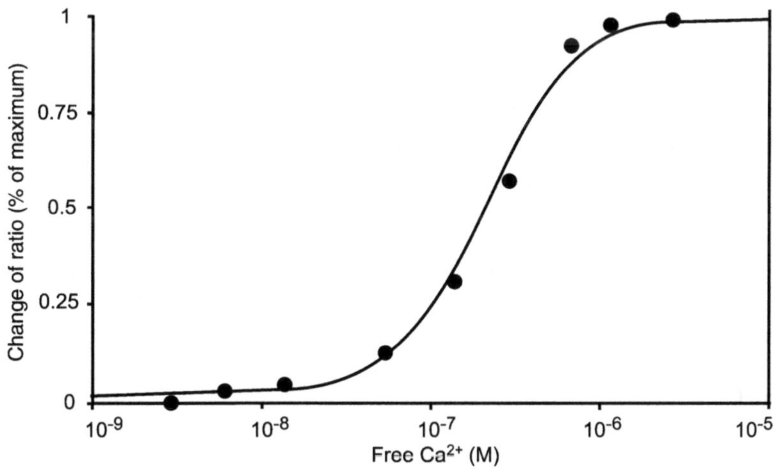

Fig. 4. Example Ca^{2+}-binding curve of YC6.1.

3. To produce the Ca^{2+}-binding curve, plot the free[Ca^{2+}] vs the change in the emission ratio. **Figure 4** shows the Ca^{2+}-binding curve of YC6.1.

3.3. Ca^{2+} Imaging With Chameleons

The following section describes a simple Ca^{2+}-imaging experiment using HeLa cells; however, with minor modifications, the method can be applied to

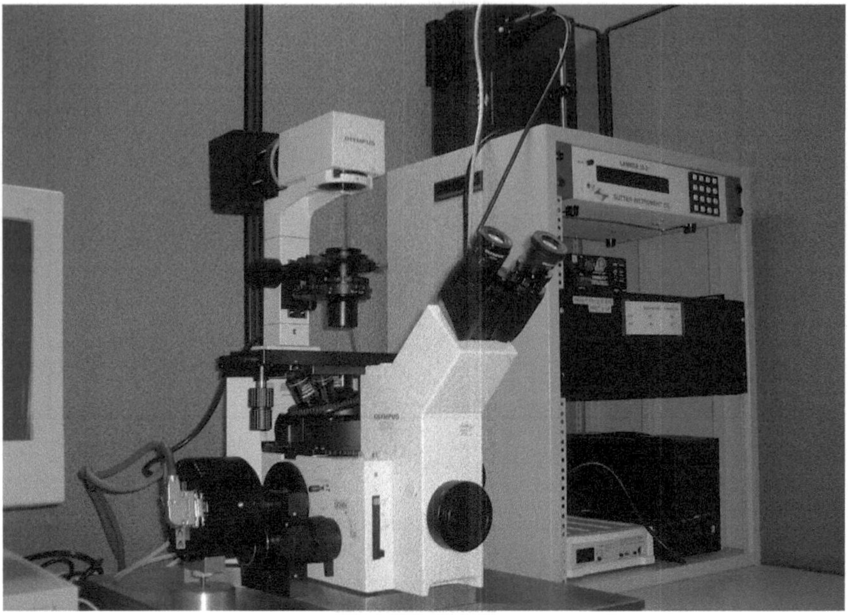

Fig. 5. Digital fluorescence microscope facilities.

other adherent cells and physiological contexts. In this section, we describe the preparation of the cell culture (*see* **Subheading 3.3.1.**) and the acquisition of data (*see* **Subheading 3.3.2.**).

3.3.1. Preparation of the Cell Culture

1. Plate HeLa cells on 35-mm-diameter glass-bottom dishes with DMEM–10% FBS media.
2. Incubate the cells at 37°C (5% CO_2) until cells are 50 to 80% confluent.
3. Transfect cells with the mammalian expression plasmid containing your chameleon, using GeneJuice (Novagen).
4. Remove the transfection mixture after 6 h and replace with 1.5 mL of DMEM–10% FBS media.
5. Incubate the cells at 37°C (5% CO_2) for 24 h. The cells are ready to perform the Ca^{2+}-imaging experiment.

3.3.2. Data Acquisition

The experiment should be performed in a dark room to reduce background light. **Figure 5** shows an image of our microscopy facilities.

1. Turn on the Xenon lamp, microscope, filter changers, and computer.
2. Remove the growth media from the culture dish.
3. Wash with 1 mL of HBSS (+$CaCl_2$).
4. Add 1 mL of fresh HBSS (+$CaCl_2$).

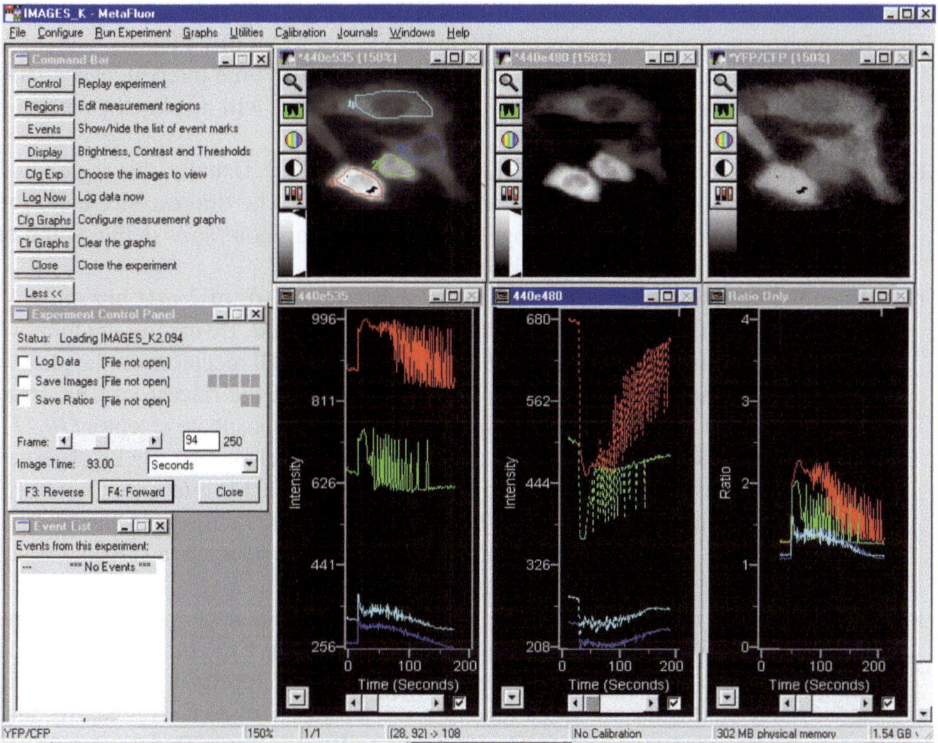

Fig. 6. Screen shot of the MetaFluor software. The regions of observation are highlighted in the 440e535 panel. The 440e535 panel and graph plot the change in YFP fluorescence as a result of CFP excitation; the 440e480 panel and graph plot the change in CFP fluorescence as a result of CFP excitation; the YFP/CFP panel and graph is the ratio of 440e535 and 440e480 panels and graphs. In this experiment, the graphs display a Ca^{2+} oscillation as a result of histamine stimulation.

5. Put the cells on the stage of microscope.
6. Start the MetaFluor software. The MetaFluor software controls the shutters, filter exchangers, and camera during data acquisition.
7. To find cells expressing the chameleon construct, screen for YFP fluorescence using the eyepiece. Turn the filter turret to the U-WNIBA band-pass mirror cube and use the ND filter in the range of 0.1 to 10% depending on the level of fluorescence emission. The U-WNIBA mirror cube is ideal for quickly finding cells with YFP fluorescence using the eyepiece, and the 10% ND filter is used to reduce photobleaching.
8. Using the ×40 oil objective, center the microscope viewing area on a cell that has a healthy morphology and displays a strong cytosolic fluorescence. **Figure 6** shows a panel with the typical fluorescence distribution of a healthy HeLa cell.
9. Acquire single images on the computer screen using MetaFluor, while adjusting the focus until you have the sharpest image.

10. Turn the filter turret to the CFP–YFP FRET mirror cube.
11. To monitor the peak emissions of CFP and YFP as a result of CFP peak excitation over time, set the data acquisition conditions as follows: time interval, every 10 s; and exposure time, 200 ms for CFP and YFP. MetaFluor will display the emission ratio over time.
12. Draw a region of interest on the field of view of the CCD. Usually, this region will outline the cell of interest. Then, start data acquisition. The fluorescence emission intensities of the CFP and YFP, together with their emission ratios in the region of interest, will be monitored over time (*see* **Note 8**).
13. When the emission ratio reaches a steady state, add 50 µL of 2 m*M* histamine to the culture dish for a final concentration of approx 100 µ*M*. Be careful not to move the culture dish in this process. The histamine binds to cell receptors on the plasma membrane that initiate a signaling cascade, resulting in the release of Ca^{2+} from the endoplasmic reticulum through the inositol-1,4,5-triphosphate receptor ***(10,11)***. This should cause a conformational change in the chameleon that can be observed by a rise in emission intensity of YFP and a decline in CFP intensity. Therefore, the emission ratio should increase. The emission ratio should return to steady-state levels when the effect of the histamine wanes. **Figure 6** shows a screen shot of MetaFluor software as it is collecting data.
14. To correlate the emission ratio to the cytosolic [Ca^{2+}], it is necessary to determine R_{min} and R_{max} so that emission ratios can be mapped to the Ca^{2+}-binding curve. Add 50 µL of 20 µ*M* ionomycin for a final concentration of approx 1 µ*M*. Ionomycin opens pores on the plasma membrane to allow permeability to Ca^{2+} ions. Because the medium is saturated with $CaCl_2$, the ratio will rise to R_{max}. To determine R_{min}, add 50 µL of 100 m*M* EGTA and 600 µ*M* BAPTA-AM for final concentrations of approx 5 m*M* and approx 30 µ*M*, respectively. The emission ratio should drop to R_{min}.
15. Stop the data acquisition.

4. Notes

1. Chameleons are not without their own limitations. In general, they have a lower signal-to-noise ratio than other techniques mentioned, but this ratio can be improved by the careful design of a new construct ***(5)*** or by finding optimal experimental conditions, as described in the text.
2. The best place to insert the CRS depends on the arrangement of the two CaM domains with respect to the CRS in the three-dimensional structure. If the N terminus of the CRS in the CaM-CRS complex is within 20 Å of the C-terminus of CaM, a CaM–CRS fusion would be effective ***(6)***. However, if the both N- and C-termini are within 10 Å of the hinge of CaM, a (N-CaM)–CRS–(C-CaM) fusion would be most effective ***(5)***.
3. If you are only interested in using CFP and YFP in your applications, the CFP–YFP FRET filter set will satisfy your needs. However, if you are using Discosoma red fluorescent protein (dsRed; **refs.** ***12*** and ***13***) for FRET, you will need the following dichoric mirrors and filters from Omega Optical or equivalents:

450-520-590TBDR for the dichoric mirror; 440DF21 for CFP excitation; 510DF23 for YFP excitation; 575DF26 for dsRed excitation; 480AF30 for CFP emission; 535AF26 for YFP emission; 600ALP for dsRed emission. This filter set will allow you to excite or acquire emission from CFP, YFP, and dsRed individually, however, with a tradeoff in efficiency.

4. In our chameleon constructs, we truncated the last 11 C-terminal amino acids of CFP (the minimal region to form GFP) to reduce the relative tumbling of the fluorophores. Additionally, glycinyl–glycine linkers were introduced between CaM and CRS fusion points to increase conformational freedom for the formation of the Ca^{2+} CaM–CRS complex.
5. When amplifying the chameleon DNA by polymerase chain reaction for insertion into PRSETB or pcDNA3, remember that the 5′ and 3′ primers for the chameleon will have internal primer sites because CFP and YFP are only different by a few point mutations. It is necessary to purify the DNA fragment of the correct size from the agarose gel.
6. If your particular chameleon degrades or is insoluble in bacterial cells, it is possible to perform the characterization using mammalian cell lysate. For harvesting, cells should be grown in 10-cm-diameter culture dishes. Twenty-four hours after transfection, cells are scraped off the culture dish and lysed in a hypotonic lysis buffer (50 m*M* HEPES, pH 7.4; 100 m*M* KCl; 5 m*M* $MgCl_2$; and 0.5% Triton X-100). Then, the mixture is centrifuged to remove cell debris. Next, the supernatant is dialyzed in 2 L of buffer (50 m*M* HEPES, pH 7.4, and 100 m*M* KCl). Finally, the sample can be used for characterization as described in **Subheadings 3.2.2.** and **3.2.3.**
7. The accuracy of the Ca^{2+}-binding curve depends on the accurate preparation of the Ca^{2+}/EGTA buffers *(9)*.
8. The emission intensities will decrease during the course of the experiment because of photobleaching; however, the emission ratio should remain constant if there is no change in FRET. To reduce photobleaching, decrease exposure time and excitation light intensity. Binning is summing the signal from multiple pixels on the CCD camera, so that less light is required, while maintaining a good signal-to-noise ratio.

Acknowledgments

K. T. acknowledges the Canadian Institutes of Health Research and Ontario Student Opportunity Trust Funds award. This work was supported by a grant from the Cancer Research Society. M. I. is a Canadian Institutes of Health Research investigator.

References

1. Takahashi, A., Camacho, P., Lechleiter, J. D., and Herman, B. (1999) Measurement of intracellular Ca^{2+}. *Physiol. Rev.* **79,** 1089–1125.
2. Miyawaki, A. and Tsien, R. Y. (2000) Monitoring protein conformations and interactions by fluorescence resonance energy transfer between mutants of green fluorescent protein. *Methods Enzymol.* **327,** 472–500.

3. Truong, K. and Ikura, M. (2001) The use of FRET imaging microscopy to detect protein-protein interactions and protein conformational changes in vivo. *Curr. Opin. Struct. Biol.* **11,** 573–578.
4. Tsien, R. Y. (1998) The green fluorescent protein. *Annu. Rev. Biochem.* **67,** 509–544.
5. Truong, K., Sawano, A., Mizuno, H., et al. (2001) FRET-based in vivo Ca^{2+} imaging by a new calmodulin-GFP fusion molecule. *Nat. Struct. Biol.* **8,** 1069–1073.
6. Miyawaki, A., Llopis, J., Heim, R., et al. (1997) Fluorescent indicators for Ca^{2+} based on green fluorescent proteins and calmodulin. *Nature* **388,** 882–887.
7. Miyawaki, A., Griesbeck, O., Heim, R., and Tsien, R.Y. (1999) Dynamic and quantitative Ca^{2+} measurements using improved chameleons. *Proc. Natl. Acad. Sci. USA* **96,** 2135–2140.
8. Sambrook, J., Fritsch, E. F., and Maniatis, T. (1989) *Molecular Cloning, A Laboratory Manual.* 2nd ed., Cold Spring Harbor Laboratory Press, New York.
9. Tsien, R. and Pozzan, T. (1989) Measurement of cytosolic free Ca^{2+} with quin2. *Methods Enzymol.* **172,** 230–262.
10. Bootman, M. D., Cheek, T. R., Moreton, R. B., Bennett, D. L., and Berridge, M. J. (1994) Smoothly graded Ca^{2+} release from inositol 1,4,5-trisphosphate-sensitive Ca^{2+} stores. *J. Biol. Chem.* **269,** 24,783–24,791.
11. Zamani, M. R. and Bristow, D. R. (1996) The histamine H1 receptor in GT1-7 neuronal cells is regulated by Ca^{2+} influx and KN-62, a putative inhibitor of Ca^{2+}/calmodulin protein kinase II. *Br. J. Pharmacol.* **118,** 1119–1126.
12. Mizuno, H., Sawano, A., Eli, P., Hama, H., and Miyawaki, A. (2001) Red fluorescent protein from Discosoma as a fusion tag and a partner for fluorescence resonance energy transfer. *Biochemistry* **40,** 2502–2510.
13. Matz, M. V., Fradkov, A. F., Labas, Y. A., et al. (1999) Fluorescent proteins from nonbioluminescent Anthozoa species. *Nat. Biotechnol.* **17**, 969–973.

5

Design and Synthesis of Artificial Zinc Finger Proteins

Wataru Nomura and Yukio Sugiura

Summary

Of the DNA-binding motifs, the $(Cys)_2(His)_2$-type zinc finger motif has great potential for manipulation. The zinc finger motif offers an attractive framework for the design of novel DNA-binding proteins. Specially, a structure-based design strategy is valuable for the creation of new artificial zinc finger proteins that have novel DNA-binding properties, namely, long-DNA recognition, DNA bending, and AT-rich sequence recognition. Herein, new strategies for the design of multi-zinc finger proteins for the recognition of a target DNA sequence, a DNA-bending zinc finger protein, a $(His)_4$-type zinc finger protein, and an AT-recognizing zinc finger protein are described based on recent experimental results.

Key Words: Zinc finger; multifinger; DNA bending; GC recognition; AT recognition; $(Cys)_2(His)_2$ type; $(His)_4$ type; artificial protein; helix substitution; protein design.

1. Introduction

Artificial zinc finger proteins have important applications in biomedical research and in gene therapy, and are novel tools for biochemical and molecular biological investigations ***(1–3)***. In particular, a $(Cys)_2(His)_2$-type zinc finger motif, one of the most common DNA-binding motifs in eukaryotes, presents a versatile and attractive framework for the design of new DNA-binding proteins. The structural analysis of some $(Cys)_2(His)_2$-type zinc finger protein–DNA complexes revealed the characteristics of this motif in DNA binding as follows:

1. The zinc finger structure is repeated by the connection with a specific linker.
2. Each zinc finger structure binds to a 3-basepair (bp) subsite of DNA with its α-helix facing toward the major groove.
3. The amino acid residues at four key positions in the α-helix of each zinc finger unit make a 1:1 contact with the DNA bases at specific positions.
4. The overall arrangement of the peptide is antiparallel to the primary interacting strand of DNA ***(4–7)***.

From: *Methods in Molecular Biology, vol. 352: Protein Engineering Protocols*
Edited by: K. M. Arndt and K. M. Müller © Humana Press Inc., Totowa, NJ

Because of these features, some artificial zinc finger proteins, for example, long sequence- and AT-sequence-binding fingers, have been created successfully. Such a designed DNA-binding protein would be expected to possess a unique binding sequence with high affinity and specificity. In addition, novel zinc finger proteins, such as a DNA-bending finger and a $(His)_4$-type finger were also designed. These artificial zinc finger proteins offer great promise as useful tools for genetic engineering and genomic-specific transcription switches in the near future.

This chapter focuses on the strategies for design and preparation of novel artificial zinc finger proteins, namely, a nine-zinc finger protein (**Subheading 3.2.**), a DNA-bending zinc finger protein (**Subheading 3.3.**), a $(His)_4$-type zinc finger protein (**Subheading 3.4.**), and an AT-recognizing zinc finger protein (**Subheading 3.5.**).

2. Materials

1. pET3b expression system (Novagen).
2. Complementary DNA-coding Sp1 zinc finger.
3. *Escherichia coli* strains DH5α and BL21 (DE3) pLysS.
4. Oligonucleotide primers, as specified in the text.
5. Restriction enzymes, T7 DNA polymerase, and T4 DNA ligase.
6. DNA sequencer (Amersham Biosciences).
7. Luria-Bertani medium: 10 g tryptone, 5 g yeast extract, 10 g NaCl; add water to a final volume of 1 L.
8. Ampicillin.
9. Isopropyl-β-D-thio-galactopyranoside (IPTG).
10. Sonication equipment.
11. Phosphate-buffered saline (PBS): 10 m*M* phosphate buffer, pH 7.6, 130 m*M* NaCl, and 2.7 m*M* KCl.
12. Sodium dodecyl sulfate polyacrylamide gel electrophoresis (SDS-PAGE) equipment.
13. Chromatography equipment.
14. High S chromatography column (Bio-Rad).
15. UNO-S1 chromatography column (Bio-Rad).
16. Superdex 75 (Amersham Biosciences).
17. 50 m*M* Tris-HCl, pH 8.0, 50 m*M* NaCl, and 1 m*M* dithiothreitol.
18. PAGE equipment.
19. DNA-sequencing equipment.
20. 9-Fluorenylmethoxycarbonyl solid-phase synthesis equipment.
21. Trifluoroacetic acid:ethanol (95:5) solution.
22. High-performance liquid chromatography equipment.
23. μBonedesphere$_5$C$_4$-300 (19 × 150-mm) column.
24. Time-of-flight mass spectrometry equipment.
25. Circular dichroism (CD) spectra equipment.
26. Ultraviolet-visible (UV-VIS) spectrophotometer.
27. Nuclear magnetic resonance (NMR) equipment.

3. Methods

The methods described in this section outline:

1. The protein expression and purification of designed zinc fingers.
2. Design strategies of artificial zinc fingers.
3. Construction procedures of artificial zinc fingers.
4. The confirmation of structure and metal coordination of zinc fingers.

3.1. Protein Expression and Purification of Artificial Zinc Fingers

The methods for protein expression and purification of designed artificial zinc finger proteins are described in **Subheadings 3.1.1.** to **3.1.6.** These include:

1. The construction of the expression vector pEV-3b.
2. Protein expression in soluble and insoluble forms.
3. Protein purification by cation exchange and gel-filtration chromatography.
4. Protein refolding of denatured structure.
5. Confirmation of the structure and metal coordination of zinc finger domains.

3.1.1. pEV-3b Expression Vector

The pEV-3b expression vector is based on a pET3b expression vector (Novagen; **Fig. 1**). This vector is converted to create the useful multicloning site for zinc finger genes.

1. Cut out the *Eco*RI/*Hin*dIII-digested fragment from the pET3b.
2. To delete the *Eco*RI and *Hin*dIII sites, insert the annealed oligonucleotides, 5′-AATTGTCATGTTTGAC-3′ and 5′-AGCTGTCAAACATGAC-3′, into the *Eco*RI/*Hin*dIII-digested pET3b. The produced plasmid is designated as pET3b′.
3. Prepare the double-stranded oligonucleotide, which includes restriction enzyme sites for *Afl*II, *Bam*HI, *Eco*RI, *Hin*dIII, and *Sma*I. This strand consists of the sequences 5′-TATGGATCCCGGGAATTCAAGCTTAAGC-3′ and 5′-TCAGCT-TAAGCTTGAATT-CCCGGGATCCA-3′.
4. Digest with *Nde*I and *Bpu*1102I, and insert this fragment into pET3b′, yielding the plasmid, pEV-3b.

The resulting plasmid, pEV-3b, is used for the expression of the designed zinc finger proteins. By inserting zinc finger genes as a *Bam*HI/*Eco*RI fragment into a similarly digested pEV-3b, the expression vectors for the artificial zinc fingers are constructed in a one-step procedure.

3.1.2. Protein Expression of Designed Artificial Zinc Fingers

1. Transform the plasmid pEV-3b maintaining the gene of the designed zinc finger proteins into the *E. coli* strain BL21(DE3) pLysS.
2. Culture the cells in Luria-Bertani medium at 37°C.
3. At an optical density of 0.6 at 600 nm, add 1 m*M* IPTG.

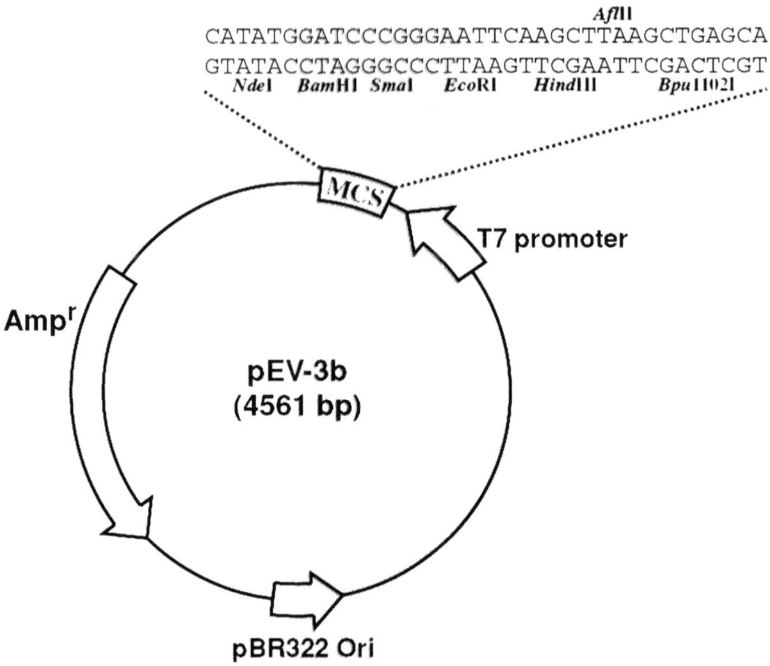

Fig. 1. Schematic drawing of our pEV-3b expression vector. The sequence of the multi-cloning site is depicted and the sites for restriction enzymes are indicated in the sequences.

4. Incubate the culture at 20°C for 8 to 12 h. This temperature is important to obtain the expressed proteins in their soluble forms.

3.1.3. Protein Purification of Designed Artificial Zinc Fingers

Purification procedures are performed at 4°C.

1. Harvest *E. coli* cells, resuspend, and lyse in PBS.
2. After centrifugation, purify the supernatant, which contains the soluble form of the zinc finger proteins, by cation exchange chromatography (High S and UNO S-1; Bio-Rad) and gel filtration (Superdex 75; Amersham Biosciences) with Tris-HCl buffer.
3. Confirm the purity of the proteins by SDS-PAGE.

For experiments of metal substitution, the insoluble form is also purified from the pellet of *E. coli* cells after centrifugation:

1. Lyse the pellet in the PBS containing 8 *M* urea and 10 m*M* chelating agents (EDTA or 1,10-phenanthrolene).
2. Purify according to the same procedure as described in **Subheading 3.1.3., step 2**.
3. Refold by heating the purified proteins at 65°C for 30 min, and by cooling gradually in 10 m*M* Tris buffer containing 125 μ*M* $ZnCl_2$, $Ni(NO_3)_2$, $CdCl_2$, $Co(NO_3)_2$, or $CuSO_4$.

3.1.4. CD Measurement

The CD spectra of the zinc finger proteins are recorded on a Jasco J-720 spectropolarimeter in Tris-HCl buffer, pH 8.0, containing 50 m*M* NaCl in a capped 1-mm path-length cell at 20°C, under nitrogen. All spectra represent the average of 8 to 16 scans. Spectra are baseline corrected and noise reduced using Jasco software.

3.1.5. UV-VIS Absorption Spectroscopy

UV-VIS absorption spectra are recorded on a Beckman Coulter DU7400 diode array spectrophotometer at 20°C in 10 m*M* Tris-HCl buffer, pH 7.5, containing 50 m*M* NaCl in a capped 1-cm path-length cell. Co(II)-substituted zinc finger complexes are obtained by titration with $CoCl_2$. The peptides are saturated with Co(II) in arbitrary conditions. All spectra are normalized by $\varepsilon = A/(l \bullet c)$, where ε is the extinction coefficient (per molar concentration per centimeter), l is the path-length of the cell (in centimeters), and c is the peptide concentration (in molar concentration).

3.1.6. NMR Experiments

In the presence of 1.5 Eq of Zn(II) ion, the complex of single finger domain and Zn(II) is prepared at a concentration of 5 m*M* in 90% H_2O/10% D_2O and D_2O (25 m*M* Tris–d_{11}, pH 5.7). All of the NMR spectra are recorded on a JOEL Lambda-600 spectrometer.

1. The nuclear Overhauser enhancement spectroscopy (NOESY) spectra are acquired with selective water presaturation (delays alternating with mutations for tailored excitation pulse) followed by the standard NOESY pulse train at 303 K, with mixing times of 100, 200, and 300 ms.
2. The total correlation spectroscopy spectra are obtained with an 80-ms MLEV-17 spin-lock duration, using water suppression by a gradient tailored excitation pulse at 303 K.

Spectra are typically acquired with 24 scans per t1 valve, for 1024 t1 valves, and 2048 complex points are collected in the direct dimension. The free-induction decay in both dimensions is multiplied by the phase-shifted sine bell apodization function, zero-filled, and Fourier transformed to yield 2048 by 2048 matrices. Sequential resonance assignments are determined by standard total correlation spectroscopy and NOESY procedures ***(8)***.

3.2. Six- and Nine-Zinc Finger Proteins

This section describes the design strategy and construction of multi-zinc fingers for the recognition of long DNA sequences.

3.2.1. Strategy of Protein Design

Novel six- and nine-zinc finger proteins, Sp1ZF6 and Sp1ZF9, were created from the three-zinc finger motifs of the transcription factor, Sp1 (**Fig. 2**). These

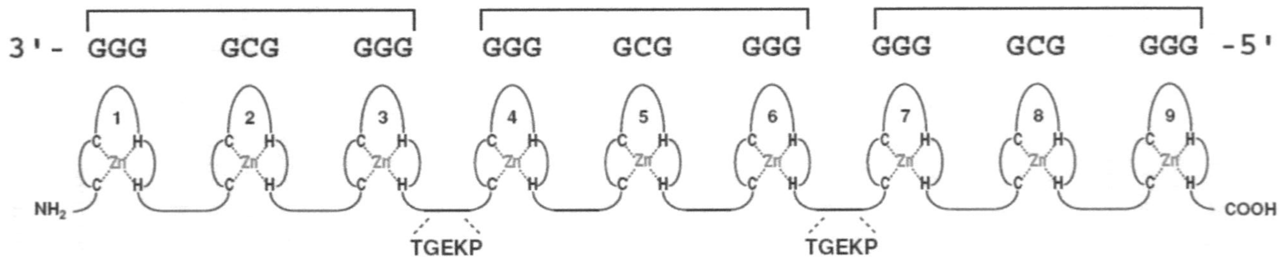

Fig. 2. Schematic representation of the artificial nine-zinc finger protein, Sp1ZF9. The DNA-binding site for the protein is shown above of the nine-zinc finger illustration (*see* **Note 2**). The Krüppel-type linker sequences and other amino acids are represented by their one-letter codes.

proteins were constructed by linking two or three Sp1 zinc finger domains with the Krüppel-type linker (*see* **Notes 1** and **2**).

3.2.2. Construction of Genes for Sp1ZF6 and Sp1ZF9

The pUC-Sp1(530-623), which codes for the Sp1 three-zinc finger region, was constructed as described previously *(9)*.

1. Prepare a synthetic oligonucleotide (84 bp) encoding the Krüppel-type linker (TGEKP) as *Bam*HI/*Sty*I fragment and insert this fragment into pUC-Sp1(530–623).
2. Cut out the *Eco*47III fragment (264 bp) and insert into similarly digested pUC-Sp1(530-623). The plasmid is renamed as pUC-Sp1ZF6.
3. Flank the middle *Sp1* gene fragments for Sp1ZF9 with *Age*I sites at the 5′ and 3′-ends by using the *Age*I site primer pair, namely, 5′-ACCGGTGAAAAACCG-CATATTTGCCACATC-3′ for the coding strand, and 5′-CGGTTTTTCACCGGT-GTGGGTCTTGATATG-3′ for the noncoding strand.
4. Ligate the resulting fragment modified with *Age*I sites into the *Age*I site of pUC-Sp1ZF6. The *Age*I enzyme site between two and three Sp1 fragments encodes the amino acids TG, which are part of the TGEKP linker peptide.
5. Confirm all sequences by DNA sequencing.
6. Cut out the DNA fragments of Sp1ZF6 and Sp1ZF9 as a *Bam*HI/*Eco*RI fragment and insert into the similarly digested plasmid, pEV-3b (*see* **Subheading 3.1.1.**; **ref.** ***10***).

3.3. DNA-Bending Zinc Finger Proteins

In **Subheadings 3.3.1.** and **3.3.2.**, the design and the construction of DNA-bending zinc fingers are described.

3.3.1. Strategy of Protein Design

By linking two three-zinc fingers of Sp1, novel six-zinc finger proteins, including polyglycine linkers, Sp1ZF6(Gly)$_4$, Sp1ZF6(Gly)$_7$ (**Fig. 3**), and Sp1ZF6(Gly)$_{10}$, were created. Sp1ZF6(Gly)$_n$ was constructed by exchanging the TGEKP linker region of Sp1ZF6 with the (Gly)$_n$ (n = 4, 7, or 10) linkers (*see* **Notes 3** and **4**).

3.3.2. Construction of Sp1ZF6(Gly)$_n$

Synthesized *Afl*II/*Mfe*I oligonucleotides containing the linker sequences were purchased from Amersham Biosciences. These oligonucleotides were annealed and inserted into pUC-Sp1ZF6. These plasmids were renamed pUC-Sp1ZF6(Gly)$_n$ (n = 4, 7, or 10). The DNA fragments coding these proteins were cut out with *Bam*HI and *Eco*RI, and inserted into the similarly digested plasmid, pEV-3b (*see* **Subheading 3.1.1.**; **ref.** ***11***).

3.4. (His)$_4$-Type Zinc Finger Protein

The procedures described in **Subheadings 3.4.1.** to **3.4.3.** are the design strategy, peptide synthesis, and gene construction of (His)$_4$-type zinc fingers.

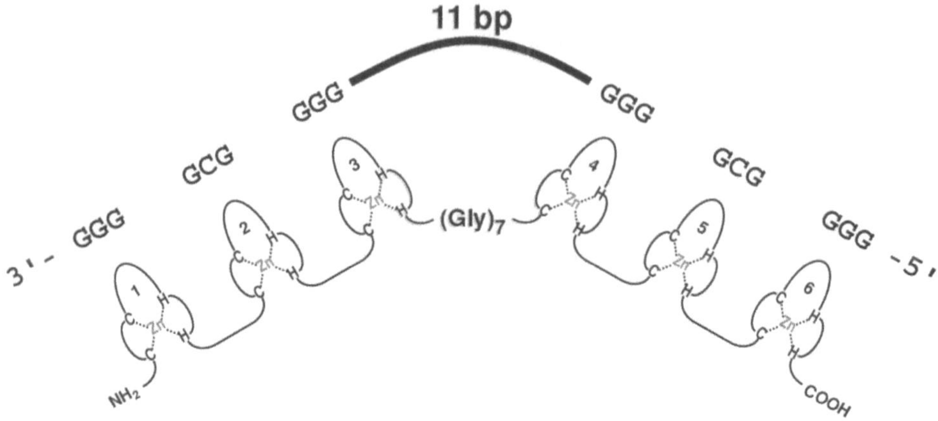

Fig. 3. Schematic drawing of Sp1ZF6(Gly)$_7$. DNA sequence bent by the six-finger protein is shown above of the figure. The amino acid residues are depicted by their one-letter codes.

3.4.1. Strategy of Protein Design

By Cys → His mutations of three (Cys)$_2$(His)$_2$-type zinc finger protein, Sp1, a novel (His)$_4$-type zinc finger protein was created (**Fig. 4**). In nature, (Cys)$_2$(His)$_2$-, (Cys)$_3$(His)-, (Cys)$_4$-, and (Cys)$_6$-type zinc finger proteins exist (*see* **Note 5**).

3.4.2. Peptide Synthesis of (His)$_4$-Type Zinc Finger Domain

The synthesis of the (His)$_4$ mutant peptide (H4Sp1f2) of the middle finger of the three-finger Sp1 was conducted by 9-fluorenylmethoxycarbonyl solid-phase synthesis on a Rink amide resin. The peptide chain was constructed with a Shimadzu PSSM-8 synthesizer, using its standard protocol with the benzotriazole-1-yloxytrispyrrolidinophosphonium hexafluorophosphate (PyBOP)–1-hydroxy-benzotriazol (HOBt)–*N*-methylmorpholine (NMM) coupling system. The protected peptide resin of H4Sp1f2 was treated with trifluoroacetic acid and ethanol (95:5) at room temperature for 2 h, followed by high-performance liquid chromatography purification on a μBondesphere$_5$C$_4$-300 (19 × 150-mm) column. The fidelity of the product was confirmed by time-of-flight mass spectrometry using a Kratos Kompact MALDI 4.

3.4.3. Construction of H$_4$Sp1

The (His)$_4$ mutant, H$_4$Sp1, was constructed by mutation of six cysteine residues to histidine residues in the three-zinc finger domain of Sp1. A DNA fragment coding H$_4$Sp1 was generated from pUC-Sp1(530-623) by means of a

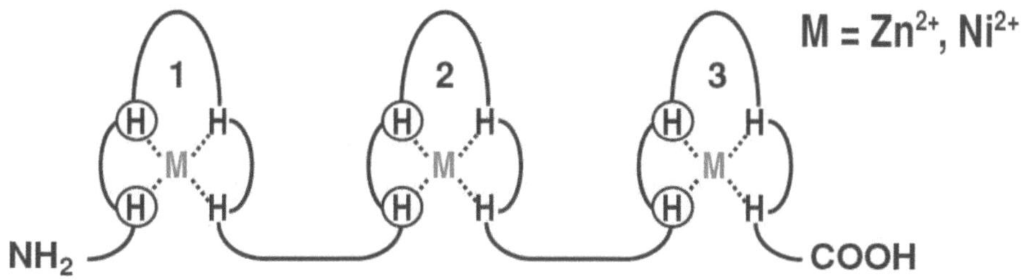

Fig. 4. Schematic drawing of H_4Sp1. The positions of Cys → His substitution are represented as circles. The amino acid residues are depicted by their one-letter codes.

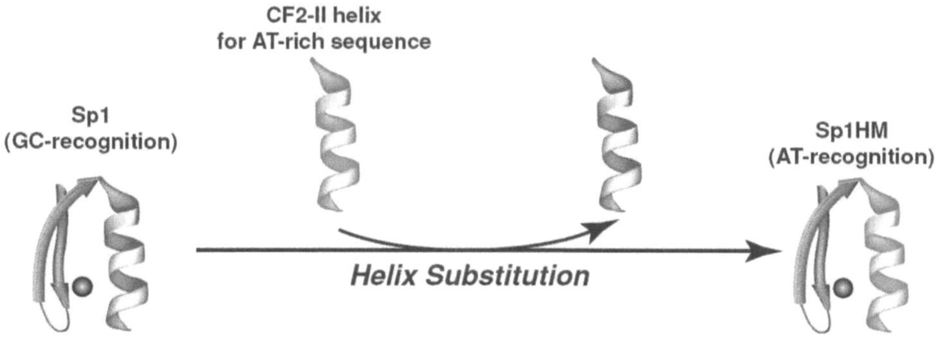

Fig. 5. Representation of the strategy of helix substitution for alternating between GC and AT recognition (*see* **Notes 6** and **7**).

polymerase chain reaction (PCR)-based site-directed mutagenesis, and the sequence was confirmed by a GeneRapid DNA sequencer (Amersham Biosciences). The amplified fragment was cut out with *Bam*HI and *Eco*RI, and inserted into a similar digested pEV-3b (*see* **Subheading 3.1.1.**; **ref. *12***).

3.5. AT-Recognizing Zinc Finger Proteins

In **Subheadings 3.5.1.** and **3.5.2.**, the design and construction of AT-recognition zinc fingers are described.

3.5.1. Strategy of Protein Design

Three fingers of the $(Cys)_2(His)_2$-type zinc finger protein, Sp1, bind to GC-rich sequences, whereas three fingers (fingers 4–6) of the six $(Cys)_2(His)_2$-type zinc finger protein, CF2-II, bind to AT-rich sequences. Novel zinc finger protein binding to AT-rich sequences was created by α-helix substitution between those two types of zinc finger proteins (**Fig. 5**).

3.5.2. Construction of Sp1HM

The oligonucleotides encoding residues of the α-helix of CF2-II were prepared. PCR-based site-directed mutagenesis was conducted by using pUC-Sp1(530–623) as the template. The PCR-amplified mutant fragment was digested with *Bam*HI and *Eco*RI, and inserted into a similarly digested plasmid, pEV-3b (*see* **Subheading 3.1.1.**; **ref.** ***13***).

4. Notes

1. The Krüppel-type linker (Thr–Gly–Glu–Lys–Pro, TGEKP) is conserved in many zinc finger proteins, and, hence, was selected for connection of the Sp1 zinc finger domains. These artificial multi-zinc finger proteins show extended sequence specificity, and their favorable sequences depend on both the number of motifs and the character of the Sp1 three-zinc finger DNA-binding domains.
2. Evidently, the footprinting analysis demonstrated that the present artificial zinc finger proteins, Sp1ZF6 and Sp1ZF9, bind to at least 18 and 27 contiguous GC-rich basepairs of DNA sequences, respectively, using all zinc finger domains.
3. The newly designed six-zinc finger proteins, Sp1ZF6$(Gly)_7$ and Sp1ZF6$(Gly)_{10}$, can induce DNA bending at the intervening region of the two distal binding sites, and the linker length between two three-zinc finger motifs has a crucial effect on the entire DNA-bending direction. The phasing assays strongly suggested that the induced DNA bending was directed toward the major groove of DNA and that Sp1ZF6$(Gly)_7$ caused the most drastic direction change in DNA bending.
4. We also created six-zinc finger proteins, Sp1ZF6$(Gly{\bullet}Arg)_4$ and Sp1ZF6$(Gly{\bullet}Glu)_4$, by connecting two DNA-binding domains of Sp1 with different charged linkers (*see* **ref.** ***14***).
5. A novel $(His)_4$-type zinc finger protein has never been observed in nature. Of special interest is the fact that the zinc finger domains of the artificial $(His)_4$-type Sp1 are folded and bind DNA similarly to the wild-type $(Cys)_2(His)_2$-type Sp1.
6. Most of the natural proteins with $(Cys)_2(His)_2$-type zinc finger proteins bind to GC-rich sequences, and AT-recognition zinc finger proteins are very rare. The creation of a zinc finger protein for the recognition of the AT-rich sequence by α-helix substitution is convenient and useful.
7. Our strategy of α-helix substitution is also effective in the creation of the zinc finger proteins binding to the sequences alternating between the AT- and GC-rich subsites.

References

1. Nagaoka, M. and Sugiura, Y. (2000) Artificial zinc finger peptides: creation, DNA recognition, and gene regulation. *J. Inorg. Biochem.* **82,** 57–63.

2. Imanishi, M., Hori, Y., Nagaoka, M., and Sugiura, Y. (2001) Design of novel zinc finger proteins: towards artificial control of specific gene expression. *Eur. J. Pharm. Sci.* **13,** 91–97.
3. Pabo, C. O., Peisach, E., and Grant, R. A. (2001) Design and selection of novel Cys_2His_2 zinc finger proteins. *Annu. Rev. Biochem.* **70,** 313–340.
4. Pavletich, N. P. and Pabo, C. O. (1991) Zinc finger-DNA recognition: crystal structure of a Zif268-DNA complex at 2.1 Å. *Science* **252,** 809–816.
5. Pavletich, N. P. and Pabo, C. O. (1993) Crystal structure of a five-finger GLI-DNA complex: new perspectives on zinc fingers. *Science* **261,** 1701–1707.
6. Kim, C. A. and Berg, J. M. (1996) A 2.2 Å resolution crystal structure of a designed zinc finger protein bound to DNA. *Nat. Struct. Biol.* **3,** 940–945.
7. Nolte, R. T., Conlin, R. M., Harrison, S. C., and Brown, R. S. (1998) Differing roles for zinc fingers in DNA recognition: structure of a six-finger transcription factor IIIA complex. *Proc. Natl. Acad. Sci. USA* **95,** 2938–2943.
8. Wütrich, K. (1986) *NMR of Proteins and Nuleic Acids.* Wiley Interscience, New York, NY.
9. Kuwahara, J. and Coleman, J. E. (1990) Role of the zinc(II) ions in the structure of the three-finger DNA binding domain of the Sp1 transcription factor. *Biochemistry* **29,** 8627–8631.
10. Kamiuchi, T., Abe, E., Imanishi, M., Kaji, T., Nagaoka, M., and Sugiura, Y. (1998) Artificial nine zinc-finger peptide with 30-base-pair binding sites. *Biochemistry* **37,** 13,827–13,834.
11. Imanishi, M., Hori, Y., Nagaoka, M., and Sugiura, Y. (2000) DNA-bending finger: artificial design of 6-zinc finger peptides with polyglycine linker and induction of DNA bending. *Biochemistry* **39,** 4383–4390.
12. Hori, Y., Suzuki, K., Okuno, Y., Nagaoka, M., Futaki, S., and Sugiura, Y. (2000) Artificial zinc finger peptide containing a novel His_4 domain. *J. Am. Chem. Soc.* **122,** 7648–7653.
13. Nagaoka, M., Doi, Y., Kuwahara, J., and Sugiura, Y. (2002) Novel strategy for the design of a new zinc finger: creation of a zinc finger for the AT-rich sequence by α-helix substitution. *J. Am. Chem. Soc.* **124,** 6526–6527.
14. Imanishi, M. and Sugiura, Y. (2002) Artificial DNA-bending six-zinc finger peptides with different charged linkers: Distinct kinetic properties of DNA binding. *Biochemistry* **41,** 1328–1334.

6

Monobodies

Antibody Mimics Based on the Scaffold of the Fibronectin Type III Domain

Akiko Koide and Shohei Koide

Summary

We developed the use of the 10th fibronectin type III domain of human fibronectin (FNfn10) as a scaffold to display multiple surface loops for target binding. We termed FNfn10 variants with novel binding function "monobodies." FNfn10 is a small (94 residues) protein with a β-sandwich structure similar to the immunoglobulin fold. It is highly stable without disulfide bonds or metal ions, and it can be expressed in the correctly folded form at a high level in bacteria. These desirable physical properties render the FNfn10 scaffold compatible with virtually any display technologies. This chapter describes methods for library construction and screening and for the production of monobodies.

Key Words: Combinatorial library; phage display; yeast two-hybrid system; overexpression.

1. Introduction

A major goal of protein engineering is to produce novel biomolecules that bind to a specified target. Such molecules would be useful in many biotechnology applications, including diagnosis, therapy, and catalysis. Although there are a vast number of natural proteins that can be exploited for such applications, the approach of engineering novel, useful biomolecules is attractive because it is not limited by the availability of natural biomolecules. The basic architecture of the antibodies, in which variable loops are presented on an essentially constant framework, and the successes of antibody engineering have inspired a number of groups to develop "molecular scaffolds" *(1,2)*. The main goal of these attempts has been to overcome the disadvantages of antibodies and antibody fragments, including the large size, heterodimeric nature (except for single immunoglobulin domains), low conformational stability, and difficulty in large-scale production,

From: *Methods in Molecular Biology, vol. 352: Protein Engineering Protocols*
Edited by: K. M. Arndt and K. M. Müller © Humana Press Inc., Totowa, NJ

while retaining the ability to present multiple peptide segments for target binding. Most of such artificial scaffolds developed to date are based on either fragments of immunoglobulins or small monomeric proteins.

We have developed the use of the fibronectin type III domain (FN3) as a molecular scaffold. Fibronectin is a large protein that plays essential roles in the formation of the extracellular matrix and cell–cell interactions. It consists of many repeats of three types (I, II, and III) of small domains ***(3)***. FN3 is a ubiquitous domain, and is estimated to occur in approx 2% of all animal proteins ***(4)***. There are 15 repeating units of FN3 in human fibronectin. We have built our system on FNfn10. FNfn10 is small (94 residues) and monomeric, and it does not contain disulfide bonds. High-level expression of correctly folded FNfn10 in bacteria is straightforward ***(5)***. The three-dimensional structure of FNfn10 ***(6,7)*** is best described as a β-sandwich similar to that of the antibody variable domain of the heavy chain domain, except that FNfn10 has seven β-strands instead of nine (**Fig. 1**). FNfn10 contains three surface loops on each end, which can potentially be used to present multiple peptide segments. These properties of FNfn10 make it an ideal candidate for a molecular scaffold.

We have demonstrated that novel binding proteins (termed monobodies) can be engineered by screening combinatorial libraries of FNfn10 in which residues in surface loops are diversified, and that the resulting monobodies retain the global structure of the scaffold and good conformational stability ***(8)***. Furthermore, because FNfn10 does not require disulfide formation for proper folding and stability, monobodies are compatible with virtually any screening methods, including phage display ***(8)***, peptide–ribonucleic acid fusion ***(9)***, and in vivo methods, such as the yeast two-hybrid system ***(10)***. We have also demonstrated that monobodies can be readily used inside living cells ***(10)***.

In this chapter, we describe methods for the construction of combinatorial libraries, for library screening using phage display and yeast two-hybrid technologies, and for bacterial production of monobodies.

2. Materials

2.1. Culture Media

1. Luria-Bertani (LB): 10 g bacto-tryptone, 5 g yeast extract, and 10 g NaCl per 1 L water, pH 7.0. For plates, add 15 g/L bacto-agar.
2. SOC: dissolve 20 g tryptone, 5 g yeast extract, and 0.5 g NaCl in approx 900 mL of water; add 10 mL of 250 m*M* KCl and 5 mL of 2 *M* $MgCl_2$; and adjust to a final volume of 1 L. Autoclave, cool, and add 20 mL of 1 *M* glucose, sterilized by filtration through a 0.2-μm filter.
3. 2xYT: 16 g bacto-tryptone, 10 g yeast extract, and 5 g NaCl per 1 L water, pH 7.0.
4. Superbroth: 12 g of bacto-tryptone, 24 g yeast extract, and 5 g glycerol per 1 L water. After autoclaving, add 50 mL of sterile solution containing 0.17 *M* KH_2PO_4 and 0.72 *M* K_2HPO_4.

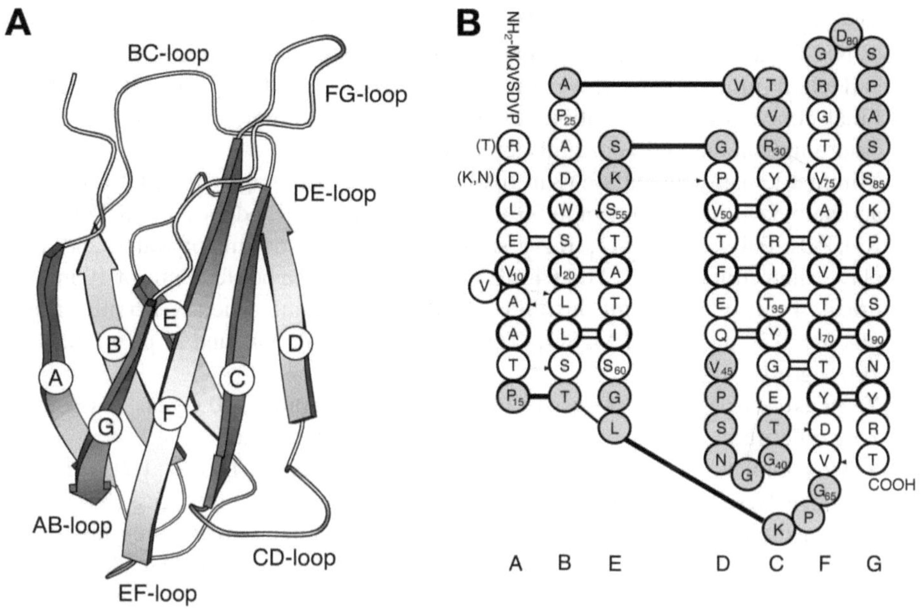

C

```
         BamHI  NdeI
  1  GGATCCCATATGCGTGTTTCTGATGTTCCGCGTGACCTGGAAGTTGTTGCTGCGACCCCG
  1  GlySerHisMetArgValSerAspValProArgAspLeuGluValValAlaAlaThrPro

                   BclI PvuII              PstI                          BsiWI
 61  ACTAGCCTGCTGATCAGCTGGGATGCTCCTGCAGTTACCGTGCGTTATTACCGTATCACG
 21  ThrSerLeuLeuIleSerTrpAspAlaProAlaValThrValArgTyrTyrArgIleThr

                                          EcoRI
121  TACGGTGAAACCGGTGGTAACTCCCCGGTTCAGGAATTCACTGTACCTGGTTCCAAGTCT
 41  TyrGlyGluThrGlyGlyAsnSerProValGlnGluPheThrValProGlySerLysSer

                                       SalI                Bst1107I
181  ACTGCTACCATCAGCGGCCTGAAACCGGGTGTCGACTATACCATCACTGTATACGCTGTT
 61  ThrAlaThrIleSerGlyLeuLysProGlyValAspTyrThrIleThrValTyrAlaVal

                                SacI
241  ACTGGCCGTGGTGACAGCCCAGCGAGCTCCAAGCCAATCTCGATTAACTACCGTACCTAG
 81  ThrGlyArgGlyAspSerProAlaSerSerLysProIleSerIleAsnTyrArgThr•••

            XhoI  BamHI
301  TAACTCGAGGATCC
```

Fig. 1. **(A)** A schematic drawing of the FNfn10 structure. The locations of β-strands A to G and loops connecting them are shown. **(B)** The amino acid sequence of FNfn10 shown in its β-strand topology. Residues in the loops have gray background. Hydrogen bonds found in the structure are shown with arrows and lines. **(C)** The nucleic acid sequence of a synthetic gene for FNfn10 that we have used in our studies.

5. M9-tryptone: Mix 6.4 g of $Na_2HPO_4 \cdot 7H_2O$, 1.5 g of KH_2PO_4, 0.25 g of NaCl, 0.5 g of NH_4Cl, and 5 g of bacto-tryptone in 500 mL water, and autoclave. Cool to 50°C and add 5 mL of 50% glucose, 1 mL of 1 *M* $MgSO_4$, and 50 µL of 1 *M* $CaCl_2$.
6. Final concentrations of antibiotics are: 100 µg/mL ampicillin (Ap); 30 µg/mL chloramphenicol; and 10 µg/mL tetracycline (Tc).
7. Minimal kanamycin sulfate (Km) plates: Mix 6 g bacto-agar in 320 mL of water and autoclave, and then cool to 50°C. In a separate bottle, add 4.2 g K_2HPO_4, 1.8 g KH_2PO_4, 0.4 g $(NH_4)_2SO_4$, 0.2 g sodium citrate, and 64 mL of water, and autoclave, then cool to 50°C. Add 64 mL of the media to the bottle, 4 mL of 10 mg/mL uracil, 4 mL of 10 mg/mL histidine, 4 mL of 10 mg/mL leucine, 1.6 mL of 50% glucose, 0.32 mL of 1 *M* $MgSO_4$, 0.66 mL of 30 mg/mL Km, and 20 µL of thiamine 100 mg/mL.
8. Yeast extract-peptone-dextrose (YPD): 20 g bacto-peptone, 10 g yeast extract, and 20 g glucose in 1 L water. For plates, add 20 g/L bacto-agar.
9. Yeast complete (YC) glucose leu– his– trp– ura–: For 1 L, mix 1.2 g of Yeast Nitrogen Base w/o amino acids w/o ammonium sulfate (Difco), 5 g $(NH_4)_2SO_4$, 10 g succinic acid, 6 g NaOH, 0.6 g complete supplement mixture-leu-his-trp-ura (Q-BIOgene), 0.1 g of cysteine, and 960 mL of water and autoclave. Add 40 mL of sterile 50% glucose and mix well. For plates, add 20 g/L bacto-agar.
10. YC Glc his– ura– trp–: add 0.1 g of leucine to YC Glc leu– his– trp– ura– media before autoclave.
11. YC Glc trp–: add 0.1 g of leucine, 0.05 g of histidine, and 0.1 g of uracil to YC Glc leu– his– trp– ura– media before autoclaving.
12. YC Glc his– ura–: add 0.1 g of leucine to YC Glc leu– his– trp– ura– media before autoclaving and use 910 mL of water instead of 960 mL. Add 50 mL of sterile 2 mg/mL tryptophan solution after autoclave.
13. YC galactose raffinose leu– his– trp– ura–: replace glucose of YC Glc leu– his– trp– ura– media with 100 mL of sterile 20% galactose and 50 mL of sterile 20% raffinose. Use 850 mL of water instead of 960 mL.
14. YC Gal Raf his– trp– ura–: add 0.1 g of leucine to YC Gal Raf leu– his– trp– ura– media before autoclaving.

2.2. Reagents

1. Polyethylene glycol (PEG)/NaCl solution: Mix 66.6 g of PEG8000, 78 g of NaCl, and 317 mL of H_2O, and autoclave.
2. HEPES buffer: 1 m*M* HEPES, pH 7.0.
3. Coating solution: 20 µg/mL ligand in 0.1 *M* $NaHCO_3$, pH 8.6.
4. Tris-buffered saline (TBS): 50 m*M* Tris-HCl and 150 m*M* NaCl, pH 7.5.
5. TBS–Tween-20 (TBST): TBS and 0.5% (v/v) Tween-20.
6. 5% sterile bovine serum albumin (BSA): filter sterilize 5% BSA in water and store at 4°C.
7. TBST–BSA: TBST with 1 mg/mL BSA.
8. Acid elution buffer: 0.1 N HCl adjusted to pH 2.2 with glycine, and 1 mg/mL BSA.
9. 2 *M* Tris-base.

10. Agarose-phosphate solution: mix 8 mL water, 2 mL 0.5 *M* potassium phosphate buffer, pH 7.0, and 0.07 g agarose per plate (multiply these amounts according to the number of plates).
11. 2X β-galactosidase assay solution: 120 m*M* Na_2HPO_4, 80 m*M* NaH_2PO_4, 20 m*M* KCl, 2 m*M* $MgSO_4$, 0.54% β-mercaptoethanol, 0.0008% sodium dodecyl sulfate, 8 mg/mL 2-nitrophenyl-β-D-galactoside, pH 7.0. Store in small aliquots at –20°C.
12. Buffer B: 20 m*M* Tris-HCl, pH 8.0, and 500 m*M* NaCl.
13. Buffer C: 20 m*M* Tris-HCl, pH 8.0, 500 m*M* NaCl, and 500 m*M* imidazole.
14. Buffer D: 50 m*M* sodium phosphate and 500 m*M* NaCl, pH 8.0.
15. Tris-HCl–EDTA (TE) buffer: 10 m*M* Tris-HCl and 1 m*M* EDTA, pH 8.0.
16. 50X Tris–acetate–EDTA buffer: dissolve 242 g Tris-base, 57.1 mL glacial acetic acid, and 18.6 g EDTA in a final volume of 1 L ddH_2O. Adjust pH to 8.3.
17. *Escherichia coli* strains: CJ236 (Bio-Rad), KC8 (Origene), XL-1 Blue (Stratagene), and SS-320 **(ref. *11*;** *see* **Note 1**).
18. Yeast strains: EGY48 and RFY206 (Origene; **refs.** ***12*** and ***13***).

2.3. Vectors

1. pAS38: a derivative of the pBluescript SK(+) phagemid (Stratagene) encoding *Fnfn10* fused to the C-terminal domain of *M13* gene III. We constructed the *FNfn10* gene using preferred codons for *E. coli* ***(8)***. This is for monovalent display of monobodies. The *Pvu*II fragment in this vector is in the opposite direction from that in pBluescript SK(+).
2. pAS45: a monobody expression plasmid derived from pET15b (Novagen). It expresses a His-tag–monobody fusion protein. The monobody gene contains an Arg6 to Thr mutation that was originally introduced to eliminate a thrombin cleavage site ***(8)***.
3. pAS47: A derivative of pAS38 containing a stop codon in the FG-loop (**Fig. 1A**). This is used as the template for mutagenesis for the construction of libraries in which residues in the FG-loop are diversified.
4. JCFN: A derivative of JC-M13-88 ***(14)*** encoding a monobody–gene VIII fusion protein. This is an M13 phage system for multivalent display of monobodies.
5. pYT45: A derivative of pYesTrp2 (Invitrogen) encoding B42 (activation protein)–monobody fusion protein ***(10,15)***. This is used to construct monobody libraries for yeast two-hybrid screening (*see* **Note 2**).
6. pSH18-34: a LacZ reporter plasmid for yeast two-hybrid screening (Origine).
7. pEG202: LexA-bait plasmid for yeast two-hybrid screening (Origene; *see* **Note 2**).

The monobody vectors and libraries are available from the corresponding author.

3. Methods

3.1. Library Construction

We used the Kunkel mutagenesis technique to construct most of our libraries ***(16)***. We used the "top" end (BC, DE, and FG loops; **Fig. 1A**) of FNfn10 as the target-binding site in most cases, although we have also shown that the AB loop can be used for target binding ***(10)***. Our method is based on the Bio-Rad Mutagene kit ***(17)***. Using *E. coli* SS320 for electrotransformation, a library containing approx 10^9 independent clones can be readily prepared.

3.1.1. Preparation of a Uracil-Single-Stranded Phagemid (for Monovalent Phage Display and Yeast Two-Hybrid Screening)

1. Transform CJ236 with pAS47, pAS38 (for monobody–gene III) or pYT45 (for B42–monobody), select colonies on LB-Ap plate.
2. Inoculate 2 mL of 2xYT-Ap-chloramphenicol with a colony; shake at 37°C overnight.
3. Inoculate prewarmed 30 mL 2xYT-Ap supplemented with 0.25 μg/mL uridine in a 250-mL baffled flask with 300 μL of the overnight CJ236 culture, and add 10^{10}/mL helper phage KO7 (Promega). Grow at 37°C for 2 h with vigorous shaking. Add Km at a final concentration of 70 μg/mL. Grow overnight at 37°C with vigorous shaking.
4. Centrifuge at 12,000*g* (10 krpm in a SS-34 [Sorvall] or equivalent rotor) for 10 min. Transfer supernatant to a new tube and centrifuge again at 12,000*g* for 10 min. Transfer supernatant to a new tube. Add 4.5 mL PEG/NaCl solution, and mix well. Incubate at 4°C overnight.
5. Centrifuge at 17,200*g* (12 krpm, SS-34 rotor) for 20 min at 4°C. Discard supernatant and place the tube upside down on a stack of paper towels for 1 min. Wipe off residual liquid with paper towel.
6. Suspend the phagemid pellet in 1 mL TBS and transfer to a microcentrifuge tube. Centrifuge for 2 min at the maximal speed in a microcentrifuge. Transfer supernatant to a new microcentrifuge tube. Add 150 μL of PEG/NaCl solution and mix well. Incubate on ice for 30 min.
7. Centrifuge for 10 min at the maximum speed at 4°C. Discard supernatant, centrifuge briefly, and remove all liquid with a pipetter. Suspend the pellet in 1 mL TBS.
8. Check the titers of the phagemid with XL-1 Blue and CJ236. Mix 100 μL of fresh *E. coli* (optical density [OD] at 600 nm [OD_{600nm}] = 0.5–1.0) and 10 μL of a serial dilution of the phagemid solution. Incubate at room temperature for 15 min and plate on LB-Ap. Incubate the plates at 37°C overnight. The titer with CJ236 should be 10^{12} to 10^{14}/mL and should be more than 10^4 higher than that with XL-1 Blue.
9. Prepare uracil-single-stranded DNA using the QIAGEN single-stranded DNA preparation kit.

3.1.2. Preparation of a Uracil-Single-Stranded Phage (for Multivalent Phage Display Vector, JCFN)

1. Transfect CJ236 with JCFN plasmid. After heat shock, mix with 30 mL prewarmed 2xYT and 500 μL of fresh CJ236 (OD_{600nm} of 0.5–1.0), and then shake at 37°C for 6 h.
2. Perform **steps 4** to **9** of **Subheading 3.1.1**. For titering, mix 3 mL of LB and 0.7% agar (melted and cooled to 45°C), 10 μL of a serial dilution of the phage solution,

and 100 µL of fresh *E. coli*. Vortex and immediately pour into an LB plate. Incubate the plates at 37°C overnight and titer.

3.1.3. Preparation of Mutagenic Oligonucleotides

1. Design an oligonucleotide that contains the 5′ and 3′ complimentary regions of approx 20 bases and approx 15 bases, respectively, depending on the GC content. We use the NNK or NNS codons for "hard" randomization, where N denotes an equimolar mixture of all nucleotides; K denotes an equimolar mixture of G and T, and S denotes an equimolar mixture of G and C.
2. Run approx 10 µg of a crude oligonucleotide dissolved in water on acrylamide gel (use ~5 cm for each 0.1-cm well), cut out a band that corresponds with the correct size, and electroelute the oligonucleotide (*see* **Note 3**).
3. Mix 2 µg purified oligonucleotide, 3 µL T4 polynucleotide kinase buffer, 5 U T4 polynucleotide kinase (New England Biolab), 1.5 µL 10 m*M* adenosine triphosphate, and water to bring up to 30 µL. Incubate at 37°C for 2 h and then at 65°C for 15 min. Store the solution at –20°C.

3.1.4. Synthesis of Double-Stranded DNA

1. Mix approx 1 µg of uracil-single-stranded DNA, 2 µL of phosphorylated oligonucleotide (**Subheading 3.1.3.**), 1 µL of 10X annealing buffer (Bio-Rad Mutagene kit), and bring up to 10 µL with H_2O (*see* **Note 4**). Prepare a control reaction without an oligonucleotide.
2. Using a polymerase chain reaction (PCR) machine, incubate at 70°C for 5 min and then decrease the temperature to 30°C, at a rate of –1°C/min. Place the tube on ice.
3. To the annealing mixture, add 1 µL of 10X synthesis buffer (Bio-Rad Mutagene kit), 1 µL of T4 DNA ligase, and 1 µL of T7 DNA polymerase (New England Biolabs). Incubate for 5 min on ice, 5 min at room temperature, 30 min at 37°C, and 15 min at 75°C, and then cool to room temperature. Run 1 µL of the sample on agarose gel. The sample with the oligonucleotide should yield products of higher molecular weights than the control without an oligonucleotide.
4. Dialyze the synthesis mixture by placing a drop of the mixture on a 0.025-µm membrane disk floating on water for 1 h, then recover the dialyzed mixture in a fresh microcentrifuge tube. Keep it on ice until use.

3.1.5. Preparation of Electrocompetent Cells

1. Grow *E. coli* SS-320 in 350 mL of superbroth containing Tc until the OD_{600nm} reaches 0.8.
2. Chill the culture in the flask quickly in ice-water bath, then keep it on ice for 15 min. Transfer the culture into a centrifuge bottle and centrifuge at 0 to 2°C at 3900*g* (4.7 krpm in a JA-10 [Beckman] or equivalent rotor) for 5 min (*see* **Note 5**).
3. Discard the supernatant and suspend cells in 10 mL HEPES buffer by knocking the centrifuge tube on ice. Add 390 mL HEPES buffer, and swirl to suspend cells. Centrifuge at 0 to 2°C at 3900*g* for 5 min.

4. Discard the supernatant. Suspend cells in 10 mL HEPES buffer and transfer the suspension to a 50-mL centrifuge tube. Rinse the first bottle with the buffer and recover the cell suspension to the 50-mL tube.
5. Centrifuge at 4300*g* (6 krpm in a SS-34 rotor) for 10 min at 0 to 2°C. Discard supernatant. Suspend cells in 10 mL of 10% glycerol by knocking the tube on ice, then add 20 mL of 10% glycerol and mix the suspension by inverting the tube several times.
6. Centrifuge at 4300*g* for 10 min at 0 to 2°C. Discard supernatant. Add 300 µL of 10% glycerol and suspend cells by knocking the tube on ice. Check the volume and add 10% glycerol to a final volume of 700 µL. Aliquot the suspension into two 350-µL suspensions in microcentrifuge tubes.

3.1.6. Electroporation for Phagemid DNA

1. Add chilled DNA (**Subheading 3.1.4., step 4**) to 350 µL of electrocompetent cells in a microcentrifuge tube and incubate on ice for 5 min. Transfer the mixture to a prechilled electroporation cuvette (2-mm gap).
2. Electroporate the cells, e.g., using BTX ECM395 electroporator in the high-voltage mode at 2.5 kV.
3. Immediately add 1 mL SOC into the cuvet and suspend the cells. Transfer the cells to a 250-mL flask. Add 1 mL of SOC into the cuvet and wash the remaining cells and transfer to the flask. Repeat this process three more times.
4. Add 20 mL SOC into the flask. Shake the flask at 180 rpm, 37°C for 30 min, and check the titer by plating a diluted portion of the suspension on LB-Ap plate.
5. Add the entire suspension to 500 mL of prewarmed 2xYT-Ap. For a plasmid preparation for the yeast two-hybrid screening, shake at 37°C overnight and prepare plasmid. For phagemid preparation, add 10^{10} pfu/mL of helper phage KO7, shake overnight at 37°C, and prepare phagemids as described in **Subheading 3.1.1.**

For transformation of phage DNA, after **step 3**, add 1.5 mL of mid-log SS-320 cells, then transfer the suspension to prewarmed 120 mL of 2xYT-Tc. Titer the phages as described in **Subheading 3.1.2.** Incubate for 4 h at 37°C. Prepare phages as described in **Subheading 3.1.1.**

3.1.7. Construction of a Yeast Library

We construct vector libraries for yeast two-hybrid screening first in *E. coli*, as described in **Subheading 3.1.** and then introduce the libraries in yeast. Transformation of yeast is based on the method of Gietz ***(18)***. We typically obtain approx 10^6 independent clones using this method.

1. Grow yeast EGY48 in 20 mL YPD with shaking at 30°C overnight. Add the culture to prewarmed 300 mL YPD in such a way that the OD_{600nm} is 0.25. Grow the cells until OD_{600nm} is approx 1.
2. Centrifuge at 3600*g* (4.5 krpm, JA-10 rotor) for 5 min at room temperature. Discard supernatant, and suspend cells in 300 mL sterile water. Repeat centrifugation and

suspension. Centrifuge again, suspend cells in 25 mL sterile water, and transfer to a 50-mL centrifuge tube. Centrifuge at 3000*g* (5 krpm, SS34 rotor) for 5 min at room temperature. Discard supernatant and suspend cells in 1.2 mL sterile water.

3. Set aside 35 µL of the cell suspension for a negative control. To the remainder of the cell suspension, add 7.2 mL of 50% PEG3350, 1.08 mL of 1 *M* lithium acetate, 500 µL of carrier single-stranded DNA (Origene), and 45 µg of DNA, in 900 µL water. Mix well and aliquot 360 µL each into 30 microcentrifuge tubes. To the tube for a negative control, add 1/30 of the above reagents, except for the DNA (i.e., PEG, lithium acetate, and carrier DNA) and mix well.
4. Incubate at 42°C for 40 min. Centrifuge at room temperature at 10 krpm in a microcentrifuge for 10 s, and discard supernatant. Suspend cells in each tube in 150 µL of water. Plate cells in each tube on a selection plate (YC Glc trp–) in a 15-cm-diameter Petri dish. Incubate at 30°C for 3 d.
5. Add 15 mL of water per plate and scrape off all of the colonies using a sterile slide glass. Combine all of the cell suspension in a 500-mL centrifuge tube, and centrifuge at 3900*g* (4.7 krpm, JA-10 rotor) for 5 min. Discard supernatant and suspend cells in 500 mL water. Centrifuge and discard supernatant again. Suspend the cells in 50 mL water and measure the OD_{600nm} to determine the cell density (1 OD_{600nm} corresponds to ~2×10^7 cells/mL). Add glycerol at a final concentration of 15% (v/v). Aliquot and store the cell suspension at –80°C.

3.2. Sorting of Phage-Display Libraries

3.2.1. Panning (11)

1. Add 50 µL of coating solution into the well (1 µg ligand/well) of an enzyme-linked immunosorbent assay plate (Nunc Maxisorp, separatable well, C-bottom). Place the plate in a sealed box with a wet paper towel inside. Keep the box at 4°C overnight or 37°C for 1 h.
2. Remove the fluid using a pipetter and wash the well twice with water. Add 360 µL of 3% BSA in TBS in the well. Incubate at 37°C for 1 h. For a later round, prepare a control well; add the buffer without ligand and block the well with BSA.
3. Remove the blocking solution from the well, and wash twice with water. Add 12 µL of 5% BSA solution and 50 µL of phage suspension (10^{11}/well). Incubate at room temperature for 2 h.
4. Remove the phage solution, and then add 360 µL of TBST. Pipet vigorously up and down. Wait 5 min, and remove the solution (one time in the first round and five times in later rounds).
5. Add 50 µL of 10 µ*M* ligand in TBST to the well. Incubate at 37°C for 1 h, pipet up and down vigorously, and recover the eluate. Alternatively, add 50 µL of acid elution buffer into the well, wait 10 min at room temperature. Pipet up and down vigorously. Recover the second eluate into a microcentrifuge tube that has 3 µL of 2 *M* Tris-base for neutralization.
6. Mix 5 mL of fresh XL-1 Blue (OD_{600nm} between 0.5 and 1; grown in LB-Tc) and eluted phage. For phage, add 100 mL of 2xYT (and isopropylthiogalactoside [IPTG]

up to 1 m*M* for higher induction of monobody) and incubate with vigorous shaking at 37°C for 6 h (longer incubation may lead to a deletion within the monobody gene). Determine the titer immediately after adding 2xYT, as described in **Subheading 3.1.2**. For phagemid, mix 5 mL of fresh XL-1 Blue and eluted phagemid, and incubate for 20 min at 37°C. Add 100 mL of 2xYT (and IPTG, if desired), 100 µg/mL Ap, and 10^{10}/mL helper phage KO7; incubate with vigorous shaking at 37°C for 2 h. Add Km at a final concentration of 70 µg/mL and incubate overnight. Determine the titer immediately after the 20 min incubation, as described in **Subheading 3.1.1**. Isolate the phage (or phagemid) as described in **Subheadings 3.1.1.** and **3.1.2.** Repeat the entire procedure until there is an increase in the eluted phage(mid) number.

3.2.2. Characterization of Individual Phage Clones

1. Prepare the phage(mid) from a single clone. For the phagemid, grow a single colony in 2 mL of LB-Ap overnight. Mix 20 µL of the culture, 2 mL of LB-Ap (and IPTG, if desired), and 10^{10}/mL helper phage KO7. Incubate overnight at 37°C with vigorous shaking. For phage clones, touch a single plaque with a sterile toothpick. Place the toothpick in 2 mL of LB-Ap (and IPTG, if desired) with 100 µL of fresh XL-1 Blue, and incubate at 37°C with vigorous shaking for 6 h. Prepare the phage(mid) as described in **Subheadings 3.1.1.** and **3.1.2.**
2. For each clone to be tested, prepare a well with an immobilized target by performing **steps 1** and **2** from **Subheading 3.2.1**. Also prepare a control well by coating with coating solution without the target (*see* **Note 6**).
3. To the ligand-coated and the control wells, add 5 µL of 5% BSA solution and 50 µL of phage suspension prepared from individual clones. Adjust the phage concentration (typically 10^8–10^{11}/well) depending on the strength of interaction. Incubate at room temperature for 2 h on a rotator.
4. Wash five times with TBST. Using a squirt bottle, add TBST to all of the wells, then shake off the liquid and slap the plate face down against a stack of paper towels.
5. To each well, add 50 µL of antibody to phage–horseradish peroxidase solution (Pharmacia; diluted 1:2000 in TBST:BSA), incubate 1 h at room temperature on a rotator. Wash five times with TBST.
6. Add 50 µL of 1step turbo-TMB (Pierce), wait approx 10 min until blue color develops, and add 50 µL of 2 *M* H_2SO_4 to stop the reaction (use filtered tips to protect the pipetter). Measure at OD_{450nm} using a plate reader. Select clones that show high binding to the ligand-coated well and low binding to the control well for further characterization (sequencing, affinity maturation, and protein production).

3.3. Yeast Two-Hybrid Screening

3.3.1. Preparation of a Target ("Bait") Plasmid

Clone the gene of a bait in pEG202 so that the bait is expressed in frame with LexA. It is important to make sure that the LexA-bait does not self-activate a reporter gene nor interact with wild-type FNfn10. This can be easily tested using the interaction mating method ***(13,19)***.

1. Transform RFY206 with a LexA-bait plasmid (or pEG202 for a negative control, and pBait [Origene] for a positive control) and pSH18-34 (β-galactosidase reporter plasmid; Origene), and select on YC Glc his– ura– plates. Also transform EGY48 with pYesTrp2 (activator B42 plasmid; Invitrogen), pTarget (positive control; Origene), or pYT45, and select on YC Glc trp– plates.
2. Streak two colonies each with a LexA plasmid and the reporter plasmid on a YC Glc his– ura– plate as parallel lines. Streak colonies with B42 plasmid on a YC Glc trp– plate in the same manner. Incubate both plates at 30°C for 1 to 2 d.
3. Replicate the plates to a single YPD plate. The lines from the two plates should be perpendicular to each other so that EGY48 cells and RFY206 cells mate at each overlap. Incubate at 30°C overnight.
4. Replicate the YPD plate to YC Glc leu– his– trp– ura–, YC Glc his– trp– ura–, and YC Gal Raf leu– his– trp– ura– plates, and incubate at 30°C for 3 d. All mated cells should grow on YC Glc his– trp– ura– plates. Growth on the YC Gal Raf leu– his– trp– ura– plate indicates interaction between the bait and prey. The other plates are for controls.

3.3.2. Yeast Two-Hybrid Library Screening

RFY206 containing a bait is mated to EGY48 containing a monobody library, and cells with a monobody that bind to the bait are selected. The use of yeast mating makes it possible to efficiently screen multiple libraries ***(13)***.

1. Grow RFY206 harboring pSH18-34 and a LexA-bait plasmid in 10 mL of YC Glc his– ura– media at 30°C for 16 h.
2. Determine the cell density at OD_{600nm} using 1.0 OD_{600nm} = 2.0×10^7 cells/mL. Spin down 10^8 bait cells in a microcentrifuge tube for 30 s at 10 krpm in a microcentrifuge, and suspend the yeast cells in 200 µL of the media.
3. Add 10^7 EGY48 cells harboring a monobody to the bait cells (**Subheading 3.1.7.**), and plate them onto a YPD plate. Allow the cells to mate for 16 h.
4. Wash the cells off the plate with 5 mL (1 mL at a time) of YPD media. Measure the OD_{600nm} of the cells after a 100-fold dilution, and plate 10^8 cells on five YC Gal Raf leu– his– trp– ura– plates. Incubate the plates for 3 to 4 d at 30°C.
5. Replicate colonies onto YC Gal Raf leu– his– trp– ura– plates and YC Glc leu– his– trp– ura– plates. Incubate the plates for 16–24 h. True positive clones should grow only on YC Gal Raf leu– his– trp– ura– plates.
6. In a chemical hood, open the lid of the plate, and add enough chloroform to cover all of the surface of the plate. After 5 min of exposure, decant the chloroform into a waste container and let the remaining chloroform on the plate evaporate for 10 min ***(20)***.
7. Heat the agarose–phosphate solution (10 mL per plate) in a microwave to dissolve the agarose, and cool to 50°C in a water bath. In a test tube, add 35 µL of 50 mg/mL X-gal solution and 10 mL agarose–phosphate solution, vortex, and pour onto the

plate. Incubate at 30°C. Observe the color. Pick the clones that show blue color for further characterization.

3.3.3. Specificity Test of Isolated Monobodies

1. Grow yeast cells of interest from library screening in 1.5 mL YPD media at 30°C for 16 h and prepare DNA from the yeast cell using Y-DER yeast DNA extraction kit (Pierce).
2. Electroporate KC8 *E. coli* cells with the yeast DNA after dialysis (*see* **Subheading 3.1.4.**) and select transformants on a LB-Ap plate.
3. Replicate the colonies on a minimal trp-Km plate, and incubate at 37°C. Grow two colonies in 2 mL LB-Ap each at 37°C for up to 10 h and extract plasmid DNA using standard methods. Using KC8 cells grown more than 10 h often yields impure plasmids. Determine the sequence of the monobody segment by standard DNA-sequencing methods (*see* **Note 7**).
4. Transform EGY48 with the B42-monobody plasmid, and test its interaction profile against various baits using the interaction mating method (**Subheading 3.3.1.**).

3.3.4. Liquid β-Galactosidase Assay

Before biochemical measurements, the affinity between a bait and a monobody can be characterized semiquantitatively by a liquid-phase β-galactosidase assay. This method is more quantitative than the plate assay described in **Subheading 3.3.2**. We have adapted this method to a 96-well plate format (*see* **Note 8**).

1. Transform a RFY206 yeast strain with pSH18-34, a LexA-bait plasmid (a derivative of pEG202), and a B42-monobody plasmid (a derivative of pYesTrp2). Inoculate 2 mL YC Glc his– ura– trp– with a colony and shake overnight at 30°C.
2. Remove the media by brief centrifugation and suspend cells in warm YC Gal Raf his– ura– trp– until the OD_{600nm} is 0.2. Shake at 30°C for 6 h. Determine the OD_{600nm} and aliquot at 350 µL each in a microcentrifuge tube. We typically measure in triplicates.
3. Mix 2X β-galactosidase assay solution with Y-Per (Pierce) at the ratio of 1 to 1 to make the working solution. Add 350 µL of the working solution into each sample. Invert several times to mix.
4. Incubate at 30°C while checking the color. When a yellow color has developed, add 300 µL of 1 N Na_2CO_3 into the tubes and mix well. Centrifuge the microcentrifuge tubes at the maximal speed for 1 min. Determine the OD_{420nm} of the supernatant.

3.4. Cloning, Expression, Purification, and Biotinylation of a Monobody

After identifying a monobody with desired properties from library selection, its gene is transferred to an *E. coli* expression vector and the monobody is expressed and purified as an isolated protein. We typically obtain 10 to 50 mg monobody per liter of culture. We also describe protocols for biotinylating a monobody for easy detection in immunochemical applications.

3.4.1. Cloning of a Monobody

1. Amplify the gene of a monobody of interest using oligonucleotides NdeMetThrFNF (5′-CGGGATCCCATATGCAGGTTTCTGATGTTCCGACCGACCTG-GAAGTTGTTGCTG-3′; this contains the Arg6 to Thr mutation) and FNGKKGKR (5′-CCGACTCGAGTTACTATTTACCTTTTTTACCGGTACGGTAGTTAATC-GAG-3′), then digest with *Nde*I and *Xho*I and clone in pET15b (Novagen) digested with the same enzymes. Confirm the gene in the expression vector by sequencing.

3.4.2. Expression of a Monobody

This protocol is for a 100 mL culture, which should yield sufficient quantities of a monobody for initial characterization.

1. Transform BL21(DE3) cells (Novagen) with a monobody expression vector (*see* **Note 9**).
2. Inoculate 10 mL M9-tryptone-Ap with transformed cells and incubate the culture with vigorous shaking at 30°C for 10 to 16 h.
3. Inoculate 2 mL of the preculture to prewarmed 100 mL M9-tryptone-Ap, and incubate with vigorous shaking at 37°C. When the OD_{600nm} reaches 0.7, add IPTG to a final concentration of 0.5 m*M*.
4. Incubate with vigorous shaking at 37°C for 3 h more, and harvest the cells by centrifugation. The cells can be stored at –20°C until use.

3.4.3. Purification of a Monobody

This protocol is for cells from 100 mL culture (**Subheading 3.4.2.**)

1. Suspend the cell pellet in 6.4 mL of 50 m*M* Tris-HCl, pH 8.0.
2. Add 64 µL of 50 mg/mL lysozyme and 128 µL of 50 m*M* PMSF. Mix and incubate at 37°C for 15 min. Sonicate the suspension until it is no longer highly viscous.
3. Add 910 µL of 4 *M* NaCl. Centrifuge at 27,000*g* (15 krpm, SS34 rotor) for 10 min at 4°C. Collect the supernatant.
4. Apply protein solution in a Hi-Trap chelating column (1 mL; Amersham-Pharmacia Biotech) that has been loaded with 0.1 *M* $NiCl_2$ and equilibrated with buffer B. Wash the column with 6 mL of buffer B, then with 6 mL of buffer B containing 60 m*M* imidazole.
5. Elute the monobody with 6 mL of buffer C. Collect 0.5 to 1 mL aliquots of the eluent in microcentrifuge tubes. Analyze the eluent with sodium dodecyl sulfate polyacrylamide gel electrophoresis.

3.4.4. Biotinylation of a Monobody

We conjugate biotin at the Lys amino group. Although this is not site specific, the Lys cluster in the C-terminal extension should be a preferred modification site. This method can be used to conjugate other moieties, e.g., a fluorescent dye.

1. Perform **steps 1** to **4** of the purification protocol, **Subheading 3.4.3.**
2. Wash the column with 10 mL buffer D.
3. Dissolve 1 mg of D-biotinoyl-ε-aminocaproic acid *N*-hydroxysuccinimide ester (Boehringer Mannheim) in 50 μL dimethylsulfoxide and dilute it in 3 mL of buffer D. Inject 1 mL of the solution into the column and incubate at room temperature for 1 h in the dark. Repeat the reaction twice.
4. Wash the column with 6 mL of buffer B and elute the monobody as in **step 5** in **Subheading 3.4.3.**

4. Notes

1. SS-320 is not commercially available, but it can be easily made by following the protocols of Sidhu et al. ***(11)***.
2. We chose pYesTrp2 as the prey plasmid mainly because of the presence of the f1 origin in this plasmid that allows Kunkel mutagenesis. We chose pEG202 as the bait plasmid because the maintenance of the plasmid in yeast was easier and more economical than pHybLex (Invitrogen), which requires zeocin.
3. Electroelution of an oligonucleotide. In dialysis tubing (MWCO 3500, 18-mm width) with one end sealed with a tubing clip, add the gel piece and 1.8 mL of TE buffer, then seal the open end with another clip, after eliminating air bubbles. Place the gel piece at an edge of the dialysis tubing and submerge the whole assembly in a gel box filled with 1X Tris–acetate–EDTA buffer. Electroelute the oligonucleotide with 200 V for 10 min. Recover the solution from the tubing. The oligonucleotide can be recovered by ethanol precipitation after phenol/chloroform extraction.
4. More than one oligonucleotide can be used in a single reaction to simultaneously diversify different loops.
5. For the preparation of electrocompetent cells, centrifuge bottles, pipets, and buffers must be chilled before use.
6. The control well should be coated with buffer without a target before blocking. Phages have significantly lower binding activity to a well that is simply blocked with BSA without a buffer-coating step.
7. A template for DNA sequencing can be generated by PCR amplification directly from a yeast colony. In this case, heat up the PCR reaction mixture at 96°C for 8 min before starting the amplification cycles.
8. The β-galactosidase assay can be done using a 96-well format. Using a deep-well plate (1.2 mL per well), perform the reaction in half as much volume. After stopping the β-galactosidase reaction and centrifuging at 2200*g* in a swing rotor for 96-well plates (3 krpm in a JS-5.3 [Beckman] or equivalent rotor) for 10 min, transfer 300 μL of the supernatant to a 96-well plate and determine the OD_{420nm} using a plate reader.
9. To obtain a high yield of a monobody, use freshly transformed BL21(DE3) cells.

References

1. Skerra, A. (2000) Engineered protein scaffolds for molecular recognition. *J. Mol. Recognit.* **13,** 167–187.

2. Nygren, P. -Å. and Uhlén, M. (1997) Scaffolds for engineering novel binding sites in proteins. *Curr. Opin. Struct. Biol.* **7,** 463–469.
3. Baron, M., Norman, D. G., and Campbell, I. D. (1991) Protein modules. *Trends Biochem. Sci.* **16,** 13–17.
4. Bork, P. and Doolittle, R. F. (1992) Proposed acquisition of an animal protein domain by bacteria. *Proc. Natl. Acad. Sci. USA* **89,** 8990–8994.
5. Plaxco, K. W., Spitzfaden, C., Campbell, I. D., and Dobson, C. M. (1996) Rapid refolding of a proline-rich all-beta-sheet fibronectin type III module. *Proc. Natl. Acad. Sci. USA* **93,** 10,703–10,706.
6. Dickinson, C. D., Veerapandian, B., Dai, X. -P., et al. (1994) Crystal structure of the tenth type III cell adhesion module of human fibronectin. *J. Mol. Biol.* **236,** 1079–1092.
7. Main, A. L., Harvey, T. S., Baron, M., Boyd, J., and Campbell, I. D. (1992) The three-dimensional structure of the tenth type III module of fibronectin: an insight into RGD-mediated interactions. *Cell* **71,** 671–678.
8. Koide, A., Bailey, C. W., Huang, X., and Koide, S. (1998) The fibronectin type III domain as a scaffold for novel binding proteins. *J. Mol. Biol.* **284,** 1141–1151.
9. Xu, L., Aha, P., Gu, K., et al. (2002) Directed evolution of high-affinity antibody mimics using mRNA display. *Chem. Biol.* **9,** 933–942.
10. Koide, A., Abbatiello, S., Rothgery, L., and Koide, S. (2002) Probing protein conformational changes by using designer binding proteins: application to the estrogen receptor. *Proc. Natl. Acad. Sci. USA* **99,** 1253–1258.
11. Sidhu, S. S., Lowman, H. B., Cunningham, B. C., and Wells, J. A. (2000) Phage display for selection of novel binding peptides. *Methods Enzymol.* **328,** 333–363.
12. Origene Technologies, Inc. (1998) *Dup-LEXA Yeast Two-Hybrid System Manual.* Rockville, MD.
13. Golemis, E. and Serebriiskii, I. (1997) Two-hybrid system/interaction trap, in *Cells: A Laboratory Manual*, CSH Laboratory Press, Cold Spring Harbor, NY, pp. 69.1–69.40.
14. Chappel, J. A., He, M., and Kang, A. S. (1998) Modulation of antibody display on M13 filamentous phage. *J. Immunol. Methods* **221,** 25–34.
15. Invitrogen. (2003) *Hybrid Hunter Manual.* Carlsbad, CA.
16. Kunkel, T. A., Roberts, J. D., and Zakour, R. A. (1987) Rapid and efficient sitc-directed mutagenesis without phenotypic selection. *Methods Enzymol.* **154,** 367–382.
17. Bio-Rad Laboratories. (1997) *Muta-Gene Mutagenesis Kit Manual.* Hercules, CA.
18. Gietz, R. D. and Woods, R. A. (2001) Genetic transformation of yeast. *Biotechniques* **30,** 816–828.
19. Finley, R. L. Jr. and Brent, R. (1994) Interaction mating reveals binary and ternary connections between Drosophila cell cycle regulators. *Proc. Natl. Acad. Sci. USA* **91,** 12,980–12,984.
20. Duttweiler, H. M. (1996) A highly sensitive and non-lethal beta-galactosidase plate assay for yeast. *Trends Genet.* **12,** 340–341.

7

Engineering Site-Specific Endonucleases

Peter Friedhoff and Alfred Pingoud

Summary

Site-specific endonucleases are involved in many aspects of the biochemistry of nucleic acids. Restriction enzymes and their relatives have become paradigms for enzymes acting on DNA. Numerous efforts have been devoted to changing their specificity by rational protein design, with, by and large, little success, presumably because the recognition process is highly redundant and recognition and catalysis are tightly coupled. This chapter describes one of the few successful examples of a change in specificity—namely the conversion of the mismatch repair nuclease, MutH, which, when stimulated by MutS and MutL, nicks hemimethylated d(GATC) sites, into a variant that cleaves fully methylated DNA as well as hemimethylated and unmethylated DNA. This chapter will describe the various steps involved in this design project, starting from the analysis of the structure and the identification of candidate amino acid residues responsible for sensing the methylation status, to the generation and characterization of MutH variants with an altered specificity toward hemimethylated d(GATC) sites.

Key Words: Restriction endonucleases; DNA mismatch repair; DNA methylation; MutH; protein engineering; site-directed mutagenesis; fluorescent labels; oligonucleotides; capillary electrophoresis.

1. Introduction

Type II restriction endonucleases are indispensable tools for the site-specific cleavage of DNA. They usually recognize double-stranded DNA sequences, 4 to 8 basepairs (bp) in length. In the presence of Mg^{2+}, they cleave the DNA within or adjacent to the recognition site in both strands to produce blunt or sticky ends with a 5′-phosphate and a 3′-OH (*see* review in **ref. *1***). Restriction enzymes are among the most specific enzymes known: sequences differing in only 1 bp are cleaved by a factor of 10^6 more slowly than the canonical recognition sequence. This extreme accuracy results from an intimate interplay between direct (interaction with the bases) and indirect readout (interaction with the sugar–phosphate

From: *Methods in Molecular Biology, vol. 352: Protein Engineering Protocols*
Edited by: K. M. Arndt and K. M. Müller © Humana Press Inc., Totowa, NJ

backbone). Typically, 15 to 20 hydrogen bonds are formed between the restriction enzyme and the bases of the recognition sequence, in addition to numerous van der Waals contacts to the bases and hydrogen bonds to the backbone, some of which are water mediated. This means that recognition is overdetermined and redundant. Any effort to change the specificity of restriction enzymes by rational protein design, therefore, is unlikely to be successful for this reason, because several contacts have to be altered and more than one amino acid residue has to be replaced ***(2)***. In addition, in most restriction enzymes, amino acid residues involved in recognition are located in very close proximity to residues responsible for catalysis. Although this guarantees a tight coupling between recognition and catalysis, it has been a great impediment for protein engineering (*see* review in **ref. *3***). Therefore, in retrospect, it is not too surprising that true changes in specificity have not been achieved by rational protein design ***(4)*** and that researchers have turned to evolutionary approaches to achieve changes in specificity of restriction enzymes, both in vitro ***(5,6)*** and in vivo ***(7,8)***. However, even with these approaches, thus far, only a reduction of specificity (e.g., BstyI, from RGATCY to preferentially AGATCT, **ref. *7***) or expansion of specificity (e.g., Eco57I from CTGAAG to CTGRAG, **ref. *8***) could be achieved. It may well be that restriction enzymes structurally are not sufficiently malleable and that multiple amino acid substitutions would be required for a true change in specificity. This could be a safeguard to prevent unwanted, in most cases deleterious, cleavage of the host genome by restriction enzyme variants that invariably are produced to a small extent in the process of translation. Using the same argument, it can be anticipated that it should be possible to generate variants of restriction enzymes with altered properties that were not counterselected during evolution. Indeed, EcoRI and EcoRV variants with simple amino acid substitutions could be produced that cleave DNA with a modified base (EcoRI Q115, **ref. *9***; EcoRV N188, **ref. *10***) or a modified backbone (EcoRV T94V, **ref. *11***) much better than the wild-type enzyme.

Here, we report how an enzyme related to restriction enzymes, namely the mismatch repair protein, MutH, from *Escherichia coli*, was converted by a single amino acid substitution into a variant with a relaxed specificity. The procedure that was used and is described in detail should be of general applicability. First, the system under study is introduced (*see* **Subheading 1.1.**) along with the structural considerations presented that led to the selection of amino acid substitutions required for the desired change in specificity (*see* **Subheading 1.2.**).

1.1. The DNA Mismatch Repair Protein, MutH

DNA mismatch repair is one of the most important processes for the correction of replication errors in almost all living organisms; it enhances the fidelity of DNA replication by up to a factor of 1000 (*see* review in **ref. *12***). In *E. coli*, DNA mismatch repair is initiated by MutS binding to the mismatch, followed by MutL-dependent activation of MutH, which nicks the unmethylated (i.e., newly replicated) strand at a hemimethylated d(GATC) site, which may be several hundred basepairs upstream or downstream from the mismatch. The nicked DNA strand is then exonucleotically degraded by one of four exonucleases, assisted by helicase II, up to and beyond the mismatch. The segment removed is subsequently replaced (and repaired) by the action of DNA polymerase III and DNA ligase.

MutH is a sequence specific Mg^{2+}-dependent DNA endonuclease, with a molecular mass of 28 kDa. It shows sequence similarity with the type II restriction endonuclease Sau3AI ***(13)***, which also recognizes the d(GATC) sequence and cleaves it 5– to the G. However, MutH cleaves only the unmethylated strand, either in a hemimethylated or unmethylated DNA, whereas Sau3AI cleaves both strands regardless of their methylation state. Different from typical restriction endonucleases and from Sau3AI, it is active as a monomer. The determination of the structure of MutH (*see* reviews in **refs. *14*** and ***15***) confirmed the relatedness between MutH and type II restriction endonucleases of the PD...D/EXK family. Based on the structural homology between MutH and PvuII ***(16,17)***, a putative DNA binding and active site of MutH was identified ***(13)*** by superposition of monomeric MutH and one subunit of the homodimeric PvuII, comprising, among other residues, Asp 70, Glu 77, and Lys 79 ***(18–21)***.

1.2. Identification of Candidate Amino Acid Residues of MutH Involved in Sensing the Methylation Status of DNA at d(GATC) Sites

We used two sources of information to identify amino acid residues in MutH that are likely candidates for sensing the methylation status of d(GATC) sites, i.e., the presence of a methyl group at N6 of adenine in one strand and the absence of such a group in the other strand.

MutH proteins have been identified in *E. coli*, *Haemophilus influenzae*, *Vibrio cholerae*, *Shewanella oneidensis*, and *Colwellia* sp.; it is likely that, in these highly homologous proteins, the amino acid residues responsible for sensing the methylation status at d(GATC) sites are conserved. MutH shares sequence similarities with restriction enzymes, e.g., Sau3AI from *Staphylococcus aureus*. These enzymes, similar to MutH, recognize d(GATC) sites. Different from MutH, which does not cleave methylated DNA (the hemimethylated DNA is only nicked in the unmethylated strand), these enzymes cleave unmethylated, hemimethylated, and fully methylated DNA in both strands. A comparison of the sequences of the MutH enzymes on one site and the restriction enzymes on the other side should allow the identification of residues important for sequence recognition and, therefore, common to both protein families, and residues required for sensing the methylation status, which should be only conserved among the MutH proteins.

The alignment of the amino acid sequence of these proteins show that these nucleases share many conserved amino acid residues, which presumably are involved in common functions, with respect to DNA binding, recognition, and cleavage. Some amino acid residues are only conserved among the MutH proteins, among them Phe94, Arg184, and Tyr212 (**Fig. 1**). To elucidate which of these amino acid residues are likely to be located in the protein–DNA interface, we superimposed the structure of MutH with the structures of restriction enzyme–DNA complexes, using the residues of the catalytic center as reference points. In the superimposed structures, the following residues in MutH turned out to be close to the nucleobases of the DNA, corresponding to the two adenine residues in the double-stranded DNA sequence: Lys48 facing the minor groove and Phe94, Arg184, and Tyr212 facing the major groove (**Fig. 2**). Because the N6 position of the adenine residues is located in the major groove, only Phe94, Arg184, and Tyr212 are good candidates to sense the methylation of N6 in one strand and the absence of methylation in the other strand. Tyr212 seemed to be of particular interest, because the superposition suggests that it is located close to the adenine residues in both strands.

1.3. Analysis of the DNA Nicking and Cleavage Activity of MutH

The analysis of the activity and specificity of MutH and MutH variants in DNA nicking and cleavage is best studied using double-stranded DNA substrates with a single d(GATC) site, which should be unmethylated, hemimethylated, or fully methylated. For detection of phosphodiester bond cleavage in the two strands, the DNA substrate should carry different labels in the two strands. Such substrates can be prepared by polymerase chain reaction (PCR) using fluorescently labeled primers. The introduction of methyl groups at d(GATC) sites requires enzymatic methylation by dam methylase. By digestion of one of the methylated strands of the fully methylated substrate by λ exonuclease and hybridization with an unmethylated complementary strand, a hemimethylated substrate is obtained. The analysis of the cleavage products obtained by incubation of these substrates with MutH can be conveniently carried out by capillary electrophoresis with laser-induced fluorescence detection using denaturing polyacrylamide gels *(19,22)*.

2. Materials

2.1. Identification of Residues Involved in DNA Binding and Recognition

The following programs are used to identify the MutH residues involved in sensing the methylation status of d(GATC) sites.

1. ClustalW (http://www.ebi.ac.uk/clustalw/).
2. ClustalX (http://bips.u-strasbg.fr/fr/Documentation/ClustalX/ [please note website is case-sensitive]).

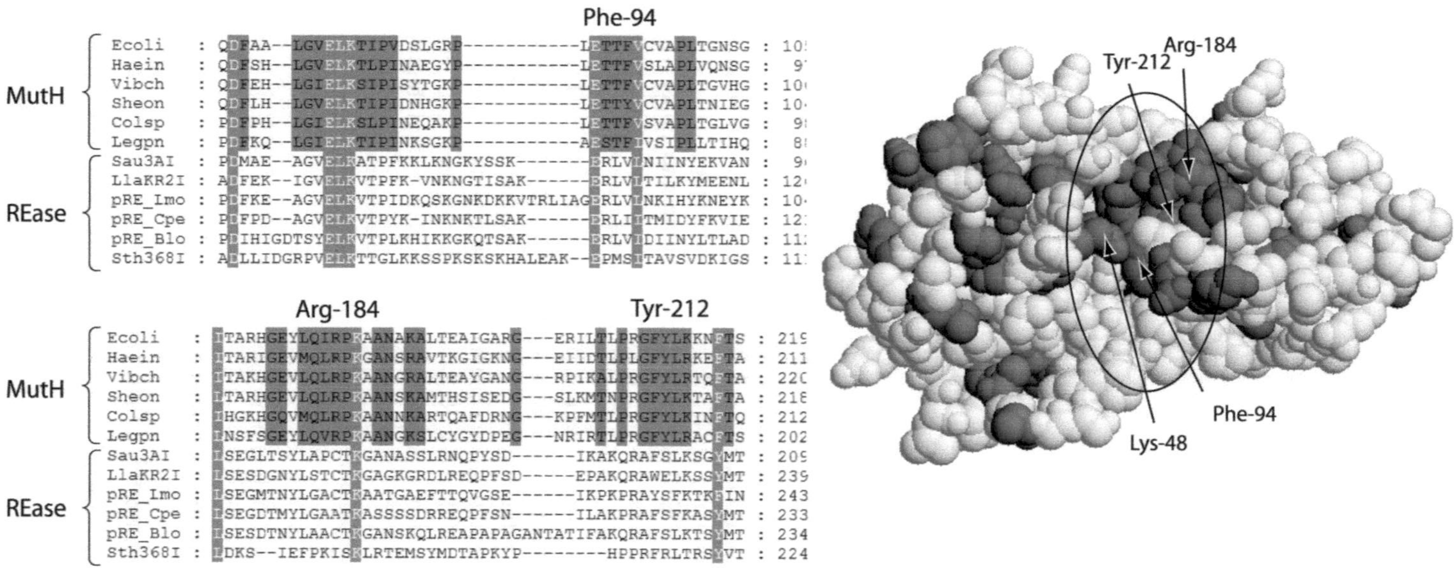

Fig. 1. Mapping group-specific residues on the structure of MutH. (**Left**) Part of the sequence alignment between the group of MutH proteins and the group of restriction endonucleases (REase). Residues conserved in both groups are shaded in gray with white lettering. Residues conserved only in the group of MutH proteins are shaded in gray with black lettering. (**Right**) In the structure of the *E. coli* MutH protein, group-specific residues are shown in gray. The positions of residues supposed to be involved in contacting the methylated bases are indicated.

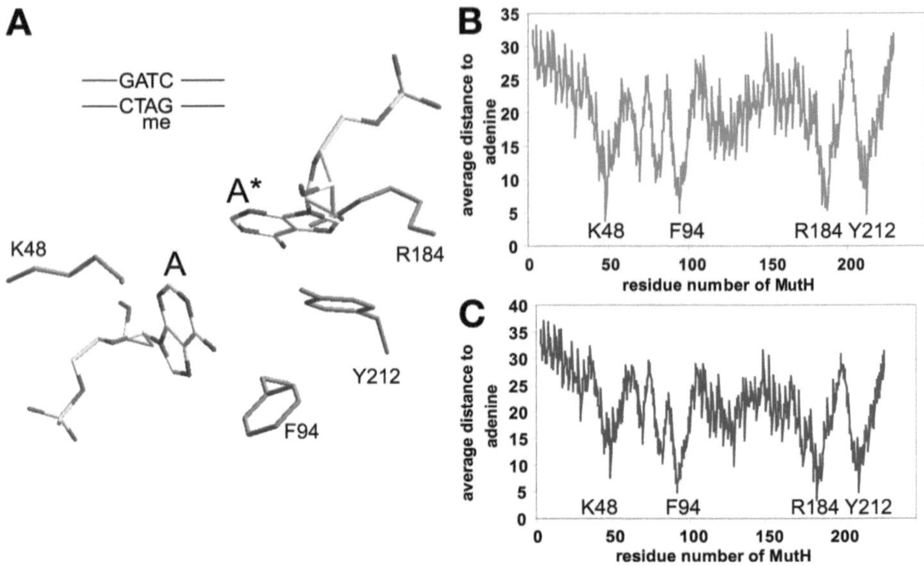

Fig. 2. Superposition of MutH with restriction endonucleases. **(A)** The two nucleotides of the MunI recognition sequence corresponding to the two adenines of the MutH recognition sequence, d(GATC), are shown. The nucleotide with the asterisks (A*) corresponds to the adenine of the MutH recognition sequence in the lower strand, which can be methylated and then directs cleavage to the upper strand. **(B,C)** Quantitative analysis of the superposition of MutH with 11 different restriction endonucleases (BamHI, BglI, BglII, BsoBI, EcoRI, EcoRV, HincII, MunI, NaeI, NgoMIV, and PvuII). Shown are the average distances of the atoms of the MutH structure to the atom N6 (adenine), O6 (guanine), N4 (cytosine), or O4 (thymine) of the nucleotide corresponding in position to the adenine in the upper strand (B) or lower strand (C), respectively, of the MutH recognition sequence. Note the minima, which indicate the proximity of residues, i.e., K48, F94, R184, and Y212, to the N6 of adenine or its equivalent in both the upper and lower strand, respectively, in all of the superimposed DNA structures.

3. GeneDoc (http://www.psc.edu/biomed/genedoc/).
4. RasMol (http://www.openrasmol.org/).
5. Swiss PDB Viewer (http://www.expasy.org/spdbv/).

2.2. Site-Directed Mutagenesis of MutH

1. pMQ402, a BAD18 derivative, is used as a bacterial expression plasmid *(23)*. Under control of an arabinose promoter for inducible protein expression, it contains the coding sequence for a His-tagged MutH protein, which can be purified by Ni-nitrilotriacetic acid (NTA) agarose affinity chromatography. The plasmid harbors an ampicillin-resistance gene for selection.

2. PCR primers for site-directed mutagenesis.
3. *Pfu* polymerase.
4. Deoxyribonucleoside triphosphates.

2.3. Overexpression, Purification, and Characterization

1. Luria-Bertani (LB) medium (per liter): 10 g bacto-tryptone, 5 g bacto yeast extract, and 5 g NaCl. Adjust the pH to 7.2 to 7.5 with NaOH, and autoclave.
2. Arabinose (Sigma).
3. Phenylmethylsulfonyl fluoride (PMSF) (Merck).
4. Ni-NTA agarose (Qiagen).
5. *E. coli* strain XL-1 Blue MRF' (Stratagene).
6. Sonicator.
7. Refrigerated centrifuge.
8. Empty columns (BioRad).
9. Binding buffer: 20 m*M* Tris-HCl, 5 m*M* imidazole, 1 *M* NaCl, and 1 m*M* PMSF, pH 7.9.
10. Washing buffer: 20 m*M* Tris-HCl, 60 m*M* imidazole, 1 *M* NaCl, and 1 m*M* PMSF, pH 7.9.
11. Elution buffer: 20 m*M* Tris-HCl, 200 m*M* imidazole, 1 *M* NaCl, and 1 m*M* PMSF, pH 7.9.
12. Dialysis buffer: 10 m*M* HEPES-KOH, 100 m*M* KCl, 1 m*M* EDTA, and 1 m*M* dithiothreitol, pH 7.9.
13. Dialysis buffer G: 10 m*M* HEPES-KOH, 100 m*M* KCl, 1 m*M* EDTA, 1 m*M* dithiothreitol, and 50% glycerol, pH 7.9.

2.4. Endonuclease Activity Assay

1. Assay buffer: 10 m*M* Tris-HCl, pH 7.5, 10 m*M* $MgCl_2$, 0.75 m*M* adenosine triphosphate, and 0.1 mg/mL bovine serum albumin.
2. *E. coli* MutL protein.
3. Fluorophore-labeled oligonucleotides or fluorescently labeled PCR products (template DNA, phosphorylated primers, fluorophore-labeled primers, *dam* methylase [NEB], λ-exonuclease [NEB], exonuclease I [NEB], PCR spin columns [Qiagen], and *Pfu* DNA polymerase [Promega]).
4. Template suppression reagent (Perkin-Elmer).
5. GeneScan-500 TAMRA size standard (Perkin-Elmer).
6. ABI PRISM 310 genetic analyzer (Perkin-Elmer).
7. 47-cm capillary (inner diameter: 50 μm) containing the POP-4 polymer supplemented with 8 *M* urea (Perkin-Elmer).
8. 1X genetic analysis buffer supplemented with 1 m*M* EDTA (Perkin Elmer).

3. Methods

The methods described here outline the selection of amino acid positions for site-directed mutagenesis (**Subheading 3.1.**), the mutagenesis of MutH proteins for expression in bacterial cells (**Subheading 3.2.**), and the characterization of the MutH variants (**Subheading 3.3.**).

3.1. Selection of Amino Acids for Site-Directed Mutagenesis

Sequences of MutH proteins and related restriction nucleases are obtained using the National Center for Biotechnology Information genomic basic local alignment search tool (BLAST) pages (http://www.ncbi.nlm.nih.gov/BLAST/) ***(24)***. Additional sequences are identified using the PSI BLAST server ***(25)***. Sequences are aligned using the Clustal X program with the PAM 250 matrix ***(26)***. The aligned sequences are analyzed using the GeneDoc program ***(27)***. Usage of the program is documented in the "Help" function within the program. Coordinates for structures are obtained from the PDB database (http://www. rcsb.org/pdb/). For structure visualization, the programs RasWin and Swiss PDB viewer were used.

3.1.1. Identification of Group Conserved Residues Involved in DNA Binding

After generating a sequence alignment (e.g., by using ClustalW or ClustalX) the following steps are performed:

1. Import alignment into the GeneDoc program (*see* **Subheading 2.1.**; **ref.** *27*).
2. Divide sequences into groups using the "Group" functionality (e.g., one group will encompass MutH proteins, the other the restriction endonucleases).
3. Identify group-specific residues (*see* **Note 1**).
4. Generate a RasMol script to map the group specific residues on the structure of the MutH protein using the "RasMol Script Dialog" function within GeneDoc.
5. Load the structure of MutH in the RasMol program (*see* **Subheading 2.1.**; **ref.** *28*) and execute the script output from GeneDoc using the "Script" command in the "RasMol Command Line" to visualize the group-specific residues.
6. Identify candidate residues located at the presumptive DNA-binding site.

3.1.2. Identification of Residues Involved in Contacting a Specific Base

1. Download sequences of available restriction endonucleases in complex with their DNA substrate (or product).
2. Align the structures of the restriction endonucleases to the structure of the target MutH with the Swiss PDB viewer using the main chain atoms of the three catalytic amino acids as a seed (e.g., D70, E77, and K79 in the case of MutH).
3. The fit can be enhanced using the "Improve fit" function.
4. Export coordinates of superimposed structure to a spreadsheet program.
5. Calculate distances between any atom of the target protein (e.g., MutH) to the bases of the superimposed structure corresponding to the base of the target DNA of the target protein.
6. Repeat the process for all superimposed structures.
7. Calculate average distances and identify residues closest to the base of interest.
8. Compare results with the analysis of group-specific residues.
9. Select promising amino acid residues for site-directed mutagenesis.

3.2. Site-Directed Mutagenesis of MutH Variants

Cloning of MutH variants is carried out using a modified QuikChange protocol (Stratagene) as described by Kirsch and Joly ***(29)*** using the plasmid pMQ402 (from Dr. M. G. Marinus, University of Massachusetts Medical School, MA) as the template and two oligodeoxynucleotides for mutagenesis suitable to generate PCR products of a length between 50 and 500 bp (*see* **Note 2**).

3.3. Purification and Characterization of MutH Variants

To be used in in vitro assays, MutH variants must be purified.

3.3.1. Purification of MutH Variants From Bacterial Cells

1. Transform XL-1 Blue MRF' cells with the bacterial plasmid using standard molecular biology methods.
2. Plate the cells on LB plates containing ampicillin and incubate overnight at 37°C.
3. Select a single colony and grow in 500 mL of LB medium containing 75 µg/mL ampicillin at 37°C.
4. At an optical density at 600 nm value of 0.8, induce with 0.2% (w/v) arabinose (final concentration) for 2.5 h at 28°C (*see* **Note 3**).
5. Centrifuge the cells for 10 min at 3000*g*.
6. Resuspend the cell pellet in 10 mL binding buffer (*see* **Subheading 2.3., item 9**).
7. Sonicate the suspension with the Branson sonifier, output level 5, 50% duty cycle, 1 min per time, five times. Cool the solution between sonication periods for 1 min each.
8. Centrifuge the cell debris for 30 min at 30,000*g* in a refrigerated centrifuge.
9. Gently mix the supernatant with 1.5 mL of a slurry of Ni-NTA agarose for 30 min at 4°C.
10. Transfer Ni-NTA agarose to an empty column.
11. Wash the column with 20 mL washing buffer (*see* **Subheading 2.3., item 10**).
12. Elute with 0.5 mL elution buffer (*see* **Subheading 2.3., item 11**). It is unnecessary to remove the His-tag because it does not interfere with the endonuclease activity (*see* **Note 4**).
13. Combine fractions containing proteins as judged by an optical density at 280 nm measurement.
14. Dialyze the sample with 500 mL dialysis buffer (*see* **Subheading 2.3., item 12**) at 4°C for at least 2 h. Change buffer twice.
15. Dialyze sample with 500 mL dialysis buffer G (*see* **Subheading 2.3., item 13**).
16. Measure the absorbance at 280 nm of a 1:10 dilution in dialysis buffer (*see* **Subheading 2.3., item 12**) to calculate the molar concentration of MutH using the theoretical extinction coefficient ***(30)***.
17. Store protein at –20°C.

3.3.2. Cleavage Analysis of MutH

Cleavage analysis of MutH and MutH variants is carried out using DNA substrates (either synthetic oligonucleotides or PCR products generated by

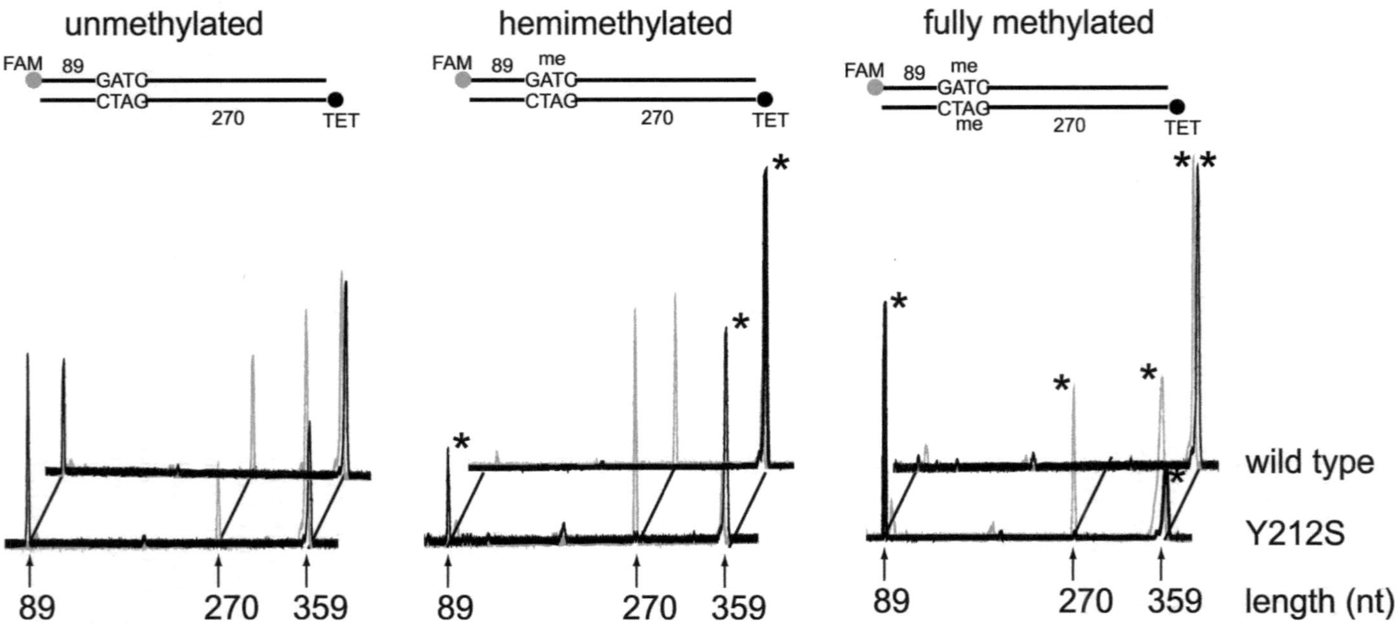

Fig. 3. Cleavage analysis of MutH wild-type and variant Y212S using unmethylated, hemimethylated or fully methylated DNA substrates. A 359-bp-long substrate containing a single d(GATC) site (unmethylated, hemimethylated, or fully methylated) labeled with FAM (gray) and TET (black) in the upper and lower strand, respectively, were incubated with wild-type MutH or the Y212S variant in the presence of MutL. Cleavage reactions were analyzed by capillary electrophoresis. The position of the substrate (359) and the product (89 and 270, respectively) strands are indicated by arrows. The methylated DNA strands are labeled with an asterisk.

procedures described elsewhere, **refs.** ***19*** and ***22***) containing a single target site, d(GATC), which is either unmethylated, hemimethylated, or fully methylated (**Fig. 3**). The substrates are labeled on the top and bottom strand with different fluorescent dyes (FAM and TET, respectively), to allow monitoring cleavage in both strands.

1. DNA substrates at 10 n*M* are incubated in 10 µL assay buffer (*see* **Note 1**), containing 500 n*M* MutL and MutH ranging from 10 to 500 n*M* (*see* **Note 5**).
2. Incubate the reaction mixture at 37°C.
3. Withdraw aliquots of the reaction mixtures after suitable time intervals (10 s to 30 min) containing 25 fmol of PCR product, mix thoroughly with 12 µL of template suppression reagent (Perkin-Elmer) and 0.5 µL of GeneScan-500 TAMRA size standard (Perkin-Elmer).
4. Heat to 95°C for 2 min and cool on ice immediately.
5. Analyze sample on an ABI PRISM 310 Genetic analyzer (Perkin-Elmer) equipped with a 47-cm capillary (inner diameter: 50 µm) containing the POP-4 polymer supplemented with 8 *M* urea (Perkin-Elmer).
6. Electroinject samples into the capillary for 5 s at 15,000 V, and perform the run at 15,000 V and 60°C for 30 min with 1X genetic analysis buffer supplemented with 1 m*M* EDTA (Perkin-Elmer) as the electrode buffer.
7. Record the amount of cleaved and uncleaved fluorescently labeled DNA (**Fig. 3**).
8. Determine the cleavage velocity.

4. Notes

1. Shading can be varied be setting the values for the "Group Cons Level" and the "PCR Maximum Level" in the "Edit Sequence Groups" dialog.
2. One primer will introduce the desired mutation whereas the other is an antisense primer to generate the so-called "megaprimer" used for site-directed mutagenesis according to the QuikChange protocol.
3. The time and temperature of induction depend on the system under investigation.
4. In some cases, it may be necessary to cleave off the His-tag. The construct used here allows cleavage at the thrombin site between the His-tag and the native N-terminal methionine.
5. Alternatively to activation by MutL, MutH endonuclease activity can be stimulated 10-fold by addition of 10% (v/v) dimethylsulfoxide.

Acknowledgments

Work in the authors' laboratory is supported by the Herbert Stolzenberg Stiftung, the Deutsche Forschungsgemeinschaft (Pi 122/12-4), and the Fonds der Chemischen Industrie.

References

1. Pingoud, A. and Jeltsch, A. (2001) Structure and function of type II restriction endonucleases. *Nucleic Acids Res.* **29,** 3705–3727.

2. Lanio, T., Jeltsch, A., and Pingoud, A. (2000) On the possibilities and limitations of rational protein design to expand the specificity of restriction enzymes: a case study employing EcoRV as the target. *Protein Eng.* **13,** 275–281.
3. Jeltsch, A., Wenz, C., Wende, W., Selent, U., and Pingoud, A. (1996) Engineering novel restriction endonucleases—principles and applications. *Trends Biotech.* **14,** 235–238.
4. Alves, J. and Vennekohl, P. (2004) in *Restriction Endonucleases* (Pingoud, A., ed.), Springer, Berlin.
5. Lanio, T., Jeltsch, A., and Pingoud, A. (1998) Towards the design of rare cutting restriction endonucleases: using directed evolution to generate variants of EcoRV differing in their substrate specificity by two orders of magnitude. *J. Mol. Biol.* **283,** 59–69.
6. Lanio, T., Jeltsch, A., and Pingoud, A. (2002) in *Directed Molecular Evolution of Proteins* (Brakmann, S. and Johnsson, K., eds.), Wiley-VCH, Weinheim, Germany, pp. 309–327.
7. Samuelson, J. C. and Xu, S. Y. (2002) Directed evolution of restriction endonuclease BstYI to achieve increased substrate specificity *J. Mol. Biol.* **319,** 673–683.
8. Rimseliene, R., Maneliene, Z., Lubys, A., and Janulaitis, A. (2003) Engineering of restriction endonucleases: using methylation activity of the bifunctional endonuclease Eco57I to select the mutant with a novel sequence specificity. *J. Mol. Biol.* **327,** 383–391.
9. Jeltsch, A., Alves, J., Oelgeschlager, T., Wolfes, H., Maass, G., and Pingoud, A. (1993) Mutational analysis of the function of Gln115 in the EcoRI restriction endonuclease, a critical amino acid for recognition of the inner thymidine residue in the sequence -GAATTC- and for coupling specific DNA binding to catalysis. *J. Mol. Biol.* **229,** 221–234.
10. Wenz, C., Selent, U., Wende, W., Jeltsch, A., Wolfes, H., and Pingoud, A. (1994) Protein engineering of the restriction endonuclease EcoRV: replacement of an amino acid residue in the DNA binding site leads to an altered selectivity towards unmodified and modified substrates. *Biochim. Biophys. Acta* **1219,** 73–80.
11. Lanio, T., Selent, U., Wenz, C., et al. (1996) EcoRV-T94C—a mutant restriction endonuclease with an altered substrate specificity towards modified oligodeoxynucleotides. *Protein Eng.* **9,** 1005–1010.
12. Modrich, P. and Lahue, R. (1996) Mismatch repair in replication fidelity, genetic recombination, and cancer biology. *Annu. Rev. Biochem.* **65,** 101–133.
13. Ban, C. and Yang, W. (1998) Structural basis for MutH activation in *E. coli* mismatch repair and relationship of MutH to restriction endonucleases. *EMBO J.* **17,** 1526–1534.
14. Yang, W. (2000) Structure and function of mismatch repair proteins. *Mutat. Res.* **460,** 245–256.
15. Nishino, T., Komori, K., Tsuchiya, D., Ishino, Y., and Morikawa, K. (2001) Crystal structure of the archaeal holliday junction resolvase Hjc and implications for DNA recognition. *Structure* **9,** 197–204.

16. Athanasiadis, A., Vlassi, M., Kotsifaki, D., Tucker, P. A., Wilson, K. S., and Kokkinidis, M. (1994) Crystal structure of PvuII endonuclease reveals extensive structural homologies to EcoRV. *Nat. Struct. Biol.* **1,** 469–475.
17. Cheng, X., Balendiran, K., Schildkraut, I., and Anderson, J. E. (1994) Structure of PvuII endonuclease with cognate DNA. *EMBO J.* **13,** 3927–3935.
18. Loh, T., Murphy, K. C., and Marinus, M. G. (2001) Mutational analysis of the MutH protein from *Escherichia coli. J. Biol. Chem.* **276,** 12,113–12,119.
19. Friedhoff, P., Thomas, E., and Pingoud, A. (2003) Tyr-212: a key residue involved in strand discrimination by the DNA mismatch repair endonuclease MutH. *J. Mol. Biol.* **325,** 285–297.
20. Friedhoff, P., Sheybani, B., Thomas, E., Merz, C., and Pingoud, A. (2002) *Haemophilus influenzae* and *Vibrio cholerae* genes for *mutH* are able to fully complement a *mutH* defect in *Escherichia coli. FEMS Microbiol. Lett.* **208,** 121–126.
21. Junop, M. S., Yang, W., Funchain, P., Clendenin, W., and Miller, J. H. (2003) In vitro and in vivo studies of MutS, MutL and MutH mutants: correlation of mismatch repair and DNA recombination. *DNA Repair (Amst.)* **2,** 387–405.
22. Thomas, E., Pingoud, A., and Friedhoff, P. (2002) An efficient method for the preparation of long heteroduplex DNA as substrate for mismatch repair by the Escherichia coli MutHLS system. *Biol. Chem.* **383,** 1459–1462.
23. Guzman, L. M., Belin, D., Carson, M. J., and Beckwith, J. (1995) Tight regulation, modulation, and high-level expression by vectors containing the arabinose PBAD promoter. *J. Bacteriol.* **177,** 4121–4130.
24. Wheeler, D. L., Church, D. M., Lash, A. E., et al. (2002) Database resources of the National Center for Biotechnology Information: 2002 update. *Nucleic Acids Res.* **30,** 13–16.
25. Altschul, S. F., Madden, T. L., Schaffer, A. A., et al. (1997) Gapped BLAST and PSI-BLAST: a new generation of protein database search programs. *Nucleic Acids Res.* **25,** 3389–3402.
26. Thompson, J. D., Gibson, T. J., Plewniak, F., Jeanmougin, F., and Higgins, D. G. (1997) The ClustalX windows interface: flexible strategies for multiple sequence alignment aided by quality analysis tools. *Nucleic Acids Research* **24,** 4876–4882.
27. Nicholas, K. B., Nicholas, H. B. J., and Deerfield, D. W. I. (1997) GeneDoc: analysis and visualization of genetic variation. *EMBnet NEWS* **4,** 14.
28. Sayle, R. A. and Milner-White, E. J. (1995) RASMOL: biomolecular graphics for all. *Trends Biochem. Sci.* **20,** 374.
29. Kirsch, R. D. and Joly, E. (1998) An improved PCR-mutagenesis strategy for two-site mutagenesis or sequence swapping between related genes. *Nucleic Acids Res.* **26,** 1848–1850.
30. Pace, C. N., Vajdos, F., Fee, L., Grimsley, G., and Gray, T. (1995) How to measure and predict the molar absorption coefficient of a protein. *Protein Sci.* **4,** 2411–2423.

II

Evolutionary Strategies for Protein Engineering

8

Protein Library Design and Screening

Working Out the Probabilities

Michel Denault and Joelle N. Pelletier

Summary

In designing protein libraries for selection, we must coordinate our capacity to create a large diversity of protein variants with the physical limitations of what we can actually screen. This chapter aims to bring the language of probabilities into the protein engineer's laboratory to answer some of our common questions: How can we most efficiently design a library? What fraction of the theoretical library diversity have we actually sampled at the end of the day? What is the probability of missing an individual of the library? Are the mutations present in the variants we have selected statistically meaningful or the product of random variation? The computation of these criteria throughout the process of experimental protein engineering will enable us to better design and evaluate the products of our libraries of protein variants.

Key Words: Library screening; experimental bias; library representation; Poisson distribution; χ^2 test of hypothesis; diversity.

1. Introduction

Natural evolution is a process of production of "variants" of a given specimen followed by selection of those best adapted to a specific environmental setting, or to a specific purpose. Natural evolution is the result of genetic mutations or rearrangements resulting in new protein variants within organisms. Antibodies well illustrate the process of evolution: the natural antibody repertoire of an organism represents a large number of variants that are sampled, and those best suited to the purpose of binding a specific ligand are selected from this natural library for further rounds of improvement. Selection of protein variants requires the coupling of genotype and phenotype—a given in natural evolution and a requirement hard worked at in the research laboratory where artificial protein evolution is undertaken. Protein engineering tools have undergone a revolution

From: *Methods in Molecular Biology, vol. 352: Protein Engineering Protocols*
Edited by: K. M. Arndt and K. M. Müller © Humana Press Inc., Totowa, NJ

in the past decade as a result of the creation of better-defined large libraries with a variety of strategies for linking genotype and phenotype, which, in turn, have allowed for the development of a host of new approaches to functional screening and selection strategies (*see* **refs.** ***1*** and ***2*** for more information).

The goal of this chapter is to present some of the main statistical and probabilistic calculations pertinent to protein library creation and to library-based screening and selection strategies in the language of the combinatorial biologist to ensure their ease of application. This chapter also aims to reveal the pertinence of these calculations to the process of library creation and screening. When designing an experiment in library-based protein engineering, one must counterbalance the scientific value of creating the most diverse protein library with the practical limitations presented by the screening or selection strategies. The statistical and probabilistic questions pertaining to library representation are important, and they vary according to the specific application. We also present tests of experimental biases, to aid in assessing the library quality and the occurrence of biases before or after selection. As we will illustrate, the information revealed by simple analyses of library-based problems can lead to important—and sometimes counter-intuitive—insights, which, in turn, allow rapid improvement of the experimental design and a stronger basis for interpretation of the results.

We first discuss some important parameters related to the design of protein libraries, such as library size, compositional bias, codon degeneracy, and encoded diversity. This is followed by the development of a series of commonly encountered problems, with discussion of the pertinence of addressing each problem and a clear mathematical development. Numerical examples are provided to clarify the application of the formulae. In cases in which the implementation of the mathematical tools is more complex or time consuming, we refer the experimenter to Excel files for input of the appropriate variables. The proposed problems are intended to be general enough to apply to a diversity of library-based systems rather than to any specific subset.

It is important to note that many results presented below are suited only when various parameters are sufficiently large or small, and can lead to erroneous conclusions if applied otherwise. Specific conditions are provided.

Complementary works on the mathematical treatment of library-based strategies include mathematical modeling of DNA shuffling, specifically treated in the important work of Moore and of Maranas ***(3–6)*** and of Sun ***(7)***, and, more recently, Blackburn and colleagues ***(8)***. The latter work also presents a mathematical treatment of library representation for libraries of equiprobable outcomes; we present the same treatment, as a prelude to the treatment of library representation for libraries of nonequiprobable outcomes. **Reference** ***8*** also presents a method for estimation of diversity in libraries generated by error-prone polymerase chain reaction and, importantly, provides simple computer programs for users.

1.1. Parameters in Library Creation and Screening

There exist a number of parameters that must necessarily be determined in the process of protein library design and screening: the library size that is desired, the library representation that is required for a given application, and the constraints that are imposed by the screening strategy. These parameters may be addressed intuitively, although this can lead to conceptual or experimental errors. The mathematics underlying these parameters are discussed in **Subheading 3.1.** to help the experimenter better plan and execute library-based experimentation.

1.1.1. Design of Library Size

Whether the combinatorial biologist relies on a naturally occurring library or a synthetic library, library size must be properly defined to later assess results of screening with a measure of accuracy. Naturally occurring libraries include natural antibody repertoires and whole repertoires of proteins from organelles and organisms. Synthetic libraries generally stem from a single protein or class of homologous proteins, although they can be more highly diversified. The techniques currently used in synthetic library generation include saturation mutagenesis; random mutagenesis by the polymerase chain reaction; "gene shuffling," in which fragments of DNA from similar origins are combined; "directed" mutagenesis, in which specific regions within the encoding DNA are targeted; as well as strategies for nonhomologous, random recombination of sequences (refer to **refs.** ***1*****,** ***2*****,** and ***9*** for nonhomologous recombination, for example). To design the desired synthetic library size, the following considerations are essential.

For a small peptide library (10 amino acids long), there are 20^{10} (= 2×10^{11}) possible amino acid combinations. The same calculation applies to a protein containing 10 randomized amino acids. However, because there are multiple coding possibilities for any 10 amino acid peptide as a result of codon degeneracy, one must create a library of DNA larger than 20^{10} to encode a reasonable proportion of the 20^{10} different peptides, which will be represented with varying distributions according to the number of coding possibilities per peptide. For example, in a random DNA library, there are six possible coding possibilities for Ser, Arg, and Leu; these are maximally redundant. Thus, the coding sequence for a deca-Ser peptide is highly redundant, with $6^{10} = 6 \times 10^7$ possible combinations, all encoding deca-Ser peptides, relative to the single coding combination that exists for the nondegenerate deca-Met peptide. The result of the codon degeneracy characterizing certain amino acids is that DNA libraries must be larger than calculated if ignoring degeneracy to carry a reasonable chance of encoding nondegenerate amino acids; these amino acids are encoded by only 1 out of the 64 possible triplet codons. Explicitly, in this example, one should encode all 64^{10} (= 1.2×10^{18}) possible DNA sequences to allow for a reasonable probability of encoding the deca-Met peptide. Unfortunately, these 1.2×10^{18} decapeptides would be encoded by approx 2 mg of DNA, a sample too large to be reasonably produced in most research contexts. It should be noted that one can consider a longer peptide or protein in which any 10 amino acids (contiguous or not) will be varied, using the same reasoning as the described decapeptide. Thus, codon degeneracy should be considered when planning a peptide or protein library.

1.1.2. Biased Libraries

It is often of interest to bias the coding DNA library. The most obvious advantages are to reduce the impact of stop codons, which render nonfunctional a large fraction of unbiased libraries (*see* **Table 1**), to skew libraries toward desired amino acid compositions, and to reduce the difference in codon representation between various amino acids, thus allowing a more-uniform representation of the amino acids in the library. Libraries with "NNC" or "NNT" repeats (where N represents any of the 4 nucleotides) serve to reduce the number of codons from 64 to 16 while eliminating the 3 stop codons (*see* **Table 1**). However, five amino acids (Met, Trp, Gln, Glu, and Lys) are not encoded, resulting in a loss of encoded diversity. A drawback in using either of these degenerate codons is that codon usage in *Escherichia coli* (the most frequently used host) is poor for certain codons ***(10)***. The "NNC/T" codon is also frequently used; it counters the problem of poor codon usage but doubles the number of codons at each position, with no increase in the number of different amino acids encoded. The advantage of having no stop codons to contend with may well be worth the loss in amino acid diversity, because, in a randomly encoded library ("NNN"), the probability of obtaining products with no stop codons drops rapidly with increasing numbers of degenerate positions (*see* **Table 1**) and can result in a low-quality library. An alternative is to encode "NNC/G" or "NNT/G," where all 20 amino acids are encoded, including a single stop codon; "NNC/G" offers better codon usage overall in *E. coli*. Here, the probability of obtaining products with no stops is significantly greater than when encoding "NNN." Thus, when multiple degenerate positions will be created, inclusion of stop codons can be very deleterious to library quality and should be avoided if the screening strategy to be used is labor-intensive or costly.

The experimenter frequently requires a specific array of amino acids at a particular position. Planning such biased oligonucleotides can be accomplished by hand, although software is available to facilitate this task. For example, the "Mixed Codon Worksheet" by T. J. Magliery (www.chemistry.ohio-state.edu/~magliery/publications.html) is an Excel file in which one lists the desired amino acids to be encoded at a position; various possibilities are computed, allowing

Table 1

Biased codon, encodes	Codons	Amino acids	Stops	Probability that no stop codon is encoded: In a protein containing 10 biased codons	In a protein containing 30 biased codons
NNN	64	20	3	0.62[a,b]	0.24
NNC or NNT	16	15	0	1	1
NNC/G or NNT/G	32	20	1	0.73	0.39

[a]The value 0.62 signifies that a peptide or protein library containing 10 fully randomized codons (NNN) is expected to contain 62% of variants that are full length, with no stop codon.

[b]Calculated as $(61/64)^{10} = 0.619$, where 61 out of the 64 codons do not encode a stop codon, for 10 codons.

the experimenter to select the mixed codon best suiting their needs. Certain oligonucleotide suppliers also offer the possibility of including different proportions of each nucleotide of interest, allowing the experimenter to bias the nucleotide distribution to favor certain amino acids relative to others. This strategy can be used to reduce or enhance differences in codon representation, as required.

In the ideal case, a biased library can be constructed from trinucleotide codons ***(11)***, in which three specific nucleotides are covalently linked into "building blocks," each encoding a single amino acid, and which can be included in oligonucleotide synthesis at specified positions and mixed in the desired proportions. This reduces the DNA library size to the protein library size that is precisely of interest, because it is nondegenerate. The advantages of trinucleotide use have been highlighted (*see* **refs. *12–14***, for example) but relatively little work has been performed with them because they have only recently become commercially available (Glen Research Sterling, VA; and Metkinen Oy, Kuusisto, Finland).

1.1.3. Limits to Library Size

The maximal physical library size that can be attained, in practice, is on the milligram scale, which corresponds to approximately 3×10^{16} molecules of double-stranded DNA for a library of 30 basepairs (bp; 10 amino acids). However, protein expression is generally undertaken via vector-based DNA. A typical expression vector may be 3000-bp long. Therefore, on the milligram scale and under ideal conditions, one can maximally produce 3×10^{14} molecules of double-stranded DNA for a vector of 3000 bp. Thus, the practical maximal representation of a library with 10 fully randomized positions (1.2×10^{18} possibilities) is but a fraction of all possibilities (more detail is provided in

Subheading 3.1.2. regarding calculation of library representation). This allows only the partial exploration of a completely randomized 10-residue peptide library (or a protein encoding 10 fully randomized residues), far from allowing for the complete exploration of an average protein. This implies that there are two courses of action in creating protein libraries. In the first course, the target protein can be "fully" randomized; in this case, the library obtained represents a subset of all of the possibilities (*see* **Subheading 1.2.**). This can be extremely informative if knowledge regarding the relative importance of the wild-type sequence for a given function is required. Those residues that are statistically less-often mutated in the selected proteins are inferred to be critical for retention of the characteristic that is selected for; this inference is analogous to that used in determining functional residues by multiple alignments of natural sequences. In the second course of action, the randomization can be limited; only specific residues (contiguous or not) are fully randomized and/or a bias is introduced into the randomization to limit the number of different possibilities. This can result in a completely defined library of known size; the size can then be kept within the constraints described in **Subheading 1.1**. The library size can also be matched to further constraints if screening of all library members, rather than sampling of a subset, is considered important.

1.2. Library Representation

Inherent to the calculation of the required library size is the consideration of the library representation that one needs to achieve for each specific application. For a library selection to be efficient, the number of independent variants encoded must be appropriate to the selection capacity of the strategy used. As described in **Subheading 1.1.**, the creation of extremely large and diverse DNA libraries is no longer a challenge. Indeed, one must generally restrict the size of the library to that which can adequately be screened. The tools for selection of a desired phenotype (while maintaining a linkage to the genotype) have been more difficult to develop than those for library creation because they are not so broadly applicable as the tools for manipulating DNA. Thus, a diversity of selections strategies must be developed to select proteins having diverse functions (refer to **ref.** ***2***).

In general, manual selection allows the experimenter to screen hundreds to thousands of protein variants; automation can increase the range by orders of magnitude. In vivo selection can allow for rapid screening of thousands to billions of variants. Cell-free systems may be the most powerful experimental technique (excluding computational approaches), allowing the researcher to screen up to 10^{14} protein variants. Clearly, the number of variants treated by any of these means is extremely limited relative to the potential variety of protein variants one could theoretically create, emphasizing the requirement for matching library design with the selection capacity. The experimenter may need to calculate the actual *representation of protein variants* within a sample; the mathematical tools to do so are provided in **Subheading 3.1.**

1.3. Determination of Experimental Biases

Library selection can yield a large amount of sequence information (at the level of nucleic acids or amino acids) that must be accurately interpreted. For example, assessment of library quality makes use of determination of biases. After the design and creation of a library, several individuals are picked *before selection* for DNA sequencing. Any deviation between the expected nucleotide distribution and the observed distribution at the mutated positions can be evaluated to assess whether a significant bias exists. A significant bias would reveal a faulty codon distribution, which could result from flawed oligonucleotide synthesis or from a positional bias in a random mutagenesis scheme, such as a bias caused by the native sequence. A bias could also indicate that a *certain amount of selection has occurred when not intended*; it may be necessary to switch to an alternate system, by changing the host cell or by switching to an in vitro system, for example. The tools for *interpretation of experimental biases*, based on the χ^2 test, are presented in **Subheading 3.2.**

A further application of the χ^2 test involves the comparison of DNA sequences *before* and *after selection* to reveal differences in codon distribution resulting from selection. One may observe a bias in the frequency of mutation of a particular residue after selection, or a bias in the occurrence of a specific amino acid at a biased position. Application of the χ^2 test will allow the experimenter to decide whether the observed biases are significant or not, and, thus, whether the selective pressure applied is stringent enough to enrich for a particular population.

2. Materials

For computation of many problems, Excel worksheets (**Figs. 1–4**) are provided for reproduction. It is also possible to access the same Excel worksheets directly through http://www.esi.umontreal.ca/~pelletjo/. The sheets presented here closely follow the structure of the paper. **Figures 1** and **2**, in the two cases of the equiprobable and nonequiprobable outcomes, are treated similarly, because they rely on the same Poisson distribution, with the same values of the parameter λ. **Figure 3** (again for both equiprobable and nonequiprobable cases) rely on a different Poisson distribution, and, hence, different values of λ. Boxed cells indicate data that the user must specify. Gray cells give answers that appear in the paper.

3. Methods

This section is divided into two main subsections. In **Subheading 3.1.**, we address problems pertaining to library design and representation. In **Subheading 3.2.**,

Problem A, equiprobable

How many of the *n* theoretical variants do we expect **not** to appear among the *m* variants chosen?

Number of variants (n)	1.00E+06
Sample size (m)	1.00E+07

Expected number of missing variants (lambda) =n*(1-1/n)^m : 45.40

Problem B, equiprobable

What is the probability that at least one variant has not been generated in the sample? That at most 50 have not been generated?

Probability that at least one variant is missing =1-EXP(-lambda) : 100.00%

Probability that at most 50 of the variants have not been generated = POISSON(50;lambda;TRUE) : 77.87%

Problem C, equiprobable

How many times can we expect a variant *i* to appear in the sample? More generally, what is the probability that it appears 10 times ? At most 10 times?

Number of variants (n)	1.00E+06
Sample size (m)	1.00E+07

Expected number of occurrences of variant "*i*" in the sample (lambda) = m/n : 10.00

Probability that variant "*i*" appears 10 times in the sample =POISSON(10;lambda;FALSE) : 12.51%

Probability that variant "*i*" appears 10 times or less in the sample =POISSON(10;lambda;TRUE): 58.30%

Fig. 1. Excel worksheet describing examples of library representation: the case of equiprobable outcomes, as presented in **Subheadings 3.1.1.1.–3.1.1.3**.

Problem A, non-equiprobable

How many of the *n* theoretical variants do we expect **not** to appear among the *m* variants chosen?

Length of peptide (Pi)	10	← use "10" for a decapeptide, for example
Number of variants (n)	1.67E+13	← for information only, not used in this sheet
Sample size (m)	1.00E+14	

Expected number of missing variants (lambda) =lambdaPeptide(Pi;m) : 5.040E+12 ← This cell calls the VBA function lambdaPeptide, which computes lambda on the basis of two parameters, the number of codons and the sample size.

Problem B, non-equiprobable

What is the probability that at least one variant has not been generated in the sample? That at most 50 have not been generated?

Probability that at least one variant is missing =1-EXP(-lambda) : 100.00%

Probability that at most 50 of the variants have not been generated =

POISSON(50;lambda;TRUE) : ERROR (see note) ← Using the Poisson distribution function

or NORMDIST(50;lambda;SQRT(lambda);TRUE) : 0.00% ← Using the Normal distribution function

Note that for the very large values of lambda obtained for many realistic experimental data (e.g. decapeptide and m=1e14 sample size), the Poisson distribution macro of Excel quickly becomes useless, returning an error message; the Normal distribution approach to the Poisson is indicated in such cases.

Fig. 2. Excel worksheet describing examples of library representation: the case of nonequiprobable outcomes, as presented in **Subheadings 3.1.2.1.** and **3.1.2.2.**

Problem C, non-equiprobable

How many times can we expect a decapeptide with 7 codons of probability 3/64 and 3 codons with probability 4/64, to appear in the sample? More generally, what is the probability that it appears 10 times ? At most 10 times?

	"1/64"	"2/64"	"3/64"	"4/64"	"6/64"
Probability:	0.015625	0.03125	0.046875	0.0625	0.09375
No. of codons	0	0	7	3	0

Probability of the decapeptide (p) = {=PRODUCT(D6:H6^D7:H7)} :	1.21E-13
Sample size (m)	1.00E+14
Expected number of occurrences of variant "*i*" in the sample (lambda) =m*p:	12.14
Probability that variant "*i*" appears 10 times in the sample = POISSON(10;lambda;FALSE) :	10.23%
Probability that variant "*i*" appears 10 times or less in the sample = POISSON(10;lambda;TRUE) :	33.27%

Note: if you modify the formula for p {=PRODUCT(D6:H6^D7:H7)}, don't forget to finish with "crtl-shift-enter", and not only "enter"! It is an array operation.

Fig. 3. Excel worksheet describing an example of library representation: the case of nonequiprobable outcomes, as presented in **Subheading 3.1.2.3**.

Tests of probability distributions (tests of experimental biases)

The hypothesis is that the experiment's results reflect their theoretical probabilities, *i.e.* the experiment is not biased.

A) 2 possible outcomes

Nb of possible outcomes	2	Nb of repeats of the experiment	100	Level of significance ("alpha")	5%
Theoretical probability of outcome 1	25%	Observed number of outcome 1	29	Value of the statistic (q) (see *)	0.85
Theoretical probability of outcome 2	75%	Observed number of outcome 2	71	Chi-square comparison value = CHIINV(L6;1)	3.84
				Reject hypothesis that experiment is unbiased? (see **)	Do not reject

B) 3 possible outcomes

Nb of possible outcomes	3	Nb of repeats of the experiment	20	Level of significance ("alpha")	5%
Theoretical probability of outcome 1	20%	Observed number of outcome 1	8	Value of the statistic (q) (see *)	5.60
Theoretical probability of outcome 2	30%	Observed number of outcome 2	6	Chi-square comparison value = CHIINV(L6;1)	5.99
Theoretical probability of outcome 3	50%	Observed number of outcome 3	6	Reject hypothesis that experiment is unbiased? (see **)	Do not reject
etc.		etc.			

* Value of the statistic = (H8-D8*H6)^2/(D8*H6)+(H9-D9*H6)^2/(D9*H6)

** Reject hypothesis? =IF(L8>L10;"Reject";"Do not reject")

Note: Clearly, this sheet can be easily adapted to more outcomes, by extending columns D and H appropriately, and by adapting the computation of the value q

Fig. 4. Excel worksheet describing two examples of tests of probability distributions, as presented in **Subheading 3.2.1**.

we analyze the significance of results, either when assessing library quality or when screening a library. In both subheadings, formulae are presented and numerical examples are provided, without full theoretical explanation, and are illustrated with an example. The full theoretical motivation of the examples is provided in the Notes (**Subheading 4.**).

3.1. Library Representation

We design a library containing n possible, different, theoretical variants. We sample m times, randomly, from this theoretical pool; we, thus, form *a sample of m variants*. The problems we will address are the following:

A. How many of the n theoretical variants do we expect **not** to appear among the m variants chosen?
B. What is the probability that *at least one* of the n theoretical variants has not been sampled, among the m variants chosen? What is the probability that at most a certain number of the theoretical variants have not been sampled?
C. How many times can we expect a specific variant i to appear in the sample? More generally, what is the probability that it appears r times?

These problems are presented as a group because their resolution relies on the same general theory, as described in **Subheading 3.1.1.** and **3.1.2**. We shall denote by p_i the probability that variant i will come up any time we sample once, randomly, from the theoretical pool.

We make a distinction between two cases, and treat them in separate subheadings below:

1. In **Subheading 3.1.1.**, the case of equiprobable outcomes, in which each of the n variants has an equal probability of being sampled. Here, $p_i = 1/n$; this case is treated in full generality.
2. In **Subheading 3.1.2.**, the case in which outcomes occur with unequal probabilities, i.e., some of the variants have a better chance of being sampled than others. By definition, it is difficult to generalize this case, because the answers to problems A, B, and C depend on the probabilities themselves. As an example, we treat one case of unequal probabilities in detail.

Equiprobable outcomes occur if there is no bias either in the expected frequency of occurrence of a given sequence nor in the encoded characteristic of interest. In the case of library selection, if the characteristic of interest is independent of other parameters, or if the impact of other parameters is negligible, we consider the outcomes to be equiprobable. Outcomes may have unequal probabilities if either considering biased codons or codon degeneracy, or else if the outcome affects the system. An example of the outcome affecting the system is the following: when screening in vivo for an increase in the catalytic activity of enzyme X, the increase in reaction product may be toxic toward the system, resulting in indirect adverse effects toward the detected catalytic activity. Such parameters are generally too complex to be systematically accounted for. However, we can take into account the use of biased codons, in which amino acids will occur with unequal probabilities, as well as other parameters which, although being unequal, can be quantified.

3.1.1. The Case of Equiprobable Outcomes

This is the simplest case, in which all outcomes are considered to have an equal probability: after sampling any one library member, all n variants have the same chance of being chosen. Formally, $p_i = 1/n$.

3.1.1.1. Problem A

How many of the n theoretical variants do we expect **not** to appear among the m variants chosen?

An easy, approximate answer is given by the parameter λ (*see* **Notes 1** and **2**), defined as:

$$\lambda = n\left(1-\frac{1}{n}\right)^m \tag{1}$$

For example, suppose we sample $m = 1 \times 10^7$ times from a theoretical pool of $n = 1 \times 10^6$ variants. The probability of any variant of coming up on any one draw is: $p_i = \frac{1}{n} = \frac{1}{1\times10^6}$

The expected number of missing theoretical variants is given by λ calculated as follows:

$$\lambda = 1\times10^6\left(1-\frac{1}{1.0\times10^6}\right)^{1\times10^7} \cong 45.4$$

Note that the expected number of missing variants need not be an integer; here, the expected number of missing variants is between 45 and 46 variants out of the 1×10^6 theoretical variants.

If instead we sample $m = 2 \times 10^7$ times from the same theoretical pool of $n = 1 \times 10^6$ variants, $\lambda = 0.00206$. Thus, the expected number of missing variants is between 0 and 1, although much closer to 0 than 1. In fact, in this example, the probabilities that, respectively, 0 or 1 variants are missing are the only probabilities that are not very small, with 0 being much more likely (>99%, as calculated in **Subheading 3.1.1.2.**).

We can conclude that sampling a 20-fold excess of items relative to the theoretical library size ($m = 20n$) results in essentially complete sampling of the theoretical library, as illustrated in **Fig. 5**. Sampling a 10-fold excess results

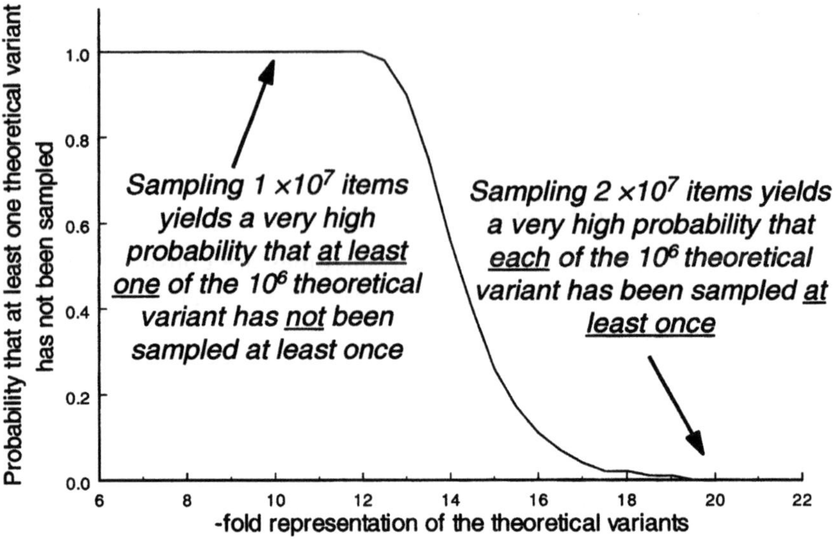

Fig. 5. Probability that at least one of the theoretical variants has not been sampled vs the fold representation of the library. The theoretical library size is given by n and the number of items sampled by m; library representation is given by the ratio n/m, where 10-fold library representation signifies that $m = 10n$. The example presented here defines the theoretical library size as $n = 1 \times 10^6$, with variable m.

in a small fraction that is not sampled (which will tend toward 0 if the theoretical library is smaller than 10^4; *see* **Table 2**); for many experimental goals, knowing that a small fraction remains unsampled meets the experimental requirements.

It is informative to note the following counterintuitive results in **Table 2**: if the sample size m is equal to the theoretical library size n, we expect that, on average, approximately 37% of the theoretical variants will not be sampled; even if $m = 2n$, the average of unsampled theoretical variants is 13.5%, far from exhaustive library sampling.

3.1.1.2. Problem B

What is the probability that at least one of the n library variants has not been sampled? What is the probability that at most 50 or else 60 of the theoretical variants have not been sampled?

The approximate probability that k variants are missing is given by:

$$\mathrm{P}(k \text{ variants are missing}) \cong \frac{e^{-\lambda}\lambda^k}{k!}, \quad k \in \{0,1,\ldots,n\} \tag{2}$$

Table 2

n	m	Ratio n/m = fold library representation	λ = expected number not picked	Expected percentage not picked	P (at least 1 missing) (%)
1000	500	0.5	606	60.6	100
	1000	1	368	36.8	100
	2000	2	135	13.5	100
	10,000	10	0.05	0.00	4.42
	20,000	20	0.00	0.00	0.00
1×10^6	5×10^5	0.5	6.07×10^5	60.7	100
	1×10^6	1	3.68×10^5	36.8	100
	2×10^6	2	1.35×10^5	13.5	100
	1×10^7	10	45.4	0.00	100
	2×10^7	20	0.00	0.00	0.21
1×10^{10}	5×10^9	0.5	6.07×10^9	60.7	100
	1×10^{10}	1	3.68×10^9	36.8	100
	2×10^{10}	2	1.35×10^9	13.5	100
	1×10^{11}	10	4.54×10^5	0.00	100
	2×10^{11}	20	20.6	0.00	100

where λ is defined as in **Eq. 1** (*see also* **Notes 1** and **3**). From there, the probability that *at least one* theoretical variant is missing is approximately $1 - e^{-\lambda}$ (*see* **Note 4**). The probability that *at most* K of the theoretical variants are missing is obtained by summing the values of the probability in **Eq. 2** for $k = 0, 1, 2, \ldots$ up to K.

To resolve problem B, we need to first compute the value λ as in **Subheading 3.1.1.1.** For our example given in problem A, in which we sample 1×10^7 times from a theoretical pool of 1×10^6 variants, we computed $\lambda \approx 45.4$. Then, the probability that *at least one* theoretical variant has not been sampled is computed as $1 - e^{-45.4} \cong 1.00$, or 100%. Therefore, it is essentially certain (100%) that at least some of the theoretical variants are missing in the sample of m variants. In practice, what does this mean? This result (and the additional examples provided in **Table 2**) indicates that if one screens a sample size m that is 10 times greater than the theoretical library size n, for n greater or equal to 10^5, it is highly probable (almost 100%) that the theoretical library n has not been completely sampled. For smaller libraries (n = 1000 and less), there is only a weak probability that sampling with $m = 10 \times n$ gives an incomplete coverage of the theoretical library (for example, the probability is ~5% for $n = 1000$, *see* **Table 2**).

By way of contrast, if we sample twice as many, i.e., $m = 2 \times 10^7$ items sampled from the same theoretical pool of $n = 1 \times 10^6$ variants, $\lambda = 0.00206$ and the probability that *at least one* theoretical variant has not been sampled is approx $1 - e^{-0.00206} = 0.00206$ or 0.2%. In this case, there is a greater than 99% probability (specifically, 100% – 0.206% = 99.794%) that all of the theoretical variants n have been sampled (the fact that λ and the probability $1 - e^{-\lambda}$ are very close numbers is not surprising, given that $1 + x \cong e^x$ for small x.).

For our main example ($n = 1 \times 10^6$; $m = 2 \times 10^7$), the probability that fewer than 50 from among the n theoretical variants have not been sampled in the m selected items is computed as the summation of the probabilities that each of 0, 1, 2, … up to 49 theoretical variants have not been sampled, yielding approx 78%. Thus:

$$\sum_{k=0}^{49} \frac{e^{-0.00206}\,0.00206^k}{k!} = 0.78$$

Similarly, the probability that fewer than 60 have not been sampled is approx 98%. For ease of computation, use the Excel computation sheets (*see* **Heading 2.** and **Fig. 1**).

3.1.1.3. Problem C

How many times can we expect a variant i to appear in the sample? What is the probability that a variant i appears 10 times? Three times? What about the probability that the variant i appears 10 times or less?

The probability that a variant appears a certain number of times is approximated through a Poisson distribution (*see* **Note 5**) with parameter:

$$\lambda = \frac{m}{n} \tag{3}$$

so that the number of times we expect a certain variant i to appear is:

$$\text{Expected number of times the variant } i \text{ occurs} = \lambda = \frac{m}{n}$$

and the probability is:

$$\text{Probability (variant } i \text{ occurs } r \text{ times in the sample of size } m) = e^{-\lambda}\frac{\lambda^r}{r!} \tag{4}$$

(*see* **Notes 2**, **3**, and **6**). Be aware that the parameter of the Poisson distribution for this problem, given in **Eq. 3**, is different from that used in Problems A and B, **Subheadings 3.1.1.1.** and **3.1.1.2.**, respectively, given in **Eq. 1**.

The probability that a variant appears a certain number of times or less can be computed by using **Eq. 4** repetitively.

For example, we take again $n = 1 \times 10^6$ and $m = 1 \times 10^7$. We expect to find variant i $\frac{1 \times 10^7}{1 \times 10^6} = 10$ times in the sample. The probability that i appears 10 times is:

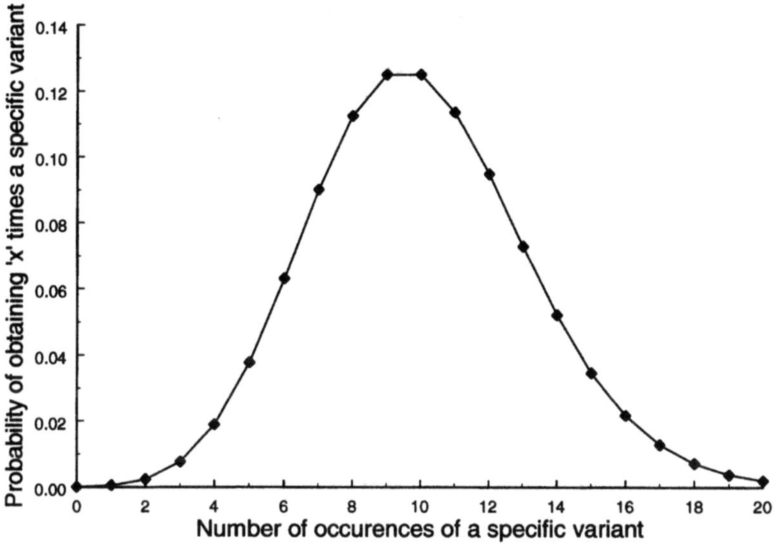

Fig. 6. Probability of obtaining "x" times a specific variant vs the number of occurrences of that variant. The example presented here uses the theoretical library size as $n = 1 \times 10^6$ and $m = 1 \times 10^7$. Thus, there is a probability greater than 10% of obtaining 8, 9, 10, or 11 times a specific variant, and a probability more than 50% of obtaining between 8 and 12 times a specific variant, using the specified values of n and m.

$$e^{-\lambda} \cdot \frac{\lambda^r}{r!} = e^{\left(-\frac{1\times10^7}{1\times10^6}\right)} \cdot \frac{\left(\frac{1\times10^7}{1\times10^6}\right)^{10}}{10!} \cong 12.5\% \quad (5)$$

This computation can be bypassed by using a built-in Poisson distribution function, available in several software programs (Excel, Matlab, Mathematica, and so on); refer to the accompanying Excel sheet (*see* **Heading 2.** and **Fig. 1**) for an example.

As illustrated in **Fig. 6**, there is a reasonable likelihood (12.5%) of obtaining 10 identical hits, and calculating for nearby values ($i = 8, 9, 11$, and 12) reveals that it is also quite likely to obtain those numbers of identical hits (11.3, 12.5, 11.4, and 9.5%, respectively). However, the probability that variant i appears three times is only:

$$e^{\left(-\frac{1\times10^7}{1\times10^6}\right)} \cdot \frac{\left(\frac{1\times10^7}{1\times10^6}\right)^{3}}{3!} \cong 0.76\%$$

that is, a less than 1% chance. Would obtaining such a result (a certain variant appearing three times) suggest that there is a bias in the system, as could result from selective pressure for example? Regarding this question, *see* **Subheading 3.2**.

Finally, the question of the variant appearing 10 times or less can clearly be computed by adding the probabilities that it appears 10 times, 9 times, and so on, down to 0 times, either manually or using the accompanying Excel worksheets (*see* **Heading 2.** and **Fig. 1**). One would, thus, find that the probability of 10 occurrences or less, for the data in this example, is approx 58.3%.

3.1.2. One Case of Outcomes With Unequal Probabilities

We present one case in which outcomes occur with unequal probabilities, i.e., some of the *n* variants have a better or worse chance of coming up than the average. As noted at the beginning of **Subheading 3.1.**, it is difficult to generalize in the case of unequal probabilities, precisely because the answers depend on what the probabilities are. The case that we provide allows some straightforward computations but also illustrates some of the difficulties one can encounter.

In the absence of selective pressure, outcomes of unequal probabilities can occur if the DNA library is biased, whether the bias was designed or whether it is accidental, as in the case of unwanted biases in oligonucleotide primer synthesis, for example. Such biases can be identified after sequencing randomly chosen library members. Unequal probabilities also occur if we define the theoretical library by the encoded protein sequences rather than the DNA sequences. For example, when encoding "NNN"-type codons, each of the six codons encoding leucine should occur with the same probability as the unique codon specifying methionine, giving a sixfold greater probability of obtaining the proteins with leucine at this position relative to methionine.

In the presence of selective pressure, it is generally not possible, at the outset of the experiment, to numerically define the selective advantage or disadvantage correlated with specific mutations. In fact, in the occurrence of selective pressure, the question is best turned around: rather than asking how frequently a variant should occur given certain sequence biases, we can observe the experimental frequency of appearance of sequences in the selected subset. We can then determine whether this frequency was expected (as in Problem C, **Subheading 3.1.1.3.**), or if it suggests that a bias is present.

3.1.2.1. Characteristics of Our Case With Unequal Probabilities

We consider sampling decapeptides. If each of the 10 amino acids within a decapeptide is randomized (NNN), there are $21^{10} \cong 1.67 \times 10^{13}$ different theoretical variants of decapeptides (we treat the stop codons as any other). Each amino acid has an unequal probability of being picked because of codon degeneracy; nonetheless, the only probabilities that occur are the five following:

$$\left\{\frac{1}{64},\frac{2}{64},\frac{3}{64},\frac{4}{64},\frac{6}{64}\right\},$$

where Met and Trp are encoded by 1 out of 64 codons each; Phe, Tyr, His, Gln, Asn, Lys, Asp, Glu, and Cys are encoded by 2 out of 64 codons each; and so on. Thus, there are, respectively, 2, 9, 2, 5, and 3 amino acids or stop codons, respectively, corresponding to each of the five probabilities given.

3.1.2.2. Problem A

How many of the n theoretical variants do we expect **not** to appear among the m variants chosen?

To resolve this problem, the principal factor to consider is the codon degeneracy. We consider that each unique decapeptide has the same probability of occurrence as any other decapeptide of the same amino acid composition (and, thus, of the same degeneracy), irrespective of the order of the amino acids. The answer is the parameter λ (*see* **Note 7**), computed as:

$$\lambda \cong \sum_{(n_1,n_2,n_3,n_4,n_5)} \left(\frac{10!}{n_1!n_2!n_3!n_4!n_5!} 2^{n_1} 9^{n_2} 2^{n_3} 5^{n_4} 3^{n_5} \right) e^{-p(n_1,n_2,n_3,n_4,n_5)m} \tag{6}$$

where the summation is over all ordered series of five numbers (n_1,n_2,n_3,n_4,n_5), which represent the five probabilities of codon degeneracy given. Each of the five numbers is taken from the set $\{0,1,\ldots,10\}$ to represent how many amino acids within the decapeptide are of each of the five degeneracies. Thus, $\sum_{i=1}^{5} n_i = 10$ because we consider 10 amino acids. For example, $(n_1,n_2,n_3,n_4,n_5) = (8,0,0,1,1)$ means that 8 positions of the decapeptide are encoded by a nondegenerate amino acid (1/64), no positions contain amino acids of degeneracy 2/64 or 3/64, and one position each is encoded by an amino acid with degeneracy 4/64 and 6/64.

$p(n_1,n_2,n_3,n_4,n_5)$ is computed as:

$$p(n_1,n_2,n_3,n_4,n_5) = \left(\frac{1}{64}\right)^{n_1} \left(\frac{2}{64}\right)^{n_2} \left(\frac{3}{64}\right)^{n_3} \left(\frac{4}{64}\right)^{n_4} \left(\frac{6}{64}\right)^{n_5} \tag{7}$$

This $p(n_1,n_2,n_3,n_4,n_5)$ is a new name for what was introduced as simply p_i at the outset of **Subheading 3.1.**, the probability that a certain variant comes up any time we sample once, randomly, from the theoretical pool.

Unfortunately, there seems to be no simpler expression for λ; the summation involved in the computation of λ must be programmed on a computer.

For example, we sample $m = 10^{14}$ decapeptides, randomly, from the theoretical pool of $n = 21^{10}$ (i.e., ~1.67×10^{13}; *see* **Fig. 2**).

The answer for these m and n is $\lambda = 5.04 \times 10^{12}$, so that we expect 5.04×10^{12} of the 1.67×10^{13} theoretical variants *not* to appear in the sample, a proportion of approx 30%.

3.1.2.3. Problem B

What is the probability that at least one of the peptide variants has not been generated in the sample? What is the probability that at most 50, or yet 60, of the peptide variants have not been sampled?

Such probabilities are found by using **Eq. 2** of **Subheading 3.1.1.** regarding equiprobable outcomes; *however*, the value of λ must be computed as in **Eq. 6**. Thus, the probability that *at least one* variant is missing is again approximately:

$$1 - e^{-\lambda} \tag{8}$$

(*see* **Note 5**). The approach to compute probabilities of "at least" or "at most" a certain number of missing variants, was also given in **Subheading 3.1.1.2.**, and is not repeated here.

For example, we again sample $m = 10^{14}$ decapeptides, randomly, from the theoretical pool of $n = 21^{10}$; and found $\lambda = 5.04 \times 10^{12}$ in **Subheading 3.1.2.2.** Clearly, with $\lambda = 5.04 \times 10^{12}$ *expected missing* variants, the probability that at least one is missing is, intuitively, bound to be very high. Using **Eq. 8**, we indeed find a probability of 100%. In other words, the probability that all variants have been generated is 0% (*see* **Fig. 2**).

In fact, the probability that at most 50 or 60 of all variants have not been generated, is also 0%. With such a high number (5.04×10^{12}) of expected missing variants, this is not surprising.

3.1.2.4. Problem C

How many times can we expect a decapeptide with seven codons of probability 3/64 and three codons of probability 4/64, to appear in the sample? What is the probability that this decapeptide appears 10 times? Three times? What about the probability that the decapeptide appears 10 times or less?

Because the answer will be the same for every peptide of the composition defined in this question, we do not need to consider the exact sequence (i.e., order or identity of the amino acids). The answers to these questions are only slightly different from those provided in **Subheading 3.1.1.3.** Because the probability of a decapeptide to occur after sampling any decapeptide is the value p (*see* **Subheading 3.1.2.2.**), then the number of times we expect that decapeptide to appear is mp, where m is the number of items sampled. Considering that any decapeptide may be sampled more than once and that this probability increases with the number of items sampled, the probability that the desired decapeptide appears a certain number of times is approximated through a Poisson distribution with parameter $\lambda = mp$ (*see* **Notes 3** and **8**), such that:

$$\text{Probability (the decapeptide occurs } r \text{ times in the sample of size } m) = e^{-\lambda}\frac{\lambda^r}{r!}$$

For example, again assuming that we sample $m = 10^{14}$ decapeptides, the (very small) probability of occurrence:

$$p(0,0,7,3,0) = \left(\frac{3}{64}\right)^7 \left(\frac{4}{64}\right)^3 = 1.21 \times 10^{-13}$$

of the decapeptide means that we expect to find $\lambda = mp = 10^{14} \times (1.21 \times 10^{13}) \cong$ 12.14 times that decapeptide in the sample. The probability that it appears 10 times is:

$$e^{-\lambda}\frac{\lambda^r}{r!} = e^{-12.14}\left(\frac{12.14^{10}}{10!}\right) \cong 10.2\%$$

Similarly, the probability that the specific decapeptide occurs three times is approx 0.16%, much less than the 10% chance of 10 appearances. Adding up all probabilities at and below 10, we find that the probability of 10 or less occurrences of the peptide in the sample is approx 33% (*see* **Fig. 3**).

3.2. Tests of Experimental Biases

This section is concerned with essentially one question, namely, the following: An experiment with two or many possible outcomes is run several times, i.e., several items are sampled. The theoretical (expected) proportion of outcomes is known, and the experimental proportion of outcomes is observed. What difference between the theoretical and experimental proportions is a significant signal that the experiment is biased?

This analysis can be routinely applied to DNA sequencing results to detect biases, before or after selection, as proposed in **Heading 1.** Application of the following χ^2 test allows the experimenter to decide whether the observed biases are significant or not.

Let the question be posed with the following notation: an experiment theoretically yields the k outcomes $A_1, A_2, \ldots, A_k$ in proportions of, respectively, $p_1, p_2, \ldots p_k$. For example, sampling a library of proteins may give $k = 2$ outcomes: functional or nonfunctional. The experiment is run n times (i.e., n proteins are randomly sampled), and results in each outcome occurring, respectively, $Y_1, Y_2, \ldots Y_k$ times.

3.2.1. The χ^2 Test

The statistic used in the χ^2 test is:

$$q = \frac{(Y_1 - np_1)^2}{np_1} + \frac{(Y_2 - np_2)^2}{np_2} + \ldots + \frac{(Y_k - np_k)^2}{np_k}$$

The hypothesis that we test, the so-called *null hypothesis*, is that the experiment was run properly, i.e., in such a way as to yield the outcomes in their theoretical proportions, on average. A large value of q suggests that there is significant discrepancy between the results of the actual experiment and the results expected, contradicting the null hypothesis.

The statistic q is compared with the χ^2 distribution of parameter $k - 1$, at any chosen level α of significance, a value that is denoted χ^2_α. The parameter α is specified by the user (*see* **Note 9** regarding the significance level). The test is as follows:

For $q < \chi^2_\alpha(k - 1)$, the hypothesis is not rejected; there is no sufficient reason to think there is a bias.

For $q > \chi^2_\alpha(k - 1)$, the hypothesis is rejected; there is sufficient reason to think there is a bias.

Values for the χ^2 distribution can be found in tables in most statistics textbooks and many software programs, including Excel.

Example 1. Two possible outcomes: an experiment is run 100 times; each time, a certain characteristic may or may not occur, thus, there are two possible outcomes and $k = 2$. The characteristic (call it "outcome 1") should theoretically occur in one sampling out of four; in the 100 items sampled, it occurred 29 times. The χ^2 test is based on the statistic:

$$q = \frac{(29 - 100 \times 0.25)^2}{100 \times 0.25} + \frac{(71 - 100 \times 0.75)^2}{100 \times 0.75} \cong 0.85$$

Let us choose a level of significance $\alpha = 0.05$. Then the test is to compare $q = 0.85$ to:

$$\chi^2_{0.05}(k - 1) = \chi^2_{0.05}(2 - 1) = \chi^2_{0.05}(1) = 3.84$$

Because:

$$q = 0.85 < 3.84 = \chi^2_{0.05}(1)$$

the hypothesis (that the experiment was run properly, without bias) is not rejected (*see* **Fig. 4**). **Figure 7** illustrates the number of outcomes with the characteristic for which the hypothesis should be kept (17 to 33 outcomes with the characteristic) or rejected (all other results).

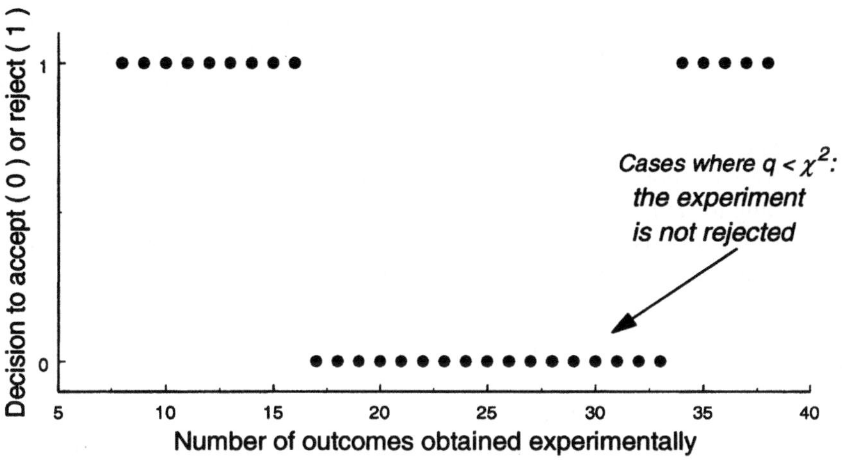

Fig. 7. Decision to accept or reject an experimental result based on the χ^2 test. The value q is calculated for an experimental outcome and is compared with a value of χ^2, chosen at the desired level (α) of significance from a table of values. If $q < \chi^2$, the outcome is assigned the value 0, do not reject. If $q > \chi^2$, the outcome is assigned the value 1, and is rejected.

Example 2. Three possible outcomes: Suppose an experiment theoretically yields outcome A, 20% of the time; outcome B, 30% of the time; and outcome C, 50% of the time. The experiment is run 20 times by a new student in the lab; the student obtains 8 outcomes A, 6 outcomes B, and only 6 outcomes C. Is this credible? Is the student's technique trustworthy?

The statistic is:

$$q = \frac{(8-20\times 0.20)^2}{20\times 0.20} + \frac{(6-20\times 0.30)^2}{20\times 0.30} + \frac{(6-20\times 0.50)^2}{20\times 0.50} = 5.6$$

At the level 5%, we have $\chi^2_{0.05}(2) = 5.991$ and: $q = 5.6 < 5.991 = \chi^2_{0.05}(2)$ indicates that the statistic q, and the discrepancy with the expected results, are not so large that we can conclude that there is a flaw in the student's procedure; the hypothesis cannot be rejected. However, if we are ready to work with $\alpha = 10\%$, then $\chi^2_{0.1}(2) = 4.605$, and the test becomes: $q = 5.6 > 4.605 = \chi^2_{0.1}(2)$ and we *are* able to reject the hypothesis that the actual experiment is the intended one.

4. Notes

1. The answers to Problems A and B (**Subheadings 3.1.1.1.** and **3.1.1.2.**) rely on the theory of Poisson approximations of sums of Bernoulli random variables, which

essentially states that if all of the λs, defined in **Eq. 9** and **Eq. 14**, are small, then the number of variants that do not occur in the sample follows approximately a Poisson distribution with parameter $\lambda = \sum_{i=1}^{n} \lambda_i$ (*see* Chapter 10 of **ref. *15***, particularly Example 10.2(B) regarding the multinomial distribution, for the mathematical details).

The Poisson approximation should be "very successful if $n \geq 20$ and $\lambda \leq 10$," according to ***(16)***, page 252.

2. The number of missing variants approximately follows a Poisson distribution with the parameter λ, defined as $\lambda = \sum_{i=1}^{n} \lambda_i$, and where λ_i is defined as:

$$\lambda_i = P(\text{variant } i \text{ is missing from the sample}) = (1 - p_i)^m \tag{9}$$

By the property of Poisson distributions, the *expected value* of the number of missing variants is the parameter λ itself.

In the case of equiprobable outcomes, because $p_i = \frac{1}{n}$, it is easy to compute λ_i and λ as:

$$\lambda_i = (1 - p_i)^m = \left(1 - \frac{1}{n}\right)^m$$

and:

$$\lambda = \sum_{i=1}^{n} \left(1 - \frac{1}{n}\right)^m = n\left(1 - \frac{1}{n}\right)^m \tag{10}$$

Note that, given the special form of λ_1, the value λ can be computed directly without computing the λ_i.

3. Computations of Poisson distributions with Excel can stall when the distribution's parameter λ is too large. In such cases, one can simply use the normal distribution with the mean equal to λ and the variance equal to λ, i.e., $N(\lambda,\lambda)$ instead. This trick is illustrated in **Fig. 2**.
4. By virtue of the Poisson distribution:

$$P(k \text{ variants are missing}) \cong \frac{e^{-\lambda}\lambda^k}{k!}, \qquad k \in \{0,1,..n\} \tag{11}$$

where λ is defined in **Eq. 1**, or, identically, in **Eq. 10**.

The probability that *none* is missing is then approximately:

$$\frac{e^{-\lambda}\lambda^0}{0!} = e^{-\lambda}$$

so that the probability that *at least one* is missing is approximately:

$$1 - e^{-\lambda} \tag{12}$$

5. The answer to Problem C relies on the Poisson approximation of the binomial distribution followed by the random variable that counts the number of times the variant is drawn. Here, the Poisson approximation should be "very successful if $m \geq 20$ and $\lambda \leq 10$," still according to ***(16)***.
6. The event of obtaining a specific variant i a certain number of times r in the sample follows a binomial distribution of parameters m (sample size) and $\frac{1}{n}$ (the probability of picking variant i anytime a variant is picked randomly). This binomial distribution is well approximated, and much easier to compute, as a Poisson distribution with parameter $\lambda = \frac{m}{n}$.
7. *See* **Note 2.** The λ for this case is computed as follows. In opposition to the equiprobable case, the λ_i are *not* identical here. At first glance, to compute λ, we would need to compute all of $n = 21^{10}$ parameters λ_i, which is enormous, but which contains many repetitions, a characteristic that we use to simplify the computations. Rather, let us identify any decapeptide with the 5-tuple (n_1,n_2,n_3,n_4,n_5) that indicates the *number of codons* of, respectively, probability $\frac{1}{64}, \frac{2}{64}, \frac{3}{64}, \frac{4}{64}$ and $\frac{6}{64}$, in the said decapeptide. Clearly, $\sum_{i=1}^{5} n_i = 10$, otherwise there are not 10 codons in the decapeptide. Also, there are:

$$\frac{10!}{n_1!n_2!n_3!n_4!n_5!}2^{n_1}9^{n_2}1^{n_3}5^{n_4}3^{n_5} \tag{13}$$

different decapeptides associated with the 5-tuple (n_1,n_2,n_3,n_4,n_5). This number accounts for all shufflings of the codons with the peptide, i.e., the order of the amino acids within the peptide needs not be explicitly defined. It also accounts for the fact that more than one amino acid displays each probability (except for Ile, the only amino acid encoded by 3/64 codons). The point we make is that the 5-tuples contain all of the information regarding the decapeptides required to answer our questions. We can now replace the computation of the $\lambda_i = (1 - p_i)^m$, which is specific to decapeptide i, by the computation of a $\lambda(n_1,n_2,n_3,n_4,n_5)$, such that:

$$\lambda_i = \lambda(n_1,n_2,n_3,n_4,n_5) = [1 - p(n_1,n_2,n_3,n_4,n_5)]^m \tag{14}$$

and which is specific to the 5-tuple (n_1,n_2,n_3,n_4,n_5) associated with decapeptide i. That is, of the 21^{10} parameters λ_i, we will compute only the *different* values that occur in the λ_i. There are:

$$\binom{10+5-1}{5-1} = \frac{14!}{10!4!} = 1001 \tag{15}$$

different values in the λ_i; here, we use the binomial coefficient notation, $\binom{r}{s} = \frac{r!}{(r-s)!s!}$

See, for example, the section of **ref. *17*** on the distribution of balls in urns, for the details on **Eq. 15**.

Because the draws follow a multinomial distribution, the probability p_i of occurrence of decapeptide i with associated 5-tuple (n_1,n_2,n_3,n_4,n_5), is:

$$p_i = p(n_1,n_2,n_3,n_4,n_5) = \left(\frac{1}{64}\right)^{n_1}\left(\frac{2}{64}\right)^{n_2}\left(\frac{3}{64}\right)^{n_3}\left(\frac{4}{64}\right)^{n_4}\left(\frac{6}{64}\right)^{n_5} \quad (16)$$

so that one can easily compute $\lambda_i = \lambda(n_1,n_2,n_3,n_4,n_5)$ through **Eq. 14**.
We can finally compute $\lambda = \sum_{i=1}^{21^{10}} \lambda_i$ using only the 1001 values for $\lambda(n_1,n_2,n_3,n_4,n_5)$ and the number of times each is repeated, given by **Eq. 13**:

$$\begin{aligned}
\lambda &= \sum_{i=1}^{21^{10}} \lambda_i \\
&= \sum_{(n_1,n_2,n_3,n_4,n_5)} \left[\frac{10!}{n_1!n_2!n_3!n_4!n_5!} 2^{n_1}9^{n_2}2^{n_3}5^{n_4}3^{n_5}\right] \lambda(n_1,n_2,n_3,n_4,n_5) \\
&= \sum_{(n_1,n_2,n_3,n_4,n_5)} \left[\frac{10!}{n_1!n_2!n_3!n_4!n_5!} 2^{n_1}9^{n_2}2^{n_3}5^{n_4}3^{n_5}\right] \left[1 - p(n_1,n_2,n_3,n_4,n_5)\right]^m \\
&\cong \sum_{(n_1,n_2,n_3,n_4,n_5)} \left[\frac{10!}{n_1!n_2!n_3!n_4!n_5!} 2^{n_1}9^{n_2}2^{n_3}5^{n_4}3^{n_5}\right] e^{-p(n_1,n_2,n_3,n_4,n_5)m}
\end{aligned}$$

where the second, third, and fourth summations are over all 5-tuples (n_1,n_2,n_3,n_4,n_5) defined in **Heading 1.** That is, they are sets of five numbers such that $\sum_{i=1}^{5} n_i = 10$, where the n_i are numbers in the set $\{0,1,\ldots,10\}$ because it is a decapeptide. Also, $p(n_1,n_2,n_3,n_4,n_5)$ is computed according to **Eq. 16**. The last equation gives an approximation of λ (justified by the fact that $1 + x \cong e^x$ for small x), which is much easier to handle than the second to last equation; too often, $1 - p(n_1,n_2,n_3,n_4,n_5)$ would round up to 1 on a computer.

8. *See* **Note 5**. The only difference here is that the binomial distribution's underlying probability is not $1/n$ (the probability in the equiprobable context), but some other probability p, which depends on the variant.
9. Recall the meaning of the level of significance $\alpha = 0.05$. It means that there is a 5% chance of a type I error, which is to reject the hypothesis although it is true. That is, there is a 1:20 chance of declaring "flawed" a perfectly well-executed experiment.

Acknowledgments

We thank Andreas Plückthun and group members for proposing problems as well as Bruno Rémillard (HEC Montréal, Canada), Nicolas Doucet (Université de

Montréal, Canada), Sabine C. Stebel (Universität Freiburg, Germany), and the editors Katja M. Arndt and Kristian M. Müller (Universität Freiburg, Germany) for their valuable comments.

This work was supported by National Sciences and Engineering Research Council grants 227853-02 (J. N. Pelletier) and 227838-00 (M. Denault).

References

1. Arnold, F. H. and Georgiou, G. (2003) *Directed Evolution Library Creation Methods and Protocols*. Methods in Molecular Biology **231,** Humana Press, Totowa, NJ.
2. Arnold, F. H. and Georgiou, G. (2003) *Directed Enzyme Evolution Screening and Selection Methods*. Methods in Molecular Biology **230,** Humana Press, Totowa, NJ.
3. Moore, J. C., Jin, H. M., Kuchner, O., and Arnold, F. H. (1997) Strategies for the in vitro evolution of protein function—enzyme evolution by random recombination of improved sequences. *J. Mol. Biol.* **272,** 336–347.
4. Moore, G. L. and Maranas, C. D. (2000) Modeling DNA mutation and recombination for directed evolution experiments. *J. Theor. Biol.* **205,** 483–503.
5. Moore, G. L., Maranas, C. D., Lutz, S., and Benkovic, S. J. (2001) Predicting crossover generation in DNA shuffling. *Proc. Natl. Acad. Sci. USA* **98,** 3226–3231.
6. Moore, G. L. and Maranas, C. D. (2002) Predicting out-of-sequence reassembly in DNA shuffling. *J. Theor. Biol.* **219,** 9–17.
7. Sun, F. (1999) Modeling DNA shuffling. *J. Comput. Biol.* **6,** 77–90.
8. Patrick, W. M., Firth, A. E., and Blackburn, J. M. (2003) User-friendly algorithms for estimating completeness and diversity in randomized protein-encoding libraries. *Protein Eng.* **16,** 451–457.
9. Bittker, J. A., Le, B. V., Liu, J. M., and Liu, D. R. (2004) Directed evolution of protein enzymes using nonhomologous random recombination. *Proc. Natl. Acad. Sci. USA* **101,** 7011–7016.
10. Gribskov, M., Devereux, J., and Burgess, R. R. (1984) The codon preference plot: graphic analysis of protein coding sequences and prediction of gene expression. *Nucleic Acids Res.* **12,** 539–549.
11. Virnekas, B., Ge, L., Plückthun, A., Schneider, K. C., Wellnhofer, G., and Moroney, S. E. (1994) Trinucleotide phosphoramidites: ideal reagents for the synthesis of mixed oligonucleotides for random mutagenesis. *Nucleic Acids Res.* **22,** 5600–5607.
12. Pelletier, J. N., Arndt, K. M., Plückthun, A., and Michnick, S. W. (1999) An in vivo library-versus-library selection of optimized protein-protein interactions. *Nature Biotechnol.* **17,** 683–690.
13. Braunagel, M. and Little, M. (1997) Construction of a semisynthetic antibody library using trinucleotide oligos. *Nucleic Acids Res.* **25,** 4690, 4691.
14. Gaytan, P., Yanez, J., Sanchez, F., and Soberon, X. (2001) Orthogonal combinatorial mutagenesis: a codon-level combinatorial mutagenesis method useful for low multiplicity and amino acid-scanning protocols. *Nucleic Acids Res.* **29,** E9.

15. Ross, S. M. (1996) *Stochastic Processes.* 2nd ed., John Wiley & Sons, New York, NY.
16. Hogg, R. V. and Tanis, E. A. (1983) *Probability and Statistical Inference*. 2nd ed., MacMillan, New York, NY.
17. Ross, S. M. (1998) *A First Course in Probability*. 5th ed., Prentice Hall, Upper Saddle River, NJ.

9

Protein Design by Binary Patterning of Polar and Nonpolar Amino Acids

Luke H. Bradley, Yinan Wei, Peter Thumfort, Christine Wurth, and Michael H. Hecht

Summary

The design of large libraries of well-folded *de novo* proteins is a powerful approach toward the ultimate goal of producing proteins with novel structures and functions for use in industry or medicine. A method for library design that incorporates both rational design and combinatorial diversity relies on the "binary patterning" of polar and nonpolar amino acids. Binary patterning is based on the premise that the appropriate arrangement of polar and nonpolar residues can direct a polypeptide chain to fold into amphipathic elements of secondary structure that anneal together to form a desired tertiary structure. A designed binary pattern exploits the periodicities inherent in protein secondary structure, and allows the identity of the side chain at each polar and nonpolar position to be varied combinatorially. This chapter provides an overview of the considerations necessary to use binary patterning to design libraries of novel proteins.

Key Words: Protein design; binary patterning; combinatorial library; *de novo* proteins; library design.

1. Introduction

Numerous studies of natural proteins have demonstrated that protein structures are remarkably tolerant to amino acid substitutions. Thus, many different amino acids can encode the information necessary to produce a given three-dimensional structure *(1–7)*.

We have taken advantage of this tolerance to develop a general strategy for protein design. The strategy—called "the binary code" strategy—is based on the premise that the appropriate patterning of polar and nonpolar residues can direct a polypeptide chain to fold into elements of secondary structure, while simultaneously allowing the burial of nonpolar amino acids in a desired tertiary structure *(8–10)*. A designed binary pattern exploits the periodicities inherent in protein

From: *Methods in Molecular Biology, vol. 352: Protein Engineering Protocols*
Edited by: K. M. Arndt and K. M. Müller © Humana Press Inc., Totowa, NJ

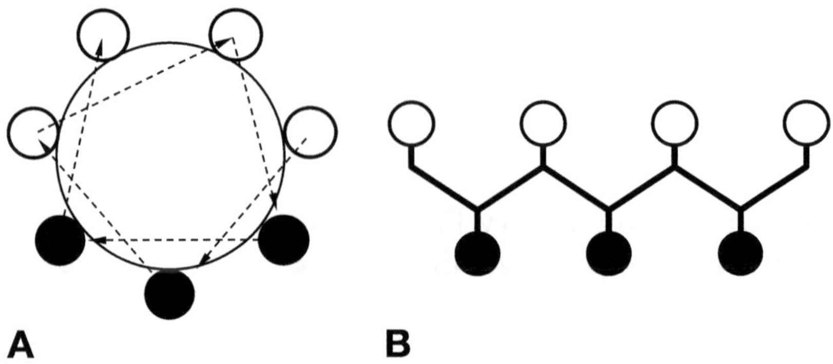

Fig. 1. The designed binary pattern of polar (○) and nonpolar (●) amino acids for α-helices **(A)** and β-strands **(B)** exploits the inherent periodicities of the respective secondary structures. (A) α-helices have a repeating periodicity of 3.6 residues per turn. By placing a nonpolar amino acid at every third or fourth position, an amphipathic helix can be encoded in which one face is polar and the opposite face is nonpolar. Note that this figure shows the positioning of seven amino acids. For longer α-helices, the binary pattern will have to be adjusted to allow the 3.6 residues per turn periodicity to maintain the amphipathic nature the entire length of the helix. (B) β-strands have an alternating periodicity of polar and nonpolar amino acids. This pattern would cause one face of the strand to be polar and the opposite face to be nonpolar.

secondary structure: α-helices have a repeating periodicity of 3.6 residues per turn, whereas β-strands have an alternating periodicity (**Fig. 1**). Thus, a binary patterned sequence designed to form amphipathic α-helices would place a nonpolar residue at every third or fourth position. In contrast, the binary pattern for an amphipathic β-strand would alternate between polar and nonpolar residues. In the binary code strategy, the precise three-dimensional packing of the side chains is not specified *a priori*. Therefore, within a library of binary patterned sequences, the identity of the side chain at each polar and nonpolar position can be varied extensively, thus, facilitating enormous combinatorial diversity.

A combinatorial library of binary patterned proteins is expressed from a combinatorial library of synthetic genes. Each gene encodes a different amino acid sequence, but all sequences within a given library have the same patterning of polar and nonpolar residues. This sequence degeneracy is made possible by the organization of the genetic code (**Fig. 2**). The degenerate codon NTN encodes nonpolar amino acids, whereas the degenerate codon NAN encodes polar amino acids. (N represents a mixture of A, G, T, and C; *see* **Subheading 2.2.** regarding codon usage.) With these degenerate codons, positions requiring a nonpolar amino acid are filled by phenylalanine, leucine, isoleucine, methionine, or valine; whereas positions requiring a polar amino acid are filled by glutamate,

First Position	Middle Position: T	C	A	G
T	TTT Phe	TCT Ser	TAT Tyr	TGT Cys
	TTC Phe	TCC Ser	TAC Tyr	TGC Cys
	TTA Leu	TCA Ser	TAA Stop	TGA Stop
	TTG Leu	TCG Ser	TAG Stop	TGG Trp
C	CTT Leu	CCT Pro	CAT His	CGT Arg
	CTC Leu	CCC Pro	CAC His	CGC Arg
	CTA Leu	CCA Pro	CAA Gln	CGA Arg
	CTG Leu	CCG Pro	CAG Gln	CGG Arg
A	ATT Ile	ACT Thr	AAT Asn	AGT Ser
	ATC Ile	ACC Thr	AAC Asn	AGC Ser
	ATA Ile	ACA Thr	AAA Lys	AGA Arg
	ATG Met	ACG Thr	AAG Lys	AGG Arg
G	GTT Val	GCT Ala	GAT Asp	GGT Gly
	GTC Val	GCC Ala	GAC Asp	GGC Gly
	GTA Val	GCA Ala	GAA Glu	GGA Gly
	GTG Val	GCG Ala	GAG Glu	GGG Gly

Fig. 2. The organization of the genetic code allows sequence degeneracy for polar and nonpolar amino acids to be incorporated into a combinatorial library of synthetic genes by defining the middle position of the codon. For a nonpolar amino acid position, the degenerate codon NTN (N represents a mixture of A, G, T, and C) would encode for Phe, Leu, Ile, Met, or Val incorporation. For positions requiring polar amino acids, the degenerate codon NAN would encode for His, Gln, Asn, Lys, Asp, or Glu incorporation. For the NAN codon, T is excluded from the first base to avoid stop codons.

aspartate, lysine, asparagine, glutamine, or histidine (**Fig. 2**). Strategies for the avoidance of stop codons are described in **Subheading 2.2.1**.

This chapter outlines the methodology for using binary patterning to design libraries of *de novo* proteins. Using examples from our laboratory, we focus on the design of all α-helical and all β-sheet proteins. For an overview of designed combinatorial libraries, *see* the reviews from our laboratory *(11,12)*.

2. Materials and Methods

2.1. Design of a Structural Template

Binary patterning can be applied to any amphipathic α-helical or β-stranded segment in a protein. Although our laboratory has focused on *de novo* proteins, the binary code strategy can also be applied to local areas of existing proteins, such as the active site, part of the core, or an interface *(13)*. For the design of *de novo* proteins, the success of the strategy depends primarily on how well the template is designed. Several factors to consider for template design are presented next.

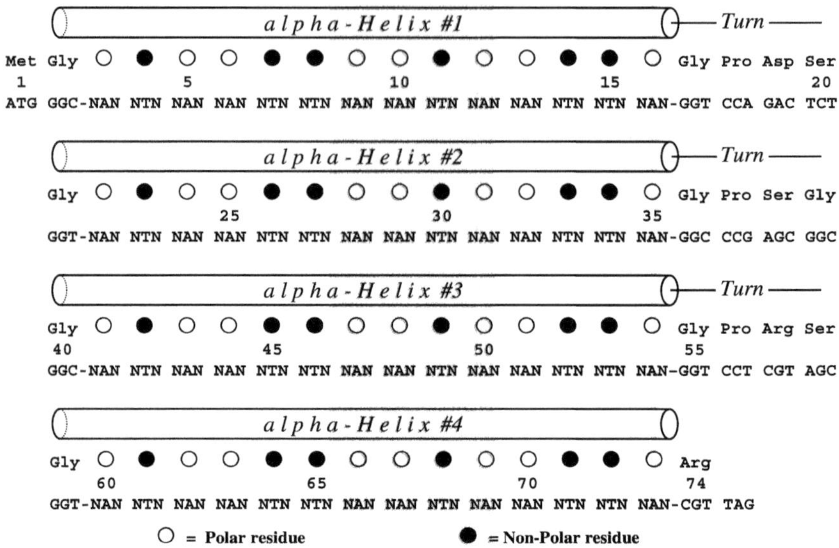

Fig. 3. The design template for the initial four-helix bundle library. Nonpolar positions (●) were encoded by the degenerate NTN codon, and polar positions (○) were encoded by the degenerate NAN codon. The defined residue positions were located at the N- and C-terminal regions as well as the interhelical turn regions of the designed protein.

2.1.1. Binary Patterned Regions

2.1.1.1. α-Helical Designs

Binary patterning exploits the periodicities inherent in secondary structures. α-Helices have a repeating periodicity of 3.6 residues per turn (**Fig. 1A**). To design an amphipathic segment of α-helical secondary structure, a binary pattern of P-N-P-P-N-N-P (P represents polar and N, nonpolar) is used. Our initial α-helical design focused on the four-helix bundle motif (**Fig. 3**). In this structure, the hydrophobic face of each of the helices is oriented toward the central core of the bundle, whereas the hydrophilic faces of the helices are exposed to the aqueous environment. The P-N-P-P-N-N-P pattern favors the formation of an amphiphilic α-helical secondary structure that can bury all nonpolar amino acids after formation of the desired tertiary structure. From our designed four-helix bundle libraries, more than 50 proteins have been purified and characterized. All have shown typical α-helical circular dichroism spectra (*see* **Note 1**). Moreover, the collection yielded several proteins with native-like properties, such as nuclear magnetic resonance chemical shift dispersion, cooperative chemical and thermal denaturations, and slow hydrogen/deutorium exchange rates *(14–18)*.

2.1.1.2. β-SHEET DESIGN

Amphiphilic β-strands have an alternating periodicity of …P-N-P-N… (**Fig. 1B**). Based on this periodicity, a combinatorial library of synthetic genes can be created to encode β-sheet structures. Polar residues will comprise one face of the β-sheet, with nonpolar residues on the opposing face. The sequences in our first library were designed to have six β-strands, with each strand having the binary pattern P-N-P-N-P-N-P *(9)*. Proteins from this library were expressed from a collection of synthetic genes cloned into *Escherichia coli*. All of the proteins studied in this collection formed a β-sheet secondary structure, having circular dichroism spectra with a characteristic minimum at 217 nm (*see* **Note 1**). The β-sheet proteins from this initial library self-assembled into amyloid-like fibrils *(9)*. The fibrils bury the nonpolar side chains in a hydrophobic core, while exposing polar side chains to solvent.

If these same β-sheet sequences are placed in a heterogeneous environment with a polar/nonpolar interface, they form a different structure. For example, at an air/water interface, these proteins self-assemble into flat β-sheet monolayers, with the nonpolar residues pointing up toward air and polar side chains pointing down toward water *(19)*. Alternatively, at an interface between water and the nonpolar surface of highly ordered pyrolytic graphite, binary patterned β-sheet sequences undergo template-directed assembly on the graphite surface to yield highly ordered structures *(20)*.

2.1.2. Fixed Regions

In practice, it is often necessary to keep part of the protein sequence fixed (i.e., not combinatorially diverse), especially if the target sequence is long. When assembling a library of synthetic genes, these constant regions serve as sites for single-stranded synthetic oligonucleotides to anneal together, and prime the enzymatic synthesis of complementary strands (**Fig. 4**; the assembly of full-length genes from single-stranded oligonucleotides is discussed in **Subheading 2.3.**).

Short or medium-length single-stranded oligonucleotides are typically used to encode the binary patterning of individual segments of secondary structure. Nondegenerate fixed regions on the 5′ and 3′ termini of these oligonucleotides are typically used to encode fixed-turn regions between elements of secondary structure (**Figs. 3** and **4**; **refs.** *8* and *9)*. The amino acid sequences chosen to occupy these turn regions are based on statistical and rational design criteria, as outlined:

1. Sequences in the turn regions are chosen based on positional preferences. For example, in the initial four-helix bundle library, glycine residues were placed at the "N-cap" and "C-cap" positions at the termini of the helices (**Fig. 3**; **ref.** *8)*. Glycine residues are frequently found at these positions in natural proteins *(21)*.

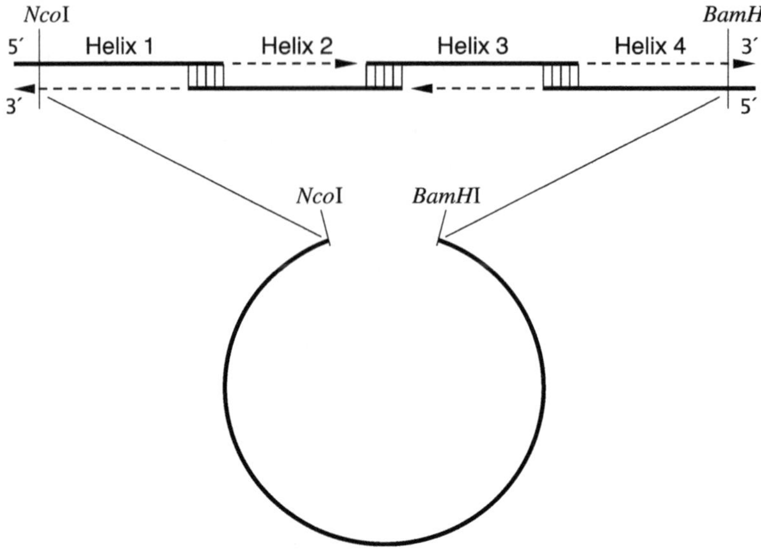

Fig. 4. Assembly of full-length genes by using four single-stranded oligonucleotides. Constant regions at the 5′- and 3′-ends of the single-stranded oligonucleotides serve as sites for annealing and for priming enzymatic synthesis (using DNA polymerase) of complementary strands. The full-length gene is then ready for insertion into an expression vector.

At the position after the C-cap, proline residues were used because they are known to be strong helix breakers. In some situations, however, proline may be undesirable because *cis/trans* isomerism could lead to multiple (rather than unique) conformations. For the β-sheet library ***(9)***, design of the turn regions was based on the "turn potentials" of the various amino acids in the known structures of natural proteins (*see* **Note 2**; **ref.** ***22***).

2. Sequences of the turns can be defined to incorporate restriction sites. These are often useful in gene assembly (*see* **Subheading 2.3.**; **ref.** ***8***).
3. The lengths of constant regions should be long enough to allow for sequence-specific annealing. Pairs of oligonucleotides with overlaps of 10 to 15 nucleotides are typically used for annealing. To further help annealing, one or two nucleotides in the codons immediately before and after the turn regions may also be held constant. For example, a synthetic oligonucleotide (5′-…NAN-NTN-NTN-NAN-<u>GGT-CCT-CGT-AGC</u>-3′) has a constant gene segment (underlined) encoding a four-amino acid-turn region that is 12 nucleotides in length. The previous codon (NAN) encodes for a polar amino acid residue. By defining the third position, for example with a G, the codon now becomes NAG—yielding two extra constant oligonucleotides (bold) for sequence-specific annealing (5′-…NAN-NTN-NTN-**NA**G-<u>GGT-CCT-CGT-AGC</u>-3′). In addition, by defining only the second and third (but not the first) positions of the codon, amino acid diversity is maintained.

In addition to the turn regions, the N and C termini of the *de novo* sequences are also held constant. Fixed sequences in these regions are typically necessary for cloning into an expression vector. Some design criteria for the termini are:

1. An initiator methionine is placed at the N terminus of the *de novo* sequence. This is required for expression in vivo.
2. An aromatic chromophore (i.e., tyrosine or tryptophan) can be incorporated at a constant site in the sequence to aid in protein purification and concentration determination ***(9,18)***. This aromatic residue could be placed either in a constant turn or at one of the chain termini. In some of our libraries, we have inserted a tyrosine immediately after the initiator methionine. This provides a chromophore and prevents methionine removal in vivo ***(23–26)***.
3. The C-terminal residue of the designed protein should contain a charged side chain. The C-terminal sequence of a protein may affect its rate of intracellular proteolysis, and the presence of a charged residue at that position can extend half-life in vivo ***(27–29)***. For the four-helix bundle libraries, an arginine residue was designed to occupy this terminal position ***(8,18)***. Moreover, positively charged side chains at the C termini of α-helices (and negatively charges residues at the N termini) can enhance stability by interacting with the helix dipole ***(30)***.

2.1.3. Considerations for the Design of Tertiary Structure

The ultimate success of a designed template is determined by the properties of the proteins it encodes. If the goal is to produce a library of well-folded globular proteins, then the designed template must not only dictate the secondary structure of the protein, but the tertiary structure as well. The template should be long enough to encode well-folded structures, but at the same time be short enough to be accessible to strategies for assembling large libraries of error-free genes. Many proteins from our first generation (74-residue) four-helix bundle library formed fluctuating structures resembling molten globule folding intermediates ***(8,14–17)***. To investigate the potential of the binary code strategy to encode collections of native-like tertiary structures, a second-generation library of binary-patterned proteins was prepared ***(18)***. This new library was based on protein 86, a pre-existing sequence from the original 74-residue library. The major change to protein 86 was the addition of six combinatorially diverse residues to each of the four helices, thereby making the second-generation proteins comparable in size to natural four-helix bundle proteins. These 24 additional helix-lengthening residues continued to follow the binary patterning.

Characterization of five sequences arbitrarily chosen from this second-generation library showed that all are substantially more stable than the parental protein, 86 ***(17)***. In addition, most of them yielded nuclear magnetic resonance spectra that are well dispersed and exhibit well-resolved nuclear Overhauser effect cross peaks, indicative of unique, well-folded tertiary structures ***(18)***. The structures of two of these proteins were determined by NMR and shown to be well-ordered four-helix bundles, as specified by design ***(12,31)***.

2.2. Codon Usage

As mentioned in **Heading 1.**, the degenerate codons NAN and NTN encode polar and nonpolar amino acids, respectively (**Fig. 2**). However, simply using an equimolar mixture of A, C, G, and T at the combinatorial N positions would introduce undesirable traits into the sequence. Most importantly, an unconstrained NAN codon would encode an unacceptably high frequency of stop codons (i.e., 2 out of 16 NAN codons ≅ 12.5%). Below is a list of considerations we have used in the design of polar and nonpolar codons.

2.2.1. The NAN (Polar) Codon

1. At the first base of the NAN codon, we use an equimolar mixture of G, C, and A. T is excluded, thereby eliminating the possibility of stop codons and tyrosine (*see* **Note 3**). If all bases were included at this first position, then the presence of thymine would give rise to the stop codons TAG or TAA.
2. The mixture of nucleotides at the third base of the NAN codon can be altered to favor some amino acids over others. An equimolar mixture of G, C, A, and T would yield an equal likelihood of histidine, glutamine, asparagine, lysine, aspartate, and glutamate. However, some of the residues of the NAN codon have higher intrinsic propensities than others to form α-helices ***(32–35)***. By omitting T at the third position, we can favor glutamine, lysine, and glutamate over histidine, asparagine, and aspartate. This matches theses residue's intrinsic propensities to form α-helices ***(32–35)***.

2.2.2. The NTN (Nonpolar) Codon

Equimolar mixtures of all four bases at the first and third N positions of the NTN codon would encode six times as many leucines as methionines (i.e., six leucine codons vs one methionine codon). In addition, an equimolar mixture would encode protein sequences in which a quarter of the hydrophobic residues in the interiors would be valine. Because valine has a relatively low α-helical propensity, this may be undesirable for some designs. By altering the molar ratio of the mixture at the first and third N positions, the relative abundance of hydrophobic residues can be altered. For example, in the initial four-helix bundle library, the first base of the NTN codon contained A:T:C:G in a molar ratio of 3:3:3:1 and the third base mixture contained equimolar concentrations of G and C ***(8)***. By biasing the mixture in this way, valine is limited to only 10% of the hydrophobic residues and leucine is represented only three times as frequently as methionine.

2.2.3. Codon Usage of the Host Expression System

The DNA sequences in both the constant and the combinatorial regions of the library should be biased to favor those codons used most frequently in the host expression system. For example, including only C and G (rather than all four bases) in the third position of a degenerate codon biases the library to favor codons preferred by *E. coli* ***(36)***. Codons that are used only rarely in the host expression system, such as CGA, AGA, and AGG (arginine); CTA (leucine); CCC (proline); and ATA (isoleucine) in *E. coli* should be avoided wherever possible, because genes containing rare codons may express poorly ***(37)***. Other (non *E. coli*) expression systems have different codon preferences, and these should be considered in the design.

2.3. Assembly of Full-Length Genes

We typically assemble full-length genes from smaller single-stranded oligonucleotides (**Fig. 4**). This is done to minimize the inherent errors (mostly deletions and frameshifts) associated with the synthesis of long degenerate oligonucleotides. Constant regions at the 5′- and 3′-ends serve as sites for the single-stranded oligonucleotides to anneal together and prime enzymatic synthesis (using DNA polymerase) of complementary strands. This strategy allows for a higher yield of error-free genes that correctly encode the desired *de novo* proteins.

When synthesizing the semirandom oligonucleotides, some are made as coding (sense) strands and others as noncoding (antisense) strands. Typically, each oligonucleotide is designed to encode an individual segment of binary patterned secondary structure. Assembly of full-length genes from such segments allows individual α-helices or β-strands to be designed and manipulated as independent modules, thereby enhancing the versatility of the binary code strategy (*see* **Note 4**).

In the design of our initial four-helix bundle library, four synthetic oligonucleotides were used to construct the full-length gene. Each oligonucleotide was designed to encode a single helix and turn. As described in **Subheading 2.1.2.**, the turn regions were precisely defined (i.e., not degenerate), thus, allowing them to serve as priming sites for DNA polymerase to synthesize the complementary strands (**Fig. 4**; **ref.** ***8***).

Various methods for assembling the genes have been used. In some cases, we have made two libraries of half genes, and then ligated them together to produce a library of genes encoding full-length proteins ***(8)***. To ensure correct head-to-tail ligation, nonpalindromic restriction sites can be designed into the constant regions to serve as the ligation sites ***(8)***. Other methods for assembling full-length genes include various polymerase chain reaction strategies (for example, overlap extension), and these have been used in constructing of several of our libraries ***(9,18)***.

3. Notes

1. Both the α-helical and β-sheet libraries are composed of the same binary-coded amino acids. Therefore, the different properties of the resulting proteins ***(8,9)*** are *not* caused by differences in amino acid composition ***(10)***. The observed differences also are *not* caused by differences in sequence length. Irrespective

of length, sequences with the P-N-P-P-N-N-P periodicity form proteins with α-helical secondary structure, whereas those with the P-N-P-N-P-N-P periodicity (examined under the same experimental conditions) form β-sheet structures. The key difference between these two libraries of sequences is the binary patterning itself.

2. The binary code codons, NAN and NTN, encode six polar amino acids (glutamate, aspartate, lysine, asparagine, glutamine, and histidine), and five nonpolar amino acids (valine, methionine, isoleucine, leucine, and phenylalanine). In addition to these 11 variable amino acids, a variety of other residues can be incorporated into the constant regions of the sequences. For example, our recent library of 102 residue four-helix bundles contains 17 of the 20 amino acids ***(18)***. Only alanine, proline, and cysteine were omitted. In natural proteins, alanine occurs both in surface and core positions. Thus, its role in the binary code as polar or nonpolar is somewhat ambiguous. Proline is a special case because its restricted ϕ angle makes it useful only in certain well-defined regions of structure. Cysteine should be used only in designs wherein a disulfide bond or metal binding is planned.
3. By excluding T from the first position of the NAN (polar) codon, tyrosine codons are avoided. This is desirable because tyrosine is not a completely polar residue and frequently occurs in the hydrophobic cores of natural proteins. Therefore, only the most polar residues (histidine, glutamine, asparagine, lysine, aspartate, and glutamate) are incorporated into the designed surface positions.
4. Polyacrylamide gel electrophoresis purification of synthetic oligonucleotides is essential. This reduces the likelihood of truncated oligonucleotides being incorporated into the library. Although this purification step reduces the quantity of DNA (and potentially the diversity), the quality of the genes and the resulting libraries is enhanced significantly.

Acknowledgments

Supported by grants from the National Institutes of Health (R01-GM62869), the Army Research Office (DAAG55-98-1-0084), the National Science Foundation (MRSEC DMR98-09483), and the Defense Advanced Research Projects Agency (n00173-01-1-0015).

References

1. Lim, W. A. and Sauer, R. T. (1989) Alternative packing arrangements in the hydrophobic core of lambda repressor. *Nature* **339,** 31–36.
2. Bowie, J. U., Reidhaar-Olson, J. F., Lim, W. A., and Sauer, R. T. (1990) Deciphering the message in protein sequences: tolerance to amino acid substitutions. *Science* **247,** 1306–1310.
3. Axe, D. D., Foster, N. W., and Fersht, A. R. (1996) Active barnase variants with completely random hydrophobic cores. *Proc. Natl. Acad. Sci. USA* **93,** 5590–5594.

4. Gassner, N. C., Baase, W. A., and Matthews, B. W. (1996) A test of the "jigsaw puzzle" model for protein folding by multiple methionine substitutions within the core of T4 lysozyme. *Proc. Natl. Acad. Sci. USA* **93**, 12,155–12,158.
5. Riddle, D. S., Santiago, J. V., Bray-Hall, S. T., et al. (1997) Functional rapidly folding proteins from simplified amino acid sequences. *Nat. Struct. Biol.* **4**, 805–809.
6. Silverman, J. A., Balakrishnan, R., and Harbury, P. B. (2001) Reverse engineering the $(\beta/\alpha)_8$ barrel fold. *Proc. Natl. Acad. Sci. USA* **98**, 3092–3097.
7. Lau, K. F. and Dill, K. A. (1990) Theory for protein mutability and biogenesis. *Proc. Natl. Acad. Sci. USA* **87**, 638–642.
8. Kamtekar, S., Schiffer, J. M., Xiong, H., Babik, J. M., and Hecht, M. H. (1993) Protein design by binary patterning of polar and nonpolar amino acids. *Science* **262**, 1680–1685.
9. West, M. W., Wang, W., Patterson, J., Mancias, J. D., Beasley, J. R., and Hecht, M. H. (1999) *De novo* amyloid proteins from designed combinatorial libraries. *Proc. Natl. Acad. Sci. USA* **96**, 11,211–11,216.
10. Xiong, H., Buckwalter, B. L., Shieh, H. M., and Hecht, M. H. (1995) Periodicity of polar and nonpolar amino acids is the major determinant of secondary structure in self-assembling oligomeric peptides. *Proc. Natl. Acad. Sci. USA* **92**, 6349–6353.
11. Moffet, D. A. and Hecht, M. H. (2001) *De novo* proteins from combinatorial libraries. *Chem. Rev.* **101**, 3191–3203.
12. Hecht, M. H., Das, A., Go, A., Bradley, L. H., and Wei, Y. (2004) De novo proteins from designed combinatorial libraries. *Protein Sci.* **13**, 1711–1723.
13. Taylor, S. V., Walter, K. U., Kast, P., and Hilvert, D. (2001) Searching sequence space for protein catalysts. *Proc. Natl. Acad. Sci. USA* **98**, 10,596–10,601.
14. Roy, S., Ratnaswamy, G., Boice, J. A., Fairman, F., McLendon, G., and Hecht, M. H. (1997) A protein designed by binary patterning of polar and nonpolar amino acids displays native-like properties. *J. Am. Chem. Soc.* **119**, 5302–5306.
15. Roy, S., Helmer, K. J., and Hecht, M. H. (1997) Detecting native-like properties in combinatorial libraries of *de novo* proteins. *Folding Des.* **2**, 89–92.
16. Roy, S. and Hecht, M. H. (2000) Cooperative thermal denaturation of proteins designed by binary patterning of polar and nonpolar amino acids. *Biochemistry* **39**, 4603–4607.
17. Rosenbaum, D. M., Roy, S., and Hecht, M. H. (1999) Screening combinatorial libraries of *de novo* proteins by hydrogen-deuterium exchange and electrospray mass spectrometry. *J. Am. Chem. Soc.* **121**, 9509–9513.
18. Wei, Y., Liu, T. I. P., Sazinsky, S. L., Moffet, D. A., and Hecht, M. H. (2003) Well folded *de novo* proteins from a designed combinatorial library. *Protein Sci.* **12**, 92–102.
19. Xu, G., Wang, W., Groves, J. T., and Hecht, M. H. (2001) Self-assembled monolayers from a designed combinatorial library of *de novo* β-sheet proteins. *Proc. Natl. Acad. Sci. USA* **98**, 3652–3657.
20. Brown, C. L., Aksay, I. A., Saville, D. A., and Hecht, M. H. (2002) Template-directed assembly of a *de novo* designed protein. *J. Am. Chem. Soc.* **124**, 6846–6848.
21. Richardson, J. S. and Richardson, D. C. (1988) Amino acid preferences for specific locations at the ends of alpha helices. *Science* **240**, 1648–1652.

22. Hutchinson, E. G. and Thornton, J. M. (1994) A revised set of potentials for β-turn formation in proteins. *Protein Sci.* **3**, 2207–2216.
23. Hirel, P. H., Schmitter, M. J., Dessen, P., Fayat, G., and Blanquet, S. (1989) Extent of N-terminal methionine excision from *Escherichia coli* proteins is governed by the side-chain length of the penultimate amino acid. *Proc. Natl. Acad. Sci. USA* **86**, 8247–8251.
24. Dalboge, H., Bayne, S., and Pedersen, J. (1990) In vivo processing of N-terminal methionine in *E. coli*. *FEBS Lett.* **266**, 1–3.
25. Tsunasawa, S., Stewart, J. W., and Sherman, F. (1985) Amino-terminal processing of mutant forms of yeast iso-1-cytochrome c. The specificities of methionine aminopeptidase and acetyltransferase. *J. Biol. Chem.* **260**, 5382–5391.
26. Huang, S., Elliott, R. C., Liu, P. S., et al. (1987) Specificity of cotranslational amino-terminal processing of proteins in yeast. *Biochemistry* **26**, 8242–8246.
27. Bowie, J. U. and Sauer, R. T. (1989) Identification of C-terminal extensions that protect proteins from intracellular proteolysis. *J. Biol. Chem.* **264**, 7596–7602.
28. Parsell, D. A., Silber, K. R., and Sauer, R. T. (1990) Carboxy-terminal determinants of intracellular protein degradation. *Genes Dev.* **4**, 277–286.
29. Milla, M. E., Brown, B. M., and Sauer, R. T. (1993) P22 Arc repressor: enhanced expression of unstable mutants by addition of polar C-terminal sequences. *Protein Sci.* **2**, 2198–2205.
30. Shoemaker, K. R., Kim, P. S., York, E. J., Stewart, J. M., and Baldwin, R. L. (1987) Tests of the helix dipole model for stabilization of alpha-helices. *Nature* **326**, 563–567.
31. Wei, Y., Kim, S., Fela, D., and Hecht, M. H. (2003) Solution structure of a de novo protein from a designed combinatorial library. *Proc. Natl. Acad. Sci. USA* **100**, 13,270–13,273.
32. Chou, P. Y. and Fasman, G. D. (1978) Empirical predictions of protein conformation. *Annu. Rev. Biochem.* **47**, 251–276.
33. Fasman, G. D. (1989) *Prediction of Protein Structure and the Principles of Protein Conformation*. Plenum, New York, NY.
34. Creighton, T. E. (1993) *Proteins: Structures and Molecular Properties*. 2nd ed., Freeman, New York, NY.
35. Pace, C. N. and Scholtz, J. M. (1998) A helix propensity scale based on experimental studies of peptides and proteins. *Biophys. J.* **75**, 422–427.
36. DeBoer, H. A. and Kastelein, R. A. (1986) in *Maximizing Gene Expression* (Rezinikoff, W. and Gold, L., eds.), Butterworth, Stoneham, MA, pp. 225–285.
37. Kane, J. F. (1995) Effects of rare codon clusters on high-level expression of heterologous proteins in *Escherichia coli*. *Curr. Opin. Biotechnol.* **6**, 494–500.

10

Versatile DNA Fragmentation and Directed Evolution With Nucleotide Exchange and Excision Technology

Sabine C. Stebel, Katja M. Arndt, and Kristian M. Müller

Summary

Mimicking natural evolution by DNA shuffling is a commonly used method for the optimization of DNA and protein properties. Here, we present an advancement of this approach whereby a gene library is amplified using a standard polymerase chain reaction (PCR), but incorporates dUTP as a fragmentation-defining exchange nucleotide, together with the four standard dNTPs. Incorporated uracil bases are excised using uracil–DNA–glycosylase, and the DNA backbone subsequently is cleaved with piperidine. This oligonucleotide pool is then reassembled with an internal primer extension procedure using a proofreading polymerase to increase yield, and, finally, is amplified by PCR. Denaturing polyacrylamide urea gels demonstrate this method to produce adjustable fragmentation size ranges dependent on the dUTP:dTTP ratios. Using the model protein, chloramphenicol acetyltransferase I, the sequencing of shuffled gene libraries based on a PCR containing 33% dUTP revealed a low mutation rate, of approx 0.1%, with an average parental fragments size of 86 bases, even without the use of a fragment-size separation. Nucleotide exchange and excision technology (NExT) DNA shuffling is, thus, reproducible and easily executed, making it superior to competing techniques. Additionally, NExT fragmentation outcome can be predicted using the computer software, NExTProg.

Key Words: DNA shuffling; directed evolution; in vitro evolution; protein engineering; UDG; dUTP; chloramphenicol acetyltransferase.

1. Introduction

There is an increasing demand for tailor-made enzymes for chemical, pharmaceutical, and food processing. Because folding, interaction, and stabilization in proteins are not fully understood, rational approaches to tailor-suit proteins until now have been complicated and have required time-consuming cycles of design and testing. Evolutionary approaches of repeated cycles of mutating, screening, and shuffling proved to be fast and effective in obtaining desired characteristics, even without knowledge of the protein structure or underlying principles.

From: *Methods in Molecular Biology, vol. 352: Protein Engineering Protocols*
Edited by: K. M. Arndt and K. M. Müller © Humana Press Inc., Totowa, NJ

Various methods are available for the in vitro recombination of gene libraries ***(1)***. To date, however, none of these methods has been without disadvantage or difficulty. In the well-established protocol of Stemmer, DNA is cut with DNase optimized to yield fragments of the desired size ***(2,3)***. For further size selection, this fragment pool is extracted from an agarose gel, the fragments are reassembled with an internal primer extension reaction, and, finally, are amplified using standard polymerase chain reaction (PCR). However, controlling such DNase digestion can be problematic and requires careful optimization of the digest conditions, e.g., amount of nuclease and DNA as well as temperature and time. Other methods, such as the staggered extension process ***(4)*** and random priming ***(5)*** are limited by the DNA composition, and matters are complicated further by a lack of controllability of the recombination and the range of fragment sizes generated. Thus, these methods require prudent optimization to ensure desired results.

In contrast to these methods, we devised a method that we are confident is both efficient and robust. The nucleotide exchange and excision technology (NExT) DNA shuffling is based on the random incorporation of so-called "exchange nucleotides." The occurrence and position of these exchange nucleotides in the DNA dictates the subsequent fragmentation without the need for further adjustment.

The NExT procedure was developed and tested by increasing the functionality of truncated mutants of chloramphenicol acetyl transferase I (CAT), which mediates resistance to the antibiotic, chloramphenicol. Directed evolution was independently applied to four sets of variants truncated at the genetic level. The first library of CAT mutants was shortened by 10 amino acids at the N-terminus (CAT_Nd10), the second library by 9 amino acids at the C terminus (CAT_Cd9), the third library by 26 amino acids at the C terminus (CAT_Nd26), and the fourth library was truncated at both ends by 10 and 9 amino acids (CAT_Nd10_Cd9), respectively. Detailed data for the NExT shuffling were obtained using test libraries with three to six members selected from error-prone PCR diversification steps ***(6)***.

A complete directed-evolution series based on error-prone PCR and NExT DNA shuffling was applied to improve the enzymatic activity of truncated CAT_Nd10, which grew on plates with up to only 25 μg/mL chloramphenicol; and CAT_Cd9, which failed to grow at all. After optimization, several clones of both libraries grew even at 400 μg/mL chloramphenicol (higher concentrations were not tested), demonstrating the efficacy of this technique. In addition, the preferred method was applied to TEM-1 β-lactamase using a dUTP fraction of 30% that was chosen based on our computer program (*see* **Subheading 3.9.**). This fragmentation and reassembly worked in the first trial, ensuring only sufficient starting material but omitting any preceding or intermediate tests or any analysis steps (data not shown).

This chapter outlines all steps for the NExT shuffling procedure in its optimized form, starting from the cloning (**Subheading 3.1.**), through the uridine-exchange PCR (**Subheading 3.2.**), the enzymatic digest and chemical cleavage (**Subheading 3.3.**), and the fragment purification (**Subheading 3.5.**), and ending with the gene reassembly and amplification (**Subheading 3.7.**). Possible variations of the protocol are discussed in the respective sections. Additionally, analytical methods for a detailed characterization of the shuffling procedure are given (**Subheadings 3.4. and 3.6.**), and the crossover rate, mean fragment length, and mutation rate are compared between different methods (**Subheading 3.8.**). Last, but not least, a computer program is introduced that we wrote to predict NExT fragmentation parameters and the respective outcomes (**Subheading 3.9.**).

Because NExT is based on a rational and predefined dUTP:dTTP ratio and is easy to perform, it is well-suited for the shuffling of short genes and large gene assemblies. Because of the robustness and simplicity of NExT shuffling, even those with little previous experience in this area should be able to apply this technique to a range of biochemical problems at the DNA or protein level.

2. Materials

2.1. Cloning Steps

1. Adequate enzymes for excising the desired gene and opening the vector. In our case, *Xba*I and *Hin*dIII were used.
2. T4 DNA ligase (from different suppliers).
3. Competent cells (*Escherichia coli* RV308 and XL-1 were tested).
4. Plasmid vector pLisc-SAFH11 *(7)*.

2.2. Uridine-Exchange PCR

1. One forward and one reverse primer, which are adequate for the gene to be amplified and contain cloning sites.
2. Standard Taq polymerase (from different suppliers).
3. 10X PCR buffer (from different suppliers), e.g., 160 m*M* $(NH_4)_2SO_4$, 670 m*M* Tris-HCl (pH 8.8 at 25°C), 15 m*M* $MgCl_2$, and 0.1% Tween-20; or 100 m*M* Tris-HCl (pH 9.0 at 25°C), 500 m*M* NaCl, 15 m*M* $MgCl_2$, and 1% Triton X-100.
4. dNTP stock solutions of dATP, dTTP, dGTP, dCTP, and dUTP (10 m*M* or 100 m*M*; from different suppliers).
5. Deionized sterile water.
6. 10X Tris–borate–EDTA buffer (TBE buffer): 1 *M* Tris, 0.83 *M* boric acid, and 1 m*M* EDTA.
7. Ethidium bromide solution: 5 mg/mL ethidium bromide in water.
8. 1% agarose gel: 1% (w/v) agarose in 0.5X TBE buffer. Boil until the agarose is completely melted. Add ethidium bromide solution (1:10,000 dilution) before pouring the gel. Run the gel in 0.5X TBE buffer.

9. Gel extraction kit (GE Healthcare or Qiagen).
10. PCR cleanup kit (GE Healthcare or Qiagen).

2.3. Enzymatic Digest and Chemical Cleavage

1. uracil–DNA–glycosylase (UDG) from *E. coli* (NEB or Peqlab Biotechnologie GmbH).
2. 99.9% piperidine (Sigma).

2.4. Denaturing Polyacrylamide Urea Gels

1. Urea (purity, at least 99.5%).
2. 30% polyacrylamide/0.8% *bis*-acrylamide (37.5:1) stock solution (Carl Roth GmbH).
3. 10X TBE buffer (*see* **Subheading 2.2.6.**).
4. 10% ammoniumperoxodisulfate in water.
5. TEMED.
6. Deionized formamide.
7. Defined oligonucleotides serving as marker or a low-range DNA ladder, e.g., a 100-basepair (bp) ladder (NEB).
8. 6X bromphenol blue loading buffer: 0.25% (w/v) in 70% (w/v) sucrose solution.
9. Ethidium bromide solution: 5 mg/mL in water.

2.5. Fragment Purification

2.5.1. Direct Purification From Piperidine Cleavage

1. QiaexII kit (Qiagen).
2. Acetate buffer: 3 *M* Na-acetate, pH 5.3.
3. Elution buffer: 10 m*M* Tris-HCl, pH 8.0.

2.5.2. Purification From Denaturing Polyacrylamide Urea Gels

1. Qiaex II purification: diffusion buffer: 0.5 *M* ammonium acetate, 10 m*M* magnesium acetate, 1 m*M* EDTA, and 0.1% sodium dodecyl sulfate, pH 8.0.
2. Water extraction: deionized sterile water.
3. Acetate buffer: 3 *M* Na-acetate, pH 5.3.
4. 1 *M* $MgCl_2$.
5. 2-Propanol.
6. 90% ethanol.
7. Elution buffer: 10 m*M* Tris-HCl, pH 8.0.

2.6. Quantification of Purified Fragments

1. 1:5000 diluted SYBR green II (Molecular Probes).

2.7. Gene Reassembly and Amplification

1. 10 m*M* mixtures of dATP, dTTP, dCTP, and dGTP.
2. Vent DNA Polymerase (NEB).
3. 25 m*M* $MgSO_4$.
4. Taq DNA polymerase (from different suppliers).

3. Methods

3.1. Cloning Steps

The priming sites of the shuffling primers, containing the restriction sites used for the NExT-shuffling procedure, should ideally be located shortly before and after the gene of interest and have a melting temperature of approx 60°C. We estimate melting temperatures using the 4 plus 2 rule: 4°C for each C and G, and 2°C for each T and A pair.

The genes we used in the NExT DNA shuffling procedures were the 657-bp *CAT* wild-type gene (CATwt; SwissProt: P00483; Protein Data Bank: 1NOC:B) and variants coding for an N-terminal 10-amino acid-truncated, C-terminal 9-amino acid-truncated, or double-truncated CAT (CAT_Nd10, CAT_Cd9, or CAT_Nd10_Cd9), or a C-terminal 26-amino acid-truncated CAT (CAT_Cd26). These genes and all shuffled genes have been cloned into the vector, pLisc-SAFH11 *(7)*, using the restriction sites *Xba*I and *Hin*dIII, thus, replacing part of the original plasmid. For diversification by error-prone PCR and the uridine-exchange PCR (*see* **Subheading 3.2.**) the cloned *CAT* genes were amplified using the primers Pr-N-shuffle (5′-ATT**TCTAGA**TAACGAGGGCAA-3′) and Pr-C-shuffle (5′-ACTTCACAGGTC**AAGCTT**TC-3′) for the wild-type and N-terminal truncated genes; and Pr-N-shuffle and Pr-Cdx-shuffle (5′-CTTCACAG GTC**AAGCTT**ATCA-3′) for the C-terminal truncated and the double-truncated genes. Plasmids are best transformed in *E. coli* using standard methods *(8)*.

3.2. Uridine-Exchange PCR

Uridine was chosen as the exchange nucleotide because dUTP is known to be incorporated into DNA by various polymerases *(9)*. The uridine-exchange PCR amplifies the gene pool of interest and additionally incorporates the exchange nucleotide, uridine. Various ratios of dUTP:dTTP can be used to obtain optimal fragmentation. This can be done either experimentally by analyzing various U:T ratios, as detailed in **Subheading 3.4.**, or using our program that was developed for this purpose (*see* **Subheading 3.9.**). No apparent difference in the amount of PCR product was observed when using dUTP fractions of up to 50% within the dUTP and dTTP pool. Only the PCR containing uridine alone was found to yield approximately a quarter of the product compared with other reactions (**Fig. 1A**).

1. For accurate pipetting, dilute 100 m*M* nucleotide stock solutions with water to 10 m*M* for dATP, dGTP, and dCTP, and to 1 m*M* for dUTP and dTTP.
2. For the uridine-exchange PCR mixture use 50 ng template (0.017 pmol of a 4340-bp plasmid), 25 pmol of each primer, 0.4 m*M* of dATP, dGTP, and dCTP each, a 0.4 m*M* mixture of dUTP:dTTP in various ratios, 5 U Taq DNA Polymerase, and 5 μL of 10X PCR buffer. Adjust the final volume to 50 μL with water (*see* **Notes 1** and **2**).
3. The thermocycler program is: one cycle of 94°C for 1 min; 25 cycles of 92°C, 30 s denaturation; 62°C, 20 s annealing; 72°C, 2 min extension; and a final incubation at

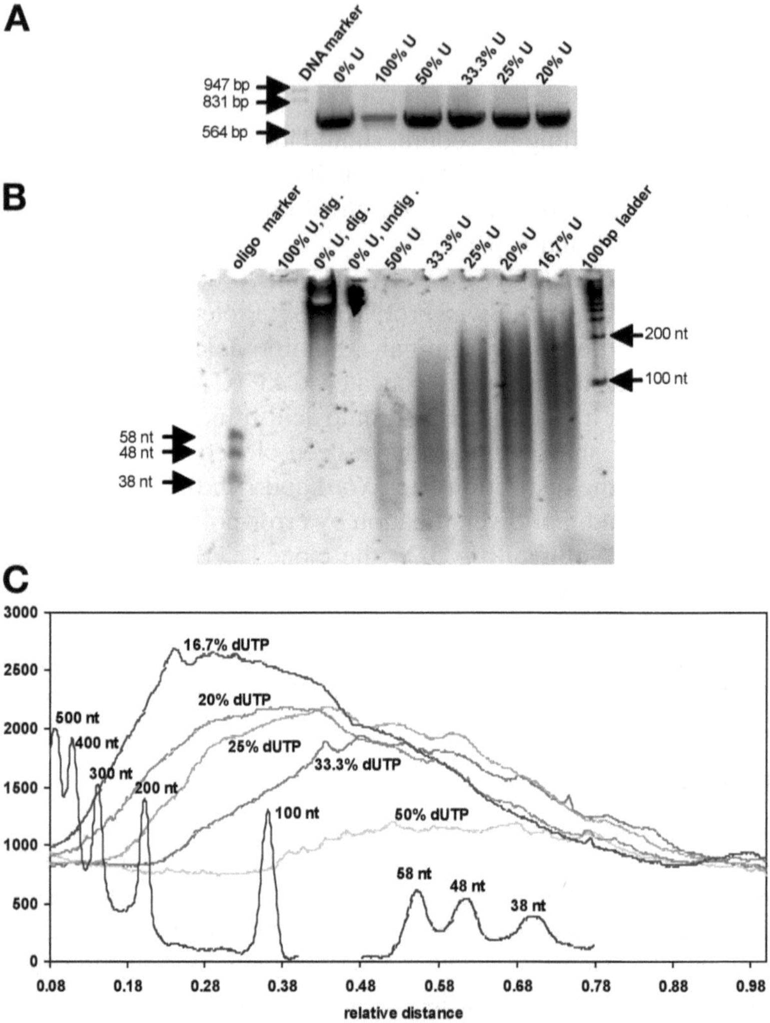

Fig. 1. Overview and quantification of NExT DNA fragmentation. (**A**) 1% agarose gel showing the uridine-exchange PCR products of CAT_Nd10 clones obtained with different amounts of uridine in the reactions. Percent U was calculated by {c(dUTP)/[c(dUTP) + c(dTTP)]} × 100. (**B**) Polyacrylamide urea gel stained with ethidium bromide showing UDG/piperidine digests of CAT_Nd10 PCR products obtained with various dUTP:dTTP ratios (1:0, 0:1, 1:1, 1:2, 1:3, 1:4, and 1:5) to determine an optimal ratio. Lane 1: oligonucleotides with 58, 48, and 36 bases as size markers; lane 2: 100% dUTP, PCR digested; lane 3: 0% dUTP, digested; lane 4: 0% dUTP, undigested; lane 5: 50% dUTP, digested; lane 6: 33.3% dUTP, digested; lane 7: 25% dUTP, digested; lane 8: 20% dUTP, digested; lane 9: 16.7% dUTP, digested; lane 10: 100-bp DNA ladder. Note that residual amounts of piperidine contribute to slightly distorted lanes. (**C**) Lane density plot of lane 1 and

72°C, 4 min. An extended elongation time of 2 min was chosen based on a test series showing that the yield was significantly improved compared with shorter times (data not shown). Depending on the efficiency of the PCR, it can be necessary do four or more 50-μL reactions to obtain enough product (optimally ~7 μg after gel extraction; *see* **Note 3**).

4. Combine the PCR samples and separate product from template using a 1% agarose gel (*see* **Note 4**). If there is enough product, the DNA can be seen in the agarose gel as a thin red line even without UV light. Purify the excised gel band using one or two spin columns (*see* **Note 5**) of a gel extraction kit and elute with 50 μL elution buffer.
5. Determine the concentration of the PCR product by taking the baseline corrected 260 nm value of an absorption spectrum from 220 to 350 nm. We usually used a 1:30 dilution and measured in a 140-μL microcuvet. The optimal amount for the NExT DNA shuffling procedure is approx 7 μg of DNA. Lower amounts might work as well, but sufficient DNA permits a trouble-free gene reassembly (*see* **Subheading 3.7.**). Ideally, use all of the DNA for the next steps.

In this study, we focused solely on uridine as the exchange nucleotide, however, the technique could equally be applied to the incorporation of several other analogs. One such example, 8-oxoguanine, can be cleaved out by 8-oxoguanine DNA glycosylase (formamidopyrimidine-DNA glycosylase; **ref.** ***10***). This base could prove useful in AT-rich regions, in which DNA cleavage by UDG is too frequent, or in GC-rich genes, in which thymidines are seldom found. Alternatively, it is feasible that a combination of several exchange nucleotides together could generate fragments of the desired range for reassembly. Additionally, incorporation of exchange nucleotides into the primers of the incorporating PCR should ensure that these regions of the gene library are also shuffled.

3.3. Enzymatic Digest and Chemical Cleavage

To fragment the gene, the exchange nucleotide is excised by incubating the product of the uridine-exchange PCR with the enzyme UDG ***(11)***. This enzyme is capable of attacking double-stranded as well as single-stranded DNA using a hydrolytic mechanism to remove, with high specificity, the uracil moiety by a nucleophilic attack at the C1′ position of deoxyuridine ***(12)***. Piperidine is used to split the backbone positions where the uracil has been removed by

Fig. 1. *(Continued)* lanes 5 to 10 of (B), detailing the fragment sizes based on the fraction of dUTP used. For the shuffling of all *CAT* variants, a uridine-exchange PCR containing 33.3% dUTP was used, producing fragments ranging from 30 to 200 bases in length (thick line). The image was acquired with a FluorS Multiimager, and the plot was generated using the Quantity One software program (Bio-Rad). For clarity of the plot, the signal of the 100-bp ladder was shifted by –750 counts and the signal of the oligonucleotides by –500 counts. Dig., digested; undig., undigested; nt, nucleotide.

UDG *(13)*. The result of such a cleavage reaction can be analyzed with high resolution on denaturing polyacrylamide urea gels (*see* **Subheading 3.4.**) for dUTP fractions ranging from 100 to 0% (**Fig. 1B**) and quantified by image analysis (**Fig. 1C**). For very small fragments, radioactive labeling can be used for better visualization (*see* **Note 2**).

1. For the enzymatic cleavage with UDG supplement, approx 45 µL (typically 7 µg DNA) of the purified PCR product from **Subheading 3.2., step 4.** with 6 µL of supplied 10X UDG buffer and 2 U of *E. coli* UDG. Adjust the volume to 60 µL with water, and incubate the digest for 1 h at 37°C (*see* **Note 6**).
2. To cleave the DNA, add piperidine to a final concentration of 10% (v/v; 6.7 µL piperidine for 60 µL enzymatic digest) and heat for 30 min at 90°C in a thermocycler with a heated lid (*see* **Note 7**).

As an alternative to piperidine with mild conditions, other endonucleases, such as endonuclease IV *(14)*, exonuclease III *(15)*, or T4-endonuclease V *(16)* can be used to cleave the DNA backbone *(11)*. We tested the latter but discontinued its use. T4-endonuclease V cleavage after UDG treatment resulted in incomplete fragmentations, as seen by the size distribution generated (**Fig. 2A**). Even more problematic was the high error rate inherent to this procedure. Applying a UDG and T4 endonuclease V fragmentation with gel extraction and a reassembly, as described for a *CAT* wild-type gene, and sequencing six clones with a total of 3930 bases gave a mutation rate of 1.75%. We propose two reasons for this finding. First, the DNA backbone is cleaved by the T4 endonuclease V with its lyase activity, which catalyses a β-elimination reaction, leaving a 3′ unsaturated aldehyde (4-hydroxy-2-pentanal) attached at the phosphate group *(11,17)*. The further chemical reaction leading to the free phosphate group is unlikely to be complete, therefore, such generated fragments are an unfavorable starting point for a polymerase. In addition, as seen by the fragment comparison, not all sites with incorporated uracil in the backbone are cleaved. However, the uracil moiety might nonetheless be missing, thus base lacking templates might lead to erroneous nucleotide incorporation. Miyazaki *(18)* proposed the use of *E. coli* endonuclease V, however, according to the data provided, cleavage was even less efficient, and short fragments were not obtained even after prolonged digests (12 h) and high dUTP fractions (75%). Further experiments could be designed to solve these problems, but were not performed because the piperidine cleavage worked well and was much more cost-effective.

As a chemical alternative to piperidine, we tested NaOH. We replaced the piperidine solution with a 0.5 *M* NaOH solution, which was added at 10% (v/v) to the DNA and treated for 30 min at 90°C. The fragmentation result of a piperidine and a NaOH cleavage started from the same uridine-exchange PCR was compared on a denaturing polyacrylamide gel with subsequent ethidium bromide staining. The intensity distribution of the two fragmentations was almost

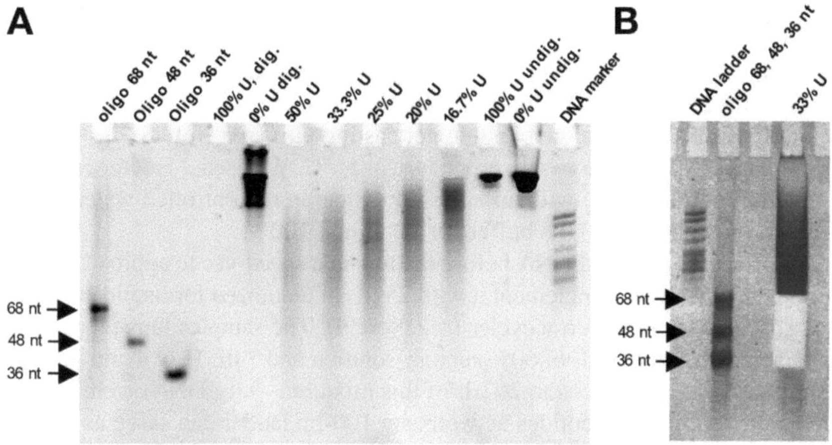

Fig. 2. Variations of the NExT DNA-shuffling procedure. **(A)** Polyacrylamide urea gel with UDG/T4 endonuclease V digests of *CAT wild-type* PCR products containing various dUTP:dTTP ratios to analyze enzymatic fragmentation. From left to right: lanes 1–3, oligonucleotides with 68, 48, and 36 bases; lanes 4–10, digests of PCR products obtained with 100, 0, 50, 33.3, 25, 20, and 16.7% dUTP; lanes 11 and 12, PCR products without digest obtained with 100 and 0% dUTP; lane 13, pBR322/*Hpa*II DNA marker. **(B)** Preparative polyacrylamide gel showing excised region of a CAT_Nd10 gene fragment library from DNA obtained with 33.3% U. Lane 1: pBR322/*Hpa*II DNA marker; lane 2: 68, 48, and 36 nucleotide oligomers used as marker; lane 3: excised preparative digest.

identical, as judged by image analysis software. A reassembly reaction run in parallel also yielded comparable results. These findings point to the fact that the main action of piperidine is the base-catalyzed β-elimination of the two phosphate groups from the sugar moiety of the DNA. Thus, if safety issues are a major concern, piperidine can be replaced by NaOH. However, we did not analyze the reassembled genes from NaOH fragmentations by sequencing. Conceptually, piperidine is better suited to also start a nucleophilic attack on the sugar to provide cleavage reactions that are more complete.

3.4. Denaturing Polyacrylamide Urea Gel

Denaturing polyacrylamide urea gels were used for analytical purposes to examine the fragmentation size distribution as well as for preparative purposes to purify gene fragments of a specific size range (*see* **Subheading 3.5.**). This step is optional and only advised for experiments requiring tight control of the fragmentation.

1. Gels are composed of 6.7 *M* urea, 11–12% polyacrylamide, and 1X TBE *(8)*. We used a size of 10 cm × 8 cm × 1 mm for analytical gels and 10 cm × 8 cm × 2 mm

for preparative gels (*see* **Subheading 3.5.2.**). Gels have to be prepared freshly because old gels do not run properly due to urea degradation. We used 31.5 g urea, 29.2 mL of 30.8% (37, 5:1) acrylamide:*bis*-acrylamide, 17.1 mL H_2O, 7.5 mL of 10X TBE, 500 µL of 10% ammoniumperoxodisulfate, and 50 µL TEMED; sufficient for nine analytical gels in one gel casting unit (*see* **Note 8**).

2. Gels are run during the whole procedure at 56°C. We used a Hoefer Mighty small basic unit (Amersham) with an attached temperature-controlled water bath.
3. Prerun the gels in 1X TBE buffer for 10 min at 100 V.
4. Concentrate the cleaved DNA before loading in a speed-vac to approx 7 µL to evaporate the piperidine, supplement it with 25 µL of deionized formamide and heat it to 80°C for 3 min in a thermocycler (*see* **Note 9**). The samples thus prepared can be supplemented with 3 µL of 60% sucrose solution and 7 µL H_2O to aid sample loading (*see* **Note 10**). Fifteen to 20 µL of this mixture (~3 µg DNA) were loaded.
5. Mixtures of oligonucleotides as well as a 100-bp ladder can serve as length standard (*see* **Note 11**). Mix at least one marker with dye-containing loading buffer.
6. Run the loaded gel at 170 V until the bromphenol blue added to the marker is approximately 0.5 cm from the end of the gel (*see* **Note 12**).
7. Stain the gel for 5 min in 30 mL of a 1 µg/mL ethidium bromide solution (1:5000 dilution) and visualize the DNA in UV light (*see* **Note 13**).

Analytical urea gels revealed that the fragmentations yielded size distributions with defined peaks, depending on the dUTP fraction. As seen in **Fig. 1B**, several length distributions were easily obtained, including pools of very small and large fragments, which is optimal for shuffling short genes or long gene clusters, respectively. Such tests were required initially to determine the optimal dUTP fraction for obtaining the desired length distribution for a given gene. Once established, the desired dUTP fraction of 33.3% for the test libraries was used and found to be highly reproducible. Thus, the gel analysis step was no longer required and omitted for subsequent rounds of directed evolution.

3.5. Fragment Purification

Gene fragments resulting from the chemical or enzymatic cleavages (*see* **Subheading 3.3.**) were cleaned either directly from the cleavage reaction solution, using a silica-based resin (**Subheading 3.5.1.**), or over a preparative denaturing polyacrylamide urea gel (**Subheading 3.5.2.**).

Initially, we thought that gel purification was crucial to permit only fragments of finite size, and to ensure that no long fragments or even full-length genes were involved in the reassembly reaction; a potential problem that could have lowered the crossover frequency. However, no full-length product could be amplified from the directly purified fragments without several cycles of the reassembly process, demonstrating a very efficient fragmentation (*see* **Subheading 3.7.**). Thus, purification from gels is only necessary when a very narrow size range is required.

3.5.1. Direct Purification From Piperidine Cleavage

1. Fragments are most easily purified directly from the piperidine cleavage (*see* **Subheading 3.3.**) using the QiaexII kit (Qiagen) according to the manufacturer's manual (*see* **Note 14**). Add the capture buffer included and neutralize (~20 µL of 3 *M* acetate buffer).
2. After two washing steps, extract the fragments from the matrix by adding 25 µL elution buffer in two steps. Pool the two elutions.
3. Centrifuge two times, transferring to a fresh tube each time (*see* **Note 15**).

3.5.2. Purification From Denaturing Polyacrylamide Urea Gels

Note that this step is optional. For the fragment extraction from urea gels, two methods were tested. Either the classic water extraction with subsequent acetate/Mg^{2+} precipitation or the procedure recommended by Qiagen for their QiaexII kit were used. Both methods suffered from loss of material not extracted from the gel, as assessed by test stainings of already extracted gel material.

1. For the extraction of fragments from preparative polyacrylamide urea gels (*see* **Subheading 3.4., step 1**), cut out the desired range (**Fig. 2B**) with a scalpel while visualizing bands on a low-intensity UV table. Crush the gel piece thoroughly in a tube and incubate it with either diffusion buffer (QiaexII extraction, **step 2**) or 1 mL water (water extraction, **step 3**) in a thermomixer (Eppendorf) at 37°C, 1000 rpm overnight.
2. Fragments extracted with diffusion buffer are purified with the QiaexII kit, as described in **Subheading 3.5.1.**
3. Precipitate the water-extracted oligonucleotides by adding 1/10 volume of acetate buffer, 1/100 volume of $MgCl_2$, and 1 volume of 2-propanol; incubate at –20°C for 1 h; and centrifuge for 15 min at 20,000*g* at 4°C. Resuspend the pellet in 50 µL 90% ethanol in a thermomixer at 37°C, 1000 rpm for 1 h; incubate for 1 h at –20°C; and centrifuge for 15 min at 20,000*g* at 4°C. Air-dry the pellet and dissolve it in 30 µL elution buffer by incubating in a thermomixer at 37°C, 1000 rpm for 1 h.

To elucidate the crossover rate with a gel extraction step, a test shuffling experiment with three parental clones (*CAT_Nd10* mutants) was set up. Two of the clones contained one and one clone contained two distinct mutations within a stretch of 100 bp. Sequencing of eight clones shuffled and assembled with Taq polymerase revealed that six clones had one crossover within the 100-bp stretch. Thus, the crossover rate is in the range of the shuffling procedure using the quick clean up (*see* **Subheadings 3.5.1.** and **3.8.**). A total of 3851 bases were sequenced, and 12 errors were found, equaling a mutation rate of 0.31%. The error rate is significantly higher than with the quick clean up, which might be explained by UV damage caused by visualization even on the weak 366-nm light source used and/or chemical modifications caused by the gel. For the preferred NExT DNA shuffling method, we omitted the gel purification step, because of the additional work without significant benefit, the loss of material, and the higher error rate.

3.6. Quantification of Purified Fragments

To analyze and compare the yield of the purified DNA fragments, we initially quantified them with SYBR Green II. Because the NExT shuffling procedure gives highly reproducible amounts, we later omitted the quantification step (*see* **Note 16**). We typically obtained a fragment concentration of 40 to 60 ng/μL.

1. Stain 2 μL samples of the purified DNA fragments (*see* **Subheading 3.4.**) in 50 μL of 1:5000 diluted SYBR Green II, which is a highly sensitive dye for single-stranded DNA (*see* **Note 17**).
2. Incubate this mixture for 5 min in a dark box and measure the fluorescence (we used a Perkin Elmer Fluorescence Spectrometer LS 50B in 96-well format; *see* **Note 18**). The dye is excited at 480 nm and the emission is measured at 515 nm.
3. Use an oligonucleotide in the size range of your fragments in different dilutions as a calibration curve to determine fragment concentration.

3.7. Gene Reassembly and Amplification

The full-length gene is reassembled from the purified gene fragments (*see* **Subheading 3.5.**) in an internal primer extension procedure with increasing annealing temperatures using, as the preferred choice, a proofreading DNA polymerase, such as Vent (**Fig. 3**). In the internal primer extension PCR, the fragments serve each other as primers and, thus, get longer with each cycle of the reaction, until full-length products are achieved. As a final step, products of the assembly reaction are amplified with a standard PCR with end primers, cloned, and sequenced. While establishing the method, the assembly reaction was monitored by agarose gel electrophoresis (**Fig. 3**). The assembly process was stopped after an increasing number of cycles, and the products obtained at theses points were subjected to the amplification PCR. The underlying principle of the final steps is the same as in other gene assembly protocols *(3)*, but there are important changes. Despite the harsh chemical cleavage conditions, the assembly works very efficiently and a proofreading polymerase was used.

1. Use approx 2 μg of the purified DNA fragments (*see* **Subheading 3.5.**) for the reassembly. In our case, even less than 1 μg of DNA fragments was sufficient when using Vent polymerase. We normally used 20 to 25 μL of purified fragments, without measuring the concentration.
2. Mix the DNA fragments with 4 μL (*see* **Note 19**) of a mixture of 10 m*M* of each dATP, dTTP, dCTP, and dGTP (800 μ*M* final), and 4 U Vent DNA Polymerase (NEB), with 1 to 4 μL of 25 m*M* $MgSO_4$ and 5 μL of the supplied 10X buffer. Adjust the volume to 50 μL with water.
3. Cycles for the reassembly are: one cycle of 94°C, 3 min; 36 cycles of 92°C, 30 s denaturation; 30°C, 60 s + 1°C per cycle (cooling ramp 1°C/s) annealing; 72°C, 1 min + 4 s per cycle extension; and final incubation at 72°C, 3 min.

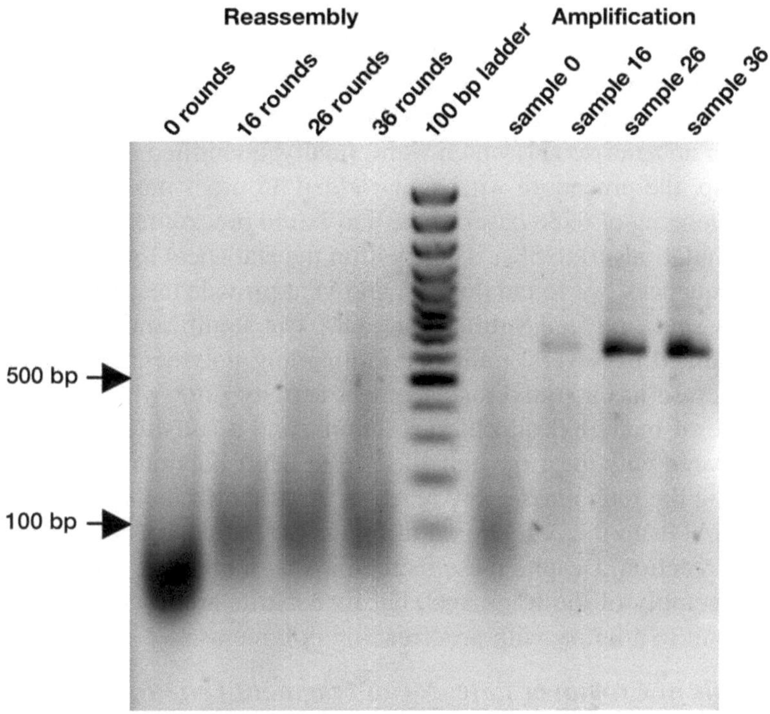

Fig. 3. Reassembly and amplification of the shuffled genes. 1% Agarose gel of *CAT_Nd10_Cd9* gene fragment libraries from DNA containing 33.3% U showing the reassembly process with Vent DNA polymerase and the amplification of reassembled genes with Taq polymerase. Lane 1: fragments without reassembly PCR; lane 2: fragments after 16 cycles of reassembly; lane 3: fragments after 26 cycles of reassembly; lane 4: fragments after 36 cycles of reassembly; lane 5: 100-bp DNA ladder; lane 6: amplification PCR of fragments without reassembly, lane 7: amplification PCR of fragments subjected to 16 reassembly cycles; lane 8: amplification PCR of fragments subjected to 26 reassembly cycles; lane 9: amplification PCR of fragments subjected to 36 reassembly cycles.

4. Amplify 10 µL of the reassembly product (*see* **Note 20**) in a standard PCR with the appropriate primers for the gene (25 pmol primer and 0.2 m*M* dNTPs; 25 cycles, 40 s elongation time).
5. Clone the amplified genes via the appropriate restriction sites, using standard methods.

To our knowledge, Taq has been the enzyme of choice for most published fragment reassembly procedures. The $3' \rightarrow 5'$ proofreading exonuclease activity of Vent DNA polymerase relies on removal of mismatched nucleotides from only the 3′ terminus of the priming strand, until polymerization can be initiated from an annealed end. This means that point mutations in nucleotides, even in close proximity to the 3′-end, will pose no problem ***(19)***. To compare these polymerases, two reassembly procedures were run in parallel with the same fragment pool, using either Taq or Vent, respectively. Aliquots of the reassembly procedures were the template for amplification PCRs with Taq resulting in one band each in an agarose gel, which were, finally, quantified by image analysis. In this set-up, the procedure with Vent yielded 35 times more product. Interestingly, sequencing of 6988 bases of the Taq-based procedure revealed only five additional mutations (0.075%). Thus, within the statistical limitations based on available sequences, we found that Taq and Vent provide the same error rate for NExT DNA shuffling (*see* **Subheading 3.8**). The significantly improved yield can be explained by several traits of proofreading polymerases, such as Vent. This polymerase has a strand displacement activity ***(20)***, which might help in the presence of many hybridization reactions, and a half-life of approx 8 h at 95°C, compared with the 1.6 h half-life of Taq DNA polymerase ***(20)***, ensuring fitness during the long reassembly reaction. Another difficulty is that Taq adds additional dATPs to the 3′ hydroxyl terminus ***(21)***, a possible hindrance in the reassembly reaction. Despite these factors, it was found that Taq was adequate for the reassembly of the fragments, but its benefits are likely limited to templates difficult to amplify with proofreading polymerases.

3.8. Analysis of Crossover Rate, Mean Fragment Length, and Mutation Rate

The NExT DNA shuffling procedure described so far has been applied to the directed evolution of a 600-bp-long *CAT* gene truncated at both ends (*CAT_Nd10_Cd9*). In the course of these experiments, a defined library of five clones with different mutation patterns between nucleotides 12 and 383 mixed

---

Fig. 4. *(Opposite page)* Analysis of crossover frequency and fragment length. **(A)** Sequencing results of a NExT DNA shuffling experiment with a CAT_Nd10_Cd9 gene mutant library with direct fragment purification from solution, and reassembly using a proofreading polymerase. Approximately 500 to 571 bases per clone were sequenced. The test shuffling was prepared with a 33.3% uridine-exchange PCR containing 26 ng (52%) truncated CAT wild-type fragments and 4.8 ng (9.6%) fragments of each mutant. The bottom panel lists the sequences of clones obtained without selection pressure, focusing on the shuffled mutations, the minimal number of parental clones that can be deduced from the mutation patterns, and the frequency of additionally introduced mutations not listed in the table. On average, the 372-bp segment analyzed is composed of 3.25 parental clones. Because of the excess of wild type, the real number of parental clones is likely to be higher than the minimal value listed. **(B)** Sequencing result of a NExT DNA shuffling experiment with four equally mixed parental clones of CAT_Cd26. Parental clones are individually shaded, and the shading is maintained to identify fragments in shuffled clones. No shading indicates that a fragment could descend from several parents. A mean fragment length of 86 bases was detected. Trunc., truncated; wt, wild type.

A

trunc. gene position	12	52	53	55	73	74	87	116	127	194	231	301	326	383	minimum parental clones	mutations / 100 bp
original nt	C	G	T	G	C	A	G	A	C	T	T	G	T	T		
clones used (CAT Nd10 Cd9)																
parent wt																
parent a		A	C			T	A	G			G	C	C	C		
parent b	A	A	C	A		T			T			C	C	C		
parent c		A	C		T	T		G			G	C	C	C		
parent d	A	A	C	A		T			T	C		C	C	C		
parent e	A	A	C	A		T						C	C	C		
clones obtained																
control 1												C	C	C	2	0.00
control 2									T						3	0.18
control 3								G						C	4	0.18
control 4	A										G	C	C		4	0.00
control 5		A	C	A		T	A		T			C	C	C	4	0.00
control 6													C		3	0.18
control 7		A	C		T	T						C	C		3	0.00
control 8		A	C	A		T									3	0.20

B

gene position	9	11	25	36	55	73	104	113	132	150	163	176	180	181	183	193	227	255	264	281	300	302	324	330	336	364	366	376	383	404	415	422	428	430	436	456	458	492	521	544	549	556	575	minimum parental clones	mutations / 100 bp
original nt	A	A	A	A	T	T	A	A	T	A	T	T	T	C	T	C	C	G	T	T	A	C	A	C	T	C	A	T	G	C	A	A	T	T	T	T	T	C	G	A	G	C	A		
clones used (CAT Cd26)																																													
parent a					A		G		C	T			A					T			T									T	T								T						
parent b						C					C								C	A								C				G			C										
parent c	G	T	T	T				G				C		T								T			C	T		G					C	A		G	C	T				T			
parent d															G	T							G				C		A											G	A		C		
clones obtained																																													
control 1							G		C	T			A				T	T												T	T													7	0.16
control 2							G		C	T			A					T			T									T	T													3	0.00
control 3								G	C	T			A								T									T	T													6	0.16
control 4				T			G		C	T			A								T																							7	0.33
control 5					A		G			T			A													T				T														10	0.00

with a truncated wild-type clone for back-crossing was shuffled based on a 33.3% uridine-exchange PCR. The fragments were directly purified from solution (*see* **Subheading 3.5.1.**) and reassembled using the proofreading polymerase, Vent (*see* **Subheading 3.7.**) Eight shuffled clones taken from control plates without selection pressure were sequenced (**Fig. 4A**). The unique mutation pattern of these clones showed that all clones tested were derived from at least two (e.g., clone 1) to four (e.g., clone 4) parental clones. Within the 372-bp stretch amenable to analysis, this resulted in one crossover per 93 to 186 bp, with a mean fragment length of 114 bp.

The mutation rate of this procedure was determined by sequencing. Within 4425 bases sequenced, 4 alterations were found (one A to G and one T to C transition, a 1-bp insertion and a 1-bp deletion) giving a mutation rate of 0.09%. This is remarkably lower than an error rate of 0.7% reported previously for DNase shuffling *(2)*. As detailed in **Subheading 3.7.** this is not a unique feature of using Vent polymerase. Because our fragment distribution and crossover rate were comparable to previous experiments, we are inclined to attribute the previously reported error rates more to the DNase digest and to the UV damage caused by gel visualization rather than the fragment size and the polymerase. A low mutation rate is particularly important when shuffling of longer DNA is envisioned, because this will avoid dilution of the gene pool with dysfunctional or undesired molecules.

In a further experiment, four parental truncated *CAT* genes (*CAT_Cd26*), containing a mutations spread from bases at positions 9 to 575 were shuffled using the same procedure as described in the first experiment, and five clones were sequenced (**Fig. 4B**). A detectable mean fragment length of 86 bases was found, including several short fragments. Mutations closely spaced by 6 bases (position 324 and 330), 12 bases (position 364 and 376), and 11 bases (position 404 and 415) were separated in clone "control 5." The mean fragment length in this experiment is smaller than in the previous experiment, because more mutations result in a better detection of the fragment length. The short fragments can be explained by the possibility that either the QiaexII kit used purified short fragments efficiently and/or that the crossover during gene assembly is a complex process that includes, e.g., PCR-based strand-switching.

3.9. NExT Fragmentation-Predicting Program

Because the dUTP incorporation and the resulting fragmentation are based on deducible principles, a computer program was developed, named NExTProg *(22)*. This program permits the prediction of the NExT fragmentation pattern of double-stranded DNA, allowing the researcher to tailor the dUTP:dTTP ratio without the need for experiments. The program was designed to read a DNA sequence file and dUTP:dTTP values and calculate all possible fragments, their likelihood of occurrence, and their relative distribution. The complementary strand for a given DNA is automatically generated and taken into account. The program displays the result in a bar chart and allows the export of all calculated data as tabulated lists for further use (**Fig. 5A**). When upper and lower ranges for the fragment size are set (e.g., because of gel purification), the program calculates the potential loss of material and adjusts the relative likelihood of the individual fragments.

3.9.1. Underlying Mathematics

1. Let the probability that a thymidine in a given DNA sequence is replaced by uridine be p.
2. A fragment between two given thymidines is generated if both are replaced by uridine, which equals the probability, $p \times p$.
3. This fragments can only exist if all n-number thymidines between the two uridines are not replaced, resulting in the overall probability of $p \times p \times (1 - p)^n$.
4. For fragments including one or both ends of the DNA, one or both p of the $p \times p$ probability are set to 1 (*see* **Note 21**).
5. For comparing fragmentation results, we normalized the fragment probabilities by dividing through the sum of all values.

Our calculation approach is distinguished from previous calculations ***(23)***, because we take into account the fact that both ends of a fragment need to be generated, and we do not face the problems of partial sequence preferences and uncertain digest conditions, which severely hamper the calculation of DNase digests.

For a gene with x number of thymidines, there is a maximum of $\sum_{i=1}^{x+1} i$ possible fragments [which is, according to Gauβ, $(x + 1) \times (x + 2)/2$ for odd x values; *see* **Note 22**]. Thus, for a typical 1000-bp gene with 250 thymidines and 240 adenines, which are calculated as thymidines in the opposite strand, in a few seconds the program calculates and generates 31,626 + 29,161 = 60,787 fragments, with their probabilities and individual sequences.

Because most users are likely interested in an overview, the program pools all fragments of the same length, sums up their probabilities, and gives the distribution as percentage of the sum of all probabilities vs the length, which is shown in the program as "%mol." To reflect the visualization on electrophoreses gels, the "mass" distribution is calculated by multiplying the probability of a certain fragment length with its length. These values are normalized, and represent the percent bases, which we termed "mass," as defined by length in basepairs, which is shown in the program as "%mass." The fragment sequence output lists all fragments by decreasing likelihood of occurrence. Identical sequences are listed once with their summed probabilities.

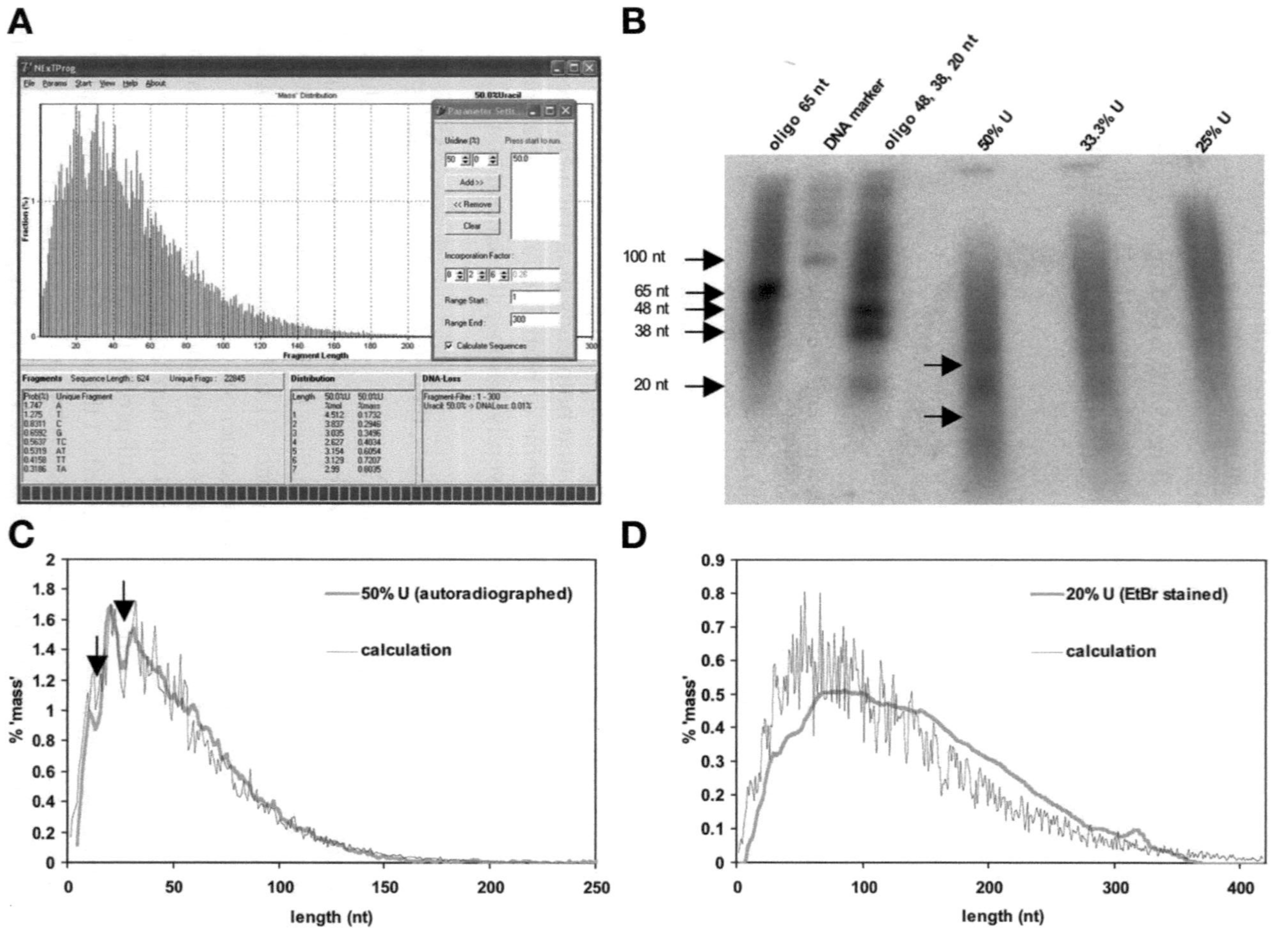
A
B
C
D
NExTProg
'Mass' Distribution
50.0%Uracil
Fraction (%)
Fragment Length
Parameter Setti...
Uridine (%)
Press start to run.
Add>>
<< Remove
Clear
Incorporation Factor:
Range Start:
Range End:
Calculate Sequences
Fragments
Distribution
DNA-Loss
oligo 65 nt
DNA marker
oligo 48, 38, 20 nt
50% U
33.3% U
25% U
100 nt
65 nt
48 nt
38 nt
20 nt
50% U (autoradiographed)
calculation
% 'mass'
length (nt)
20% U (EtBr stained)
calculation
% 'mass'
length (nt)

3.9.2. Calibration of the Program and Comparison With Experimental Results

Before one can compare the measured and calculated values, there is one important point to consider: the incorporation rate of uridine vs thymidine by the polymerase in the exchange PCR. This value might not only depend on the ratio of dUTP:dTTP, but also on the type of polymerase used, the absolute concentration of nucleotides, and the buffer. Thus, this value needs to be set when using the program. The uridine incorporation could be measured directly using radioactively labeled dUTP and counting the radioactivity of PCR products or, more indirectly, by quantitatively analyzing the fragmentation pattern. We opted for the latter approach, because it also provided the basis for the comparison with our software. However, one should note that this twofold use of the data is only valid if the fragmentation is complete.

For quantitative analyses of fragmentation experiments, we relied on widely used conditions (standard Taq polymerase *[24]* and 200 μ*M* dNTPs).

Fig. 5. *(Opposite page)* Comparison of calculated NExT fragmentations with radioactively labeled and ethidium bromide-stained fragmentation experiments. **(A)** Graphical front end of NExTProg 1.0 *(22)*. This program reads DNA sequences and calculates all possible fragments. The program exports lists of fragment length vs normalized fraction of molecules or mass, as defined by number of nucleotides, respectively. The sequences of all fragments can be generated, whereby identical sequences are combined, and exported for subsequent assembly calculations. **(B)** Denaturing polyacrylamide urea gel of radioactively labeled DNA samples. Lanes 1–3: marker oligonucleotides kinased with ^{32}P-ATP; lanes 4–6: fragments of a gene (*CAT_Cd26*, 624 bp), based on the indicated amount of uridine and ^{32}P-CTP in the exchange PCR (each lane contained 0.3 μCi). The gel was autoradiographed with a phosphor screen (Kodak) and read with a phosphoimager (BioRad Fx). Note the inhomogeneities in the lower third of the lanes, which correspond to sequence-specific peaks in the fragmentation. **(C)** Measured and calculated fragment distributions used to determine the incorporation rate of uridine vs thymidine in the exchange PCR. The gray line represents a line-density plot of the radioactive 50% U lane in (B), which was converted from relative migration distance to nucleotide length based on the marker nucleotides, set to integer numbers by averaging the respective values, and normalized. The black line represents the calculation of NExTProg for the fragment "mass" distribution for the same gene with 50% uridine and an incorporation rate of 0.26, which provided the lowest mean root square deviation. **(D)** The gray line is the line density plot of a 20% U reaction (**Fig. 1B,C**) stained with ethidium bromide, which was converted to fragment length and normalized. The black line is the calculation of NExTProg with 20% uridine and an incorporation rate of 0.26. Note that the staining of short single-stranded oligonucleotides with ethidium bromide is inefficient and, consequently, longer fragments are overrepresented in the normalized plot. EtBr, ethidium bromide.

1. Label DNA radioactively by adding ^{32}P-dCTP in the uridine-incorporating PCR (*see* **Subheading 3.2.** and **Note 2**).
2. Run a denaturing polyacrylamide gel (*see* **Subheading 3.4.**).
3. Autoradiograph the gel with an Imager (*see* **Note 23**). This avoids signal distortions caused by inefficient staining of short fragments.
4. Create line-density plots for each lane with an image analysis software program (e.g., QuantityOne (BioRad) or NIH Image/Scion Image/ImageJ).
5. Calculate the relative migration for the fragment smear and the radioactively labeled markers from the line-density plots.
6. Fit the equation $rel.distance = a \times \ln(\text{length in basepairs}) + b$, with the variables a and b being the relative distance and length of the markers.
7. Convert the relative distance of the fragment smear to nucleotide length using the reverse of the above equation.
8. Convert the intensity signal vs a continuous length distribution to a intensity signal vs integer numbers (only discrete oligonucleotide lengths can be present) by combining rounded nucleotide length values and by averaging the respective intensity values.
9. Normalize the distribution by dividing each intensity value through the sum of all values.

An autoradiograph of such a denaturing gel is shown in **Fig. 5B**. The normalized intensities taken from the gel are shown as "%mass" in **Fig. 5C**. This experiment was compared with program calculations to determine the relative uridine vs thymidine incorporation value. For the calculations, incorporation values from 0.2. to 0.3, in 0.01 increments were used. Setting a value of 0.26 resulted in the best agreement of measurement and prediction based on a root mean square analysis of the fragment size range between 10 and 150 bases, where the length calculation is likely reliable and between 4 and 200 bases (**Fig. 5C**). In the case of using the range between the markers (20 and 100 bases), a factor of 0.25 scored marginally better. It is remarkable how well these plots overlay, and how well peaks and dips within the fragment smear (indicated by arrows in **Fig. 5B,C**) can be explained. Importantly, the scaling of the *y*-axis falls into place without further adjustment. We are very confident that the uridine incorporation rate has been determined and not a factor accounting for incomplete fragmentation, because of the nearly perfect agreement of calculation and experiment, and the many experimental conditions tested for fragmentation. Measurements of the radioactive 33.3% uridine fragmentation and calculations agreed equally well. The radioactive 25% dUTP fragmentation produced significant amounts of fragments beyond the 100-nucleotide marker, and showed some deviations caused by scaling problems. However, using the same incorporation rate value of 0.26 and comparing the results with the ethidium bromide-stained fragmentation results, which were more accurate for longer fragments because of the available markers, demonstrated a good agreement (**Fig. 5D**) considering the length dependency of the gel staining. Because the uridine-incorporating PCRs for the radioactive and ethidium bromide-stained gels were carried out with two different buffers (*see* **Subheading 2.2.**), the addition of PCR enhancers, such as ammonium sulfate, Tween-20, or Triton X-100 have no significant effect on the dUTP vs dTTP incorporation rate.

4. Notes

1. Although the overall error rate of the described NExT DNA shuffling is already quite low, the uridine-exchange PCR could, in addition, be performed with a proofreading polymerase. However, the polymerase should be selected carefully to be able to sufficiently incorporate uridine and not to stall at uracil sites in the template. Vent exo$^+$ (New England Biolabs) and nonuracil-stalling mutants of *Pfu* polymerase (Stratagene) have been reported to incorporate uridine.
2. Very small fragments can be visualized radioactively. In this case, add 0.5 µL of a 3.3 µ*M* ^{32}P-dCTP solution (approx 0.5 µCi) to the uridine-exchange PCR mixture. For radioactive experiments, a PCR cleanup kit was used without the agarose gel step (*see* **Subheading 3.2.4.**), because the nonradioactive template does not disturb visualization of radioactive fragments.
3. We used four 50-µL uridine-exchange PCR vials because our thermocycler works best with this volume.
4. It is necessary to separate the uracil-containing products from nonuracil-containing template, to make sure that no template is carried over.
5. We preferred to use two columns in parallel to save time. In this case, we did the enzymatic and chemical cleavage of the uridine-exchange PCR (*see* **Subheading 3.3.**) separately for the two tubes to keep the volume in a reasonable range for our thermocycler. The samples can be combined for the fragment purification either at **Subheading 3.5.1, step 2.** or **Subheading 3.4., step 4.**
6. Digests of up to 2 h or more yielded equivalent results, indicating a selective and consistent digest.
7. Piperidine is toxic and should be handled in a fume hood. All other solutions containing piperidine, e.g., the used capture buffer of the Qiaex II kit (*see* **Subheading 3.5.1.**), should be handled with care, and discharged in a closed tube.
8. We used gels up to a maximum of 3 d after preparation. The urea powder adds to the volume, which is taken into account in the final concentrations. The volume for nine gels was used to pour eight gels.
9. Piperidine influences the running properties of fragments and, thus, has to be evaporated as much as possible.
10. For the radioactive experiments, only 7 µL of the DNA–formamide sample were used, supplemented with 3 µL of 60% sucrose solution and 7 µL H_2O; 15 µL were loaded onto the gel.
11. For radioactive experiments, oligonucleotides and DNA ladder were kinased with ^{32}P-γATP and purified by size exclusion to remove excessive ^{32}P-γATP. We used Sephadex G-50 material, but any nucleotide removal kit should serve this purpose.

12. We used bromphenol blue only for the marker, but not for our samples, to avoid overlaying of the dye with the DNA fragments.
13. Incubation times should preferably not exceed 5 min because smaller fragments start to elude from the gel. For this reason, it is difficult to stain the gel with SYBR green II, because this dye needs to be incubated for 5 to 10 min. Note also that urea gels bleach fast in UV light.
14. To avoid loss of small fragments, we found the QiaexII kit to be best suited. The manufacture recommends the kits only for fragments larger than 40 bp, however, we were able to recover fragments down to approx 20 bases in low amounts.
15. The two centrifugation steps are essential to ensure that the DNA fragments are not contaminated with residual amounts of matrix because the matrix will hinder the reassembly reaction by binding DNA.
16. Normally, it is not necessary to measure the concentration of the fragments if the procedure is started with 7 μg of uridine-exchange PCR product or more.
17. The supplier recommends a 1:10,000 dilution, however, we achieved better results using a 1:5000 dilution for small fragments and low concentration.
18. Ideally, use white 96-well plates because these enhance fluorescence.
19. We used higher amounts of dNTPs because they might degrade during the approx 4-h-long reassembly procedure.
20. This volume was chosen to ensure diversity.
21. The sum for all possible patterns of the uridine incorporation in the gene is 1, but the sum of the probabilities for the fragments is larger, because each pattern results in several fragments.
22. This calculation includes fragments with a length of zero bases, which are "generated" when two uracil moieties are next to each other.
23. Gels can also be autoradiographed with a film. However, modern imagers provide a larger linear detection range.

Acknowledgment

We thank Susanne Knall for help with the experiments, Gregor Zipf for coding NExTProg, Hubert Bernanser for fruitful discussions, and Jody Mason for critical reading of the manuscript. K. M. A. was funded by the Deutsche Forschungsgemeinschaft, Emmy Noether-Programm (Ar373).

References

1. Neylon, C. (2004) Chemical and biochemical strategies for the randomization of protein encoding DNA sequences: library construction methods for directed evolution. *Nucleic Acids Res* **32**, 1448–1459.
2. Stemmer, W. P. (1994) DNA shuffling by random fragmentation and reassembly: in vitro recombination for molecular evolution. *Proc. Natl. Acad. Sci. USA* **91**, 10,747–10,751.
3. Stemmer, W. P. (1994) Rapid evolution of a protein in vitro by DNA shuffling. *Nature* **370**, 389–391.

4. Zhao, H., Giver, L., Shao, Z., Affholter, J. A., and Arnold, F. H. (1998) Molecular evolution by staggered extension process (StEP) in vitro recombination. *Nat. Biotechnol.* **16**, 258–261.
5. Shao, Z., Zhao, H., Giver, L., and Arnold, F. H. (1998) Random-priming in vitro recombination: an effective tool for directed evolution. *Nucleic Acids Res.* **26**, 681–683.
6. Cadwell, R. C. and Joyce, G. F. (1992) Randomization of genes by PCR mutagenesis. *PCR Methods Appl.* **2**, 28–33.
7. Arndt, K. M., Müller, K. M., and Plückthun, A. (2001) Helix-stabilized Fv (hsFv) antibody fragments: substituting the constant domains of a Fab fragment for a heterodimeric coiled-coil domain. *J. Mol. Biol.* **312**, 221–228.
8. Sambrook, J. and Russel, D. W. (2001) *Molecular Cloning: A Laboratory Manual.* 3rd ed., Cold Spring Harbor Laboratory Press, Cold Spring Harbor, NY.
9. Patel, P. H. and Loeb, L. A. (2000) Multiple amino acid substitutions allow DNA polymerases to synthesize RNA. *J. Biol. Chem.* **275**, 40,266–40,272.
10. Boiteux, S., O'Connor, T. R., and Laval, J. (1987) Formamidopyrimidine-DNA glycosylase of Escherichia coli: cloning and sequencing of the fpg structural gene and overproduction of the protein. *Embo J.* **6**, 3177–3183.
11. Friedberg, E. C., Walker, G. C., and Siede, W. (1995) *DNA Repair and Mutagenesis.* American Society of Microbiology, Washington, DC.
12. Drohat, A. C., Jagadeesh, J., Ferguson, E., and Stivers, J. T. (1999) Role of electrophilic and general base catalysis in the mechanism of Escherichia coli uracil DNA glycosylase. *Biochemistry* **38**, 11,866–11,875.
13. Maxam, A. M. and Gilbert, W. (1980) Sequencing end-labeled DNA with base-specific chemical cleavages. *Methods Enzymol.* **65**, 499–560.
14. Ljungquist, S. (1977) A new endonuclease from Escherichia coli acting at apurinic sites in DNA. *J. Biol. Chem.* **252**, 2808–2814.
15. Richardson, C. C. and Kornberg, A. (1964) A deoxyribonucleic acid phosphatase-exonuclease from Escherichia coli. I. Purification of the enzyme and characterization of the phosphatase activity. *J. Biol. Chem.* **239**, 242–250.
16. Dodson, M. L., Schrock, R. D., 3rd, and Lloyd, R. S. (1993) Evidence for an imino intermediate in the T4 endonuclease V reaction. *Biochemistry* **32**, 8284–8290.
17. Levin, J. D. and Demple, B. (1990) Analysis of class II (hydrolytic) and class I (beta-lyase) apurinic/apyrimidinic endonucleases with a synthetic DNA substrate. *Nucleic Acids Res.* **18**, 5069–5075.
18. Miyazaki, K. (2002) Random DNA fragmentation with endonuclease V: application to DNA shuffling. *Nucleic Acids Res.* **30**, e139.
19. Mattila, P., Korpela, J., Tenkanen, T., and Pitkanen, K. (1991) Fidelity of DNA synthesis by the Thermococcus litoralis DNA polymerase—an extremely heat stable enzyme with proofreading activity. *Nucleic Acids Res.* **19**, 4967–4973.
20. Kong, H., Kucera, R. B., and Jack, W. E. (1993) Characterization of a DNA polymerase from the hyperthermophile archaea Thermococcus litoralis. Vent DNA polymerase, steady state kinetics, thermal stability, processivity, strand displacement, and exonuclease activities. *J. Biol. Chem.* **268**, 1965–1975.

21. Clark, J. M. (1988) Novel non-templated nucleotide addition reactions catalyzed by procaryotic and eucaryotic DNA polymerases. *Nucleic Acids Res.* **16**, 9677–9686.
22. Müller, K. M. and Zipf, G. (2004) NExTProg 1.0, download available at http://www.molbiotech.uni-freiburg.de/next or http://www.ATG-biosynthetics.com.
23. Moore, G. L. and Maranas, C. D. (2000) Modeling DNA mutation and recombination for directed evolution experiments. *J. Theor. Biol.* **205**, 483–503.
24. Kaledin, A. S., Sliusarenko, A. G., and Gorodetskii, S. I. (1980) [Isolation and properties of DNA polymerase from extreme thermophylic bacteria Thermus aquaticus YT-1]. *Biokhimiia* **45**, 644–651.

11

Degenerate Oligonucleotide Gene Shuffling

Peter L. Bergquist and Moreland D. Gibbs

Summary

Improvement of the biochemical characteristics of enzymes has been aided by misincorporation mutagenesis and DNA shuffling. Shuffling techniques can be used on a collection of mutants of the same gene, or related families of genes can be shuffled to produce mutants encoding chimeric gene products. One difficulty with current shuffling procedures is the predominance of unshuffled ("parental") molecules in the pool of mutants. We describe a procedure for gene shuffling using degenerate primers that allows control of the relative levels of recombination between the genes that are shuffled and reduces the regeneration of unshuffled parental genes. This procedure has the advantage of avoiding the use of endonucleases for gene fragmentation before shuffling and allows the use of random mutagenesis of selected segments of the gene as part of the procedure. We illustrate the use of the technique with a diverse family of β-xylanase genes that possess widely different G and C contents.

Key Words: Polymerase chain reaction; primer extension; degenerate primers; in vitro evolution; gene shuffling; complementary degenerate-end primers.

1. Introduction

Until recently, the most popular methods of creating combinatorial libraries were recursive strategies that sought to evolve sequences by the addition of point mutations. For in vitro evolution, inclusion of recombinant polymerase chain reaction (PCR; gene shuffling) offers practical and theoretical advantages over simple recursive mutagenesis methods ***(1–3)***. It will rapidly fine-tune the mutational load in several parts of the protein by recombining point mutations and wild-type sequences. Family shuffling is usually achieved by fragmentation of the genes to be shuffled, followed by PCR. It relies on homologous recombination during the PCR reassembly step. Most methods require relatively high levels of sequence similarity between the genes to be shuffled, because "crossover points" seem to occur in these regions.

From: *Methods in Molecular Biology, vol. 352: Protein Engineering Protocols*
Edited by: K. M. Arndt and K. M. Müller © Humana Press Inc., Totowa, NJ

If sequence similarity is low between the input genes, the majority of products tend to be the reassembled parental genes, and extensive searches need to be performed to find the chimeric recombinants ***(4,5)***. Kichuchi et al. described a method for gene shuffling that makes use of unique restriction enzyme sites in the sequences of the parental molecules ***(5)***. The complete restriction enzyme digestion of parental genes ensured that subsequent overlap extension gave rise to hybrid genes at high frequencies.

We isolated a gene coding for a thermophilic β-xylanase that had superior performance in the bleaching of paper pulp ***(6)***. We wished to investigate the possibility of obtaining mutant derivatives that had enhanced stability and an altered pH optimum. Experiments using error-prone PCR and misincorporation mutagenesis followed by gene shuffling allowed the identification of mutant genes that coded for a limited sample of the variations in sequence space but required extensive screening for their identification. Gene shuffling after DNase I fragmentation of related genes (family shuffling) overwhelmingly yielded wild-type parental sequences as the major products. After several trials of methods designed to reduce the background, we devised a technique that allows shuffling of genes that differ widely in sequence similarity and G and C content, and greatly reduces the regeneration of wild-type genes. Furthermore, the primer extension conditions may be modified to bias the resulting progeny genes toward any one (or more) of the parental input genes. We term this procedure degenerate oligonucleotide gene shuffling (DOGS; **ref. *7***), and note its compatibility with other recursive point mutation techniques.

2. Materials

1. Source of genes: Family 11 xylanase genes were obtained from the following bacterial strains: *Dictyoglomus thermophilum* strain Rt46B.1 *xynB* ***(8)***; *Clostridium stercorarium xynB* ***(9)***; *Bacillus* sp. strain V1-4 ***(10)***; *Caldicellulosiruptor* sp. strain Rt69B.1 *xynD* ***(11)***; *Clostridium thermocellum xynV* ***(12)***; and *Streptomyces roseiscleroticus xyl3* ***(13)***. Each gene was PCR amplified from genomic DNA using the respective gene-specific primers.
2. Platinum *Pfx* polymerase (Invitrogen, Victoria, Australia).
3. Plasmid pBSII KS– (Stratagene, San Diego, CA).
4. Shrimp alkaline phosphatase (Roche Diagnostics Australia, NSW, Australia).
5. *Escherichia coli* strain DH5α.
6. Universal buffer: 50 m*M* phosphoric acid, 50 m*M* boric acid, 50 m*M* acetic acid. Adjust pH with NaOH.
7. Overlay solution: 0.5% birchwood xylan (Sigma-Aldrich, Sydney, Australia), 0.5% agarose in 120 m*M* Universal buffer, pH 6.5. Mix, then autoclave to completely solubilize the xylan and agarose. Allow mixture to cool to approx 50°C.
8. Congo Red solution: 1% Congo Red (Sigma-Aldrich, Sydney, Australia) in water. Ensure the pH is slightly alkaline by addition of NaOH to approx 5 m*M*.

9. Destaining solution: 1 *M* NaCl, 5 m*M* NaOH.
10. Protein extraction reagent BPER II (Pierce Chemical Company, Rockford, IL).
11. Birchwood xylan substrate solution: Add 0.5 g Birchwood xylan to 100 mL of 120 m*M* Universal buffer, pH 6.5. Stir 5 min until homogeneous, then autoclave to completely solubilize the xylan.
12. PAHBAH stock solution: Add 5 mL of each of 0.5 *M* Na_3 citrate, 1.0 *M* Na_2SO_3, 0.2 *M* $CaCl_2$, and 5.0 *M* NaOH to 25 mL water, mixing after each addition. Add 0.76 g *p*-hydroxybenzoic acid hydrazide (Sigma-Aldrich), mix, and bring volume up to 50 mL in water.

3. Methods

As an example of the DOGS procedure, we give examples from the shuffling of the *D. thermophilum* Rt46B.1 xylanase gene, *xynB*, with five related xylanase genes, reported by us previously *(7)*. **Figure 1** shows schematically the steps involved in the design of primers and the subsequent primer extension steps to create the shuffled segments that are incorporated into the final amplified and cloned chimeric genes (*see* **Note 1**).

3.1. Complementary Degenerate-End Primer Pairs for Efficient PCR Amplification of Gene Segments and Overlap-Extension of Adjacent Segments

The most commonly used strategy to isolate distantly related sequences by PCR has been to design degenerate primers that bind to highly conserved regions of DNA sequences. The difficulty with this method is that, as the primer degeneracy increases to accommodate more divergent genes, the number of primer molecules in a PCR that can correctly prime synthesis drops, and these primers may be used up in the first few cycles of the reaction. Nonspecific amplification may then occur because of the abundance of primers that do not participate in amplification of the targeted gene, and, therefore, are available to prime nonspecific synthesis, especially because low stringency annealing conditions are usually needed to detect mismatched templates.

Rose et al. ***(13)*** described a strategy that overcomes problems of degenerate methods for primer design, called consensus–degenerate hybrid oligonucleotide primers (CODEHOP). CODEHOP primers consist of a relatively short 3′ degenerate end and a 5′ nondegenerate consensus clamp. Reducing the length of the 3′-end to a minimum decreases the total number of individual primers in the degenerate primer pool. Hybridization of the 3′ degenerate end with the target template is stabilized by the 5′ nondegenerate consensus clamp, which allows higher annealing temperatures without increasing the degeneracy of the pool. Although potential mismatches may occur between the 5′ consensus clamp of the primer and the target sequence during the initial PCR cycles, they

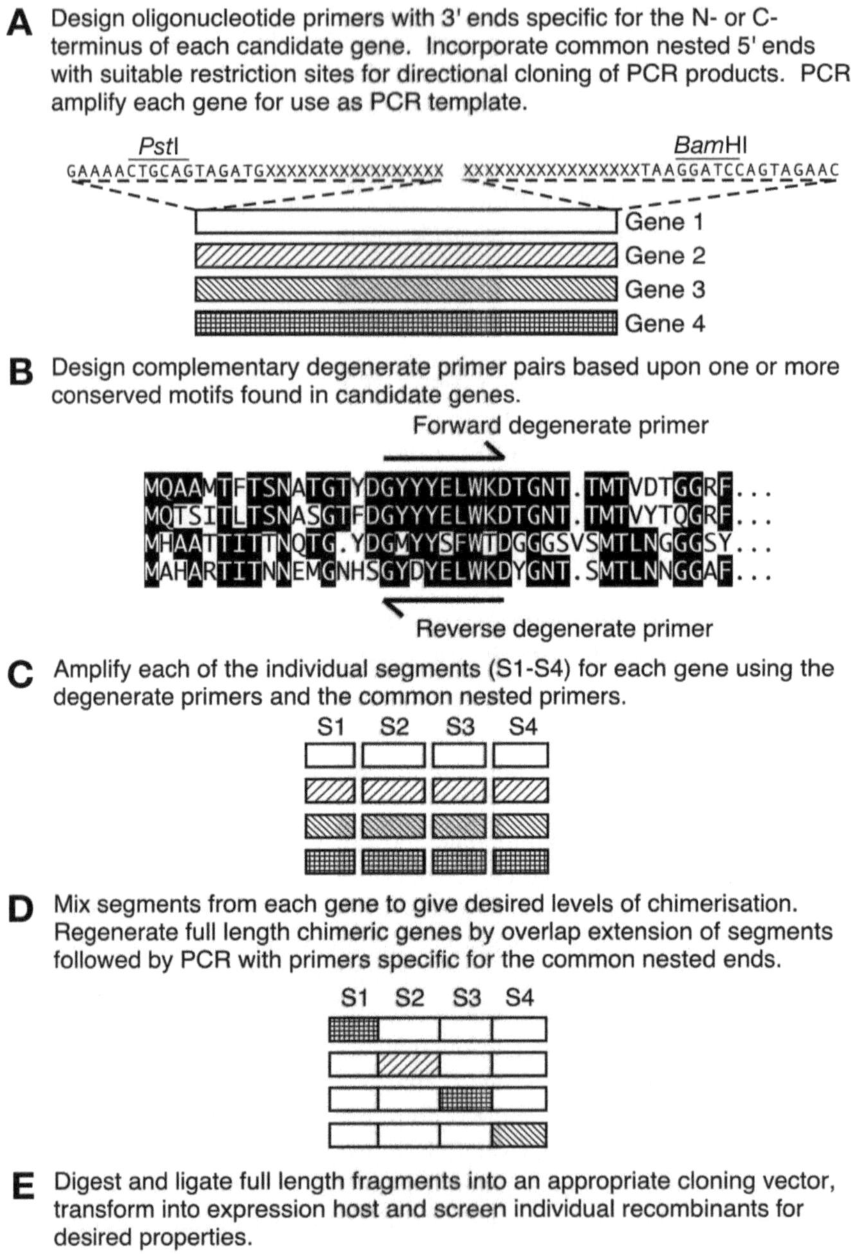

Fig. 1. Schematic diagram of DOGS procedure.

are situated away from the 3′ hydroxyl extension site, and, thus, mismatches between the primer and the target are less disruptive to priming of polymerase extension. Further amplification of primed PCR products during subsequent rounds of primer hybridization and extension is enhanced by the sequence similarity of all primers in the pool; this potentially allows use of all primers in the reaction. A modification of the CODEHOP method allows efficient amplification of overlapping segments of related genes, and subsequent overlap-extension of adjacent segments from different genes resulting in the formation of chimeric gene fragments. The careful design of the primer sequences is the most important step in the DOGS procedure.

The modification of the CODEHOP technique entails the design of perfectly complementary pairs of primers. Each primer has a nondegenerate core flanked by both 5′ and 3′ degenerate ends, referred to herein as complementary degenerate-end (CDE) primers. As with the CODEHOP primers, the 3′ degenerate end gives each CDE primer their template-binding specificity, whereas the nondegenerate region acts as a stabilizing clamp in subsequent rounds of the PCR. The 5′ degenerate end is not required to contribute to the binding efficiency of the CDE primer during PCR, however, it plays a pivotal role in allowing efficient binding and subsequent overlap-extension of separate PCR products (gene segments) generated using, respectively, the forward or the reverse CDE primers.

The nondegenerate core of individual CDE primers is generally based on the corresponding coding sequence of one gene, designated the parental gene for shuffling. This results in the formation of chimeric fragments that retain the parental sequence at the points of segment overlap.

3.1.1. Design and Use of Gene-Specific Nested End Primers

1. Design and synthesize forward and reverse primer pairs suitable for PCR amplification of each gene to be shuffled. Each primer should comprise a 17- to 20-nucleotide gene-specific 3′-end and a 17- to 20-nucleotide common 5′-end. We incorporated restriction sites into the common ends to facilitate directional ligation of PCR products into pBSII KS–.
2. Synthesize two nested primers with sequences corresponding directly to the forward and reverse common ends of the gene-specific nested primers. The nested primers will be used in combination with CDE primers to amplify the first and last segments of each gene.
3. Use the gene specific primers to amplify each gene from a suitable source of template DNA, usually either genomic DNA or, preferably, a cloned example of the gene.

3.1.2. Summary of the Features of CDE Primers

CDE primers allow efficient and specific amplification of portions (referred to herein as gene segments) of related but divergent genes. The 5′ degenerate end of CDE primers ensures that separate PCR products generated with the respective forward or reverse complementary CDE primers will anneal efficiently in subsequent overlap-extension PCR steps. Furthermore, multiple (one or more) pairs of CDE primers allow the generation of consecutive PCR products (gene segments) with complementary ends suitable for overlap extension and PCR,

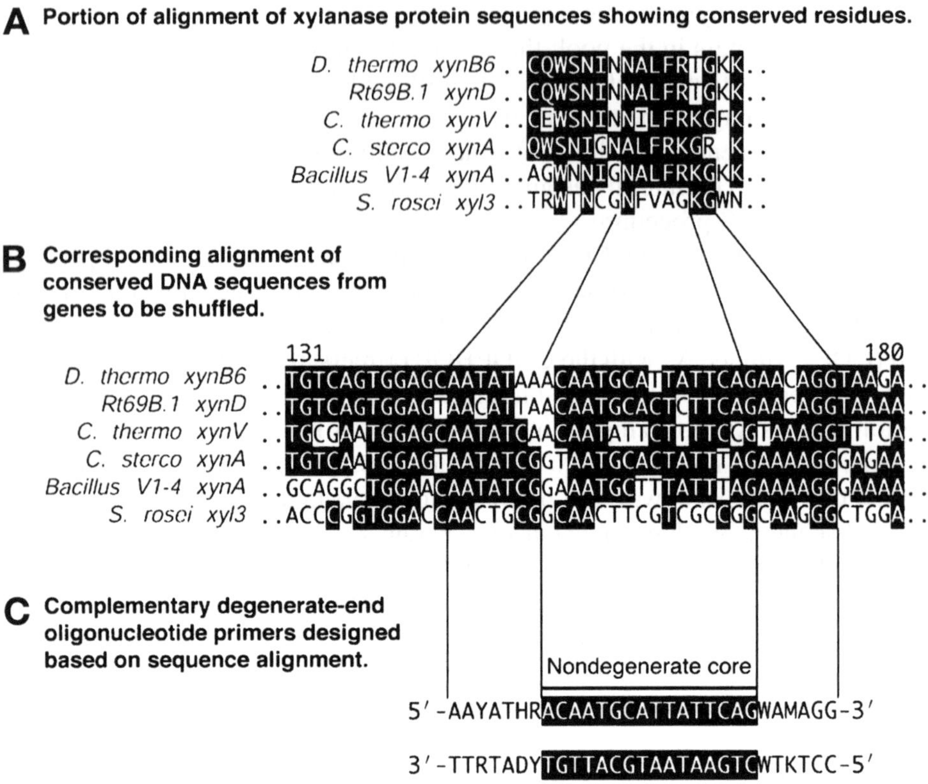

Fig. 2. An example of the design of CDE primers based on the DNA-coding sequence of conserved amino acids. **(A)** Portion of alignment of xylanase protein sequences. **(B)** DNA sequence alignment corresponding to amino acid sequences. **(C)** CDE primers based on conserved sequence. The degenerate ends correspond to conserved residues and the corresponding conserved DNA sequence. The nondegenerate core sequence (shown in reverse text) is designed to match the sequence of the selected parental gene (in the example, *D. thermophilum xynB6*).

resulting in the generation of recombined segments. An example of the design strategy for making complementary oligonucleotide pairs suitable for the amplification of gene segments from related genes, and for the subsequent overlap extension of segments to generate chimeric genes is given in **Fig. 2**.

CDE primers may also be used in combination with complementary degenerate primers that do not have a nondegenerate core, to generate consecutive PCR products (gene segments) with complementary ends suitable for overlap extension and PCR, resulting in the generation of recombined segments. The mixing of segments amplified from related genes followed by overlap extension and PCR results in the efficient generation of chimeric gene fragments. The nondegenerate core of each complementary CDE primer set may be (but does not have to be) based on the gene designated as the parental gene. Finally, gene segments amplified from related genes can be mixed in unequal amounts, allowing control of the level of incorporation of each segment into resultant chimeric gene fragments (**Note 2**). An overview of the gene segment amplification and overlap-extension with CDE primers is depicted in **Fig. 3**.

3.1.3. Design of CDE Primer Pairs for Efficient PCR Amplification of Gene Segments and Overlap-Extension of Adjacent Segments

1. Align the amino acids sequences of the respective proteins using a suitable sequence alignment software program, such as ClustalX ***(14)***.
2. Identify blocks of conserved amino acid motifs in the proteins. In our example, the genes coding for the xylanases were divided into eight segments based on the relative positions of observed conserved regions.
3. Align the nucleotide sequences of the corresponding genes with a suitable sequence alignment software program, such as Tranalign ***(15)***.
4. Design CDE forward and reverse primers for the amplification of the DNA from the defined conserved sites, allowing the amplification of gene segments when combined, as appropriate, with the nested 5′ and 3′ common primers (*see* **Fig. 2**).
5. Amplify each segment of each gene using combinations of adjacent CDE primers and nested 5′ and 3′ common primers. In the example, the PCR conditions were as follow: 1 cycle of 95°C for 1 min; then 35 cycles at 95°C (denaturation) for 30 s; annealing at 35°C for 20 s; extension at 72°C for 40 s; and a final incubation at 72°C for 5 min. We used the archaeal DNA polymerase, Platinum *Pfx* (*see* **Notes 3** and **4**).
6. Individually gel-purify each PCR product.

3.1.4. Overlap-Extension of Gene Segments

1. The individual segments of each gene should be mixed in appropriate ratios to give the desired level of gene chimerization. For example, using the six candidate genes G1 to G6, where G1 is the *D. thermophilum* Rt46B.1 *xynB* gene, and deciding that this sequence should predominate in the shuffled progeny, the pooled PCR segments for each gene were mixed in the ratio of 8.75 G1 to 1:1:1:1:1 to yield, on average, chimeras with 5/8 segments from Rt46B.1 *xynB* (a simple formula for calculating suitable ratios is given in **Note 5**).
2. Next, 50 to 100 ng of the mixed segments were used as templates for overlap extension ***(16)***, using the following conditions: 1 cycle of 95°C for 1 min; then 35 cycles at 95°C (denaturation) for 30 s; annealing at 35°C for 20 s; extension at 72°C for 40 s; and a final incubation at 72°C for 5 min. We used the archaeal DNA polymerase, Platinum *Pfx* (*see* **Note 3**).

3.1.5. Amplification of Full-Length Chimeric Genes

Chimeric fragments were recombined into complete genes by using the overlap-extended products (50–100 ng; *see* **Subheading 3.1.4.**) as a template for PCR

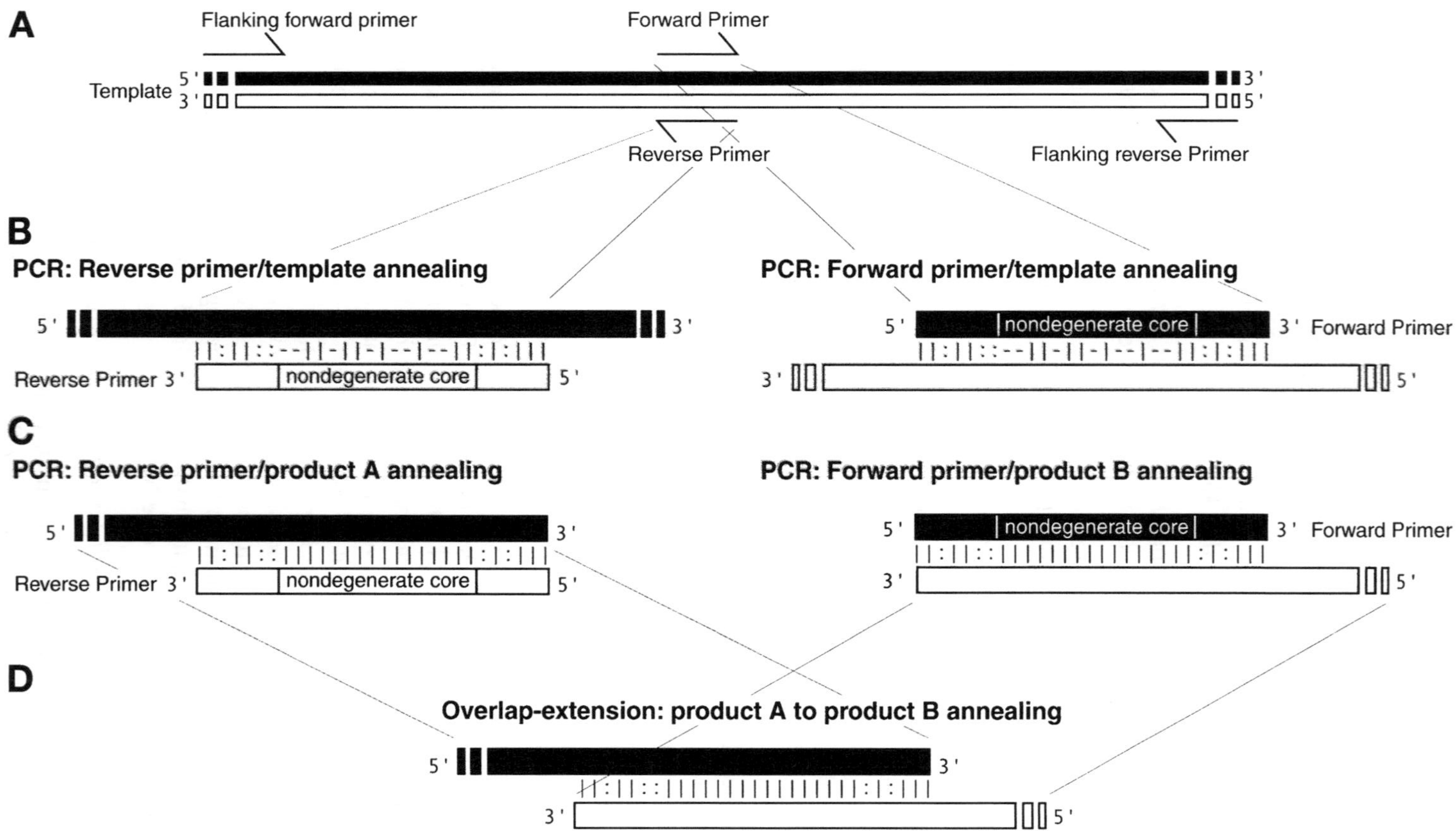

A
Flanking forward primer
Forward Primer
Template
5'
3'
3'
5'
Reverse Primer
Flanking reverse Primer
B
PCR: Reverse primer/template annealing
PCR: Forward primer/template annealing
nondegenerate core
Reverse Primer
Forward Primer
C
PCR: Reverse primer/product A annealing
PCR: Forward primer/product B annealing
D
Overlap-extension: product A to product B annealing

using the common flanking nested 5′ and 3′ primers under the following conditions: 1 cycle of 95°C for 1 min to activate the enzyme; then 20 cycles at 95°C (denaturation) for 30 s; annealing at 50°C for 20 s; extension at 72°C for 40 s; and a final incubation at 72°C for 5 min, using Platinum *Pfx* DNA polymerase.

3.1.6. Cloning of Shuffled Products

1. Digest DOGS PCR products with the restriction enzymes *Bam*HI and *Hin*dIII.
2. Digest pBSII KS– with the same restriction enzymes and treat with shrimp alkaline phosphatase.
3. Ligate the PCR product in the vector pBSII KS–.
4. Transform the ligated plasmid, which also carries an ampicillin resistance gene into the *E. coli* strain DH5α and spread onto LB plates containing 100 μg/mL ampicillin and 5 m*M* isopropylthiogalactoside (IPTG).
5. Pick individual colonies, patch in duplicate onto new plates containing 100 μg/mL ampicillin and 5 m*M* IPTG, and screen for the expression of xylanase activity by the Congo Red overlay method ***(17)*** as detailed in **Subheading 3.1.7.**

3.1.7. Plate Assays Using the Congo Red Method

1. Gently pipet 4 mL of overlay solution cooled to approx 50°C onto plates with patched colonies. The plates should be prewarmed to 37°C to ensure that the overlay solution does not set before forming an even layer.

Fig. 3. *(Opposite page)* CDE primers for PCR and overlap extension. **(A)** A diagrammatic representation of double-stranded template DNA and the relative binding positions of the CDE forward and reverse primers. In separate PCR amplifications, the forward CDE primer is used combination with the reverse flanking primer, whereas the reverse CDE primer is used in combination with the forward flanking primer. **(B)** A diagrammatic representation showing the correct binding of each of the CDE forward and reverse primers to the DNA template. A thin vertical line (|) indicates correct primer/template pairing of adjacent nucleotides; a colon (:) indicates potential matching of adjacent nucleotides caused by degeneracy in the primer pool; whereas a dash (–) indicates a nucleotide mismatch. As depicted here, in the first round of PCR, the nondegenerate core does not contribute to primer binding, and primer binding specificity is attained by the 3′ degenerate end of each primer. **(C)** A diagrammatic representation showing the binding of each of the CDE forward and reverse primers to products generated in early rounds of PCR amplification. A thin vertical line (|) indicates correct primer/template pairing of adjacent nucleotides; a colon (:) indicates potential matching of adjacent nucleotides caused by degeneracy in the primer pool. The nondegenerate core now acts as a clamp, ensuring efficient use of the degenerate primer pool in amplification of a gene segment. **(D)** A diagrammatic representation showing the complementarity of two gene segments generated by PCR using, respectively, the forward or the reverse CDE primer. This complementarity allows for efficient polymerase-mediated overlap extension, resulting in the regeneration of a single DNA fragment comprised of both DNA segments. If the two PCR products originated from different genes, a chimeric fragment will be generated.

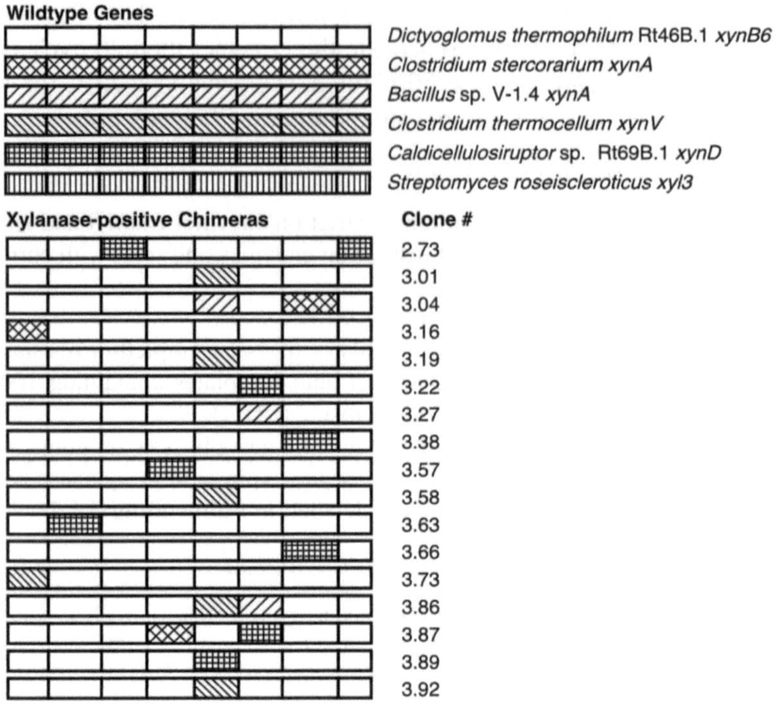

Fig. 4. An example of functional chimeric xylanase genes generated using the DOGS procedure.

2. Allow the overlay solution to set, then place the plates "lid side up" in a "zip-lock" bag. Seal the bag and place in a 70°C incubator for 3 h.
3. Remove plates from bag and allow to cool to room temperature.
4. Add approx 5 mL of the Congo Red solution to each plate, ensuring that the overlay layer is completely covered. Incubate for 5 to 10 min.
5. Pour off excess Congo Red solution, then destain plates by soaking plates in a destaining solution.
6. Xylanase-positive colonies are detected by an unstained circular zone (halo) around the colony.

Figure 4 summarizes some results from one round of DOGS of the six xylanase genes (*see* **Note 6**).

3.1.8. Enzyme Assays for Xylanase Activity (PAHBAH Method)

Whole-cell extracts containing recombinant xylanase were prepared for enzyme assays using the protein extraction reagent, BPER II. Xylanase activity was determined in a liquid assay using the method of Lever ***(18)***, with Birchwood xylan, as substrate. The standard assay reaction mixture consisted of 0.5% (w/v) xylan in 120 m*M* universal buffer ***(19)***, pH 6.5, and enzyme, to give a final volume of 0.03 mL. The reaction mixture was incubated at 60°C for 20 min.

3.1.8.1. Cell Lysis and Enzyme Extraction

1. Grow positive "plate assay" transformants overnight in 2 mL Luria broth containing 100 μg/mL ampicillin and IPTG.
2. Pellet cells from 1.5 mL of culture by centrifugation at 11,000*g* for 30 s, discard supernatant.
3. Completely resuspend pellet by vortexing, then add 150 μL of BPER II solution to resuspended cells.
4. Lyse cells by vortexing mixture for 1 min and clarify cell lysates by centrifugation at 11,000*g* for 1 min.
5. Transfer supernatant to a new tube and store at 4°C.

3.1.8.2. The Xylanase Assay

1. On ice, in a 0.2-mL PCR tube, mix 5 μL of appropriately diluted enzyme extract with Birchwood xylan substrate solution to a final volume of 50 μL.
2. Incubate tubes in a "hot-top" PCR block at an appropriate assay temperature for 10 to 30 min.
3. Place in ice water, add 100 μL PAHBAH stock solution to stop enzymatic reaction, and mix.
4. Heat the assay mixture to 99°C on a PCR block for 5 min to develop color. Ensure that the "hot-top" is in position to prevent the caps from opening. Set the PCR block to cool immediately to 4°C to halt color development.
5. Mix tube contents, then transfer 100 μL to a flat-bottomed microtiter dish. Measure absorbance at *A*420 in a suitable spectrophotometric plate reader.

4. Notes

1. The DOGS procedure that we have described demonstrates that it is possible to shuffle members of a gene family that are not particularly closely related and still generate chimeric molecules at a high enough frequency that comprehensive and time-consuming screens are not necessary. The use of CDE primers has allowed the reliable PCR amplification and shuffling of equivalent gene segments from a diverse range of genes with low overall sequence homology. Although in the example we have given, we used broadly related xylanase genes of Family 11 for shuffling, the procedure can be used with a single gene from which a misincorporation mutagenesis library can be produced, and the most promising mutants can take the place of the different gene families used in our example. The individual mutants are amplified by the PCR, using the CDE primers, and are mixed in appropriate ratios and subjected to primer extension as in the DOGS procedure described to shuffle the genes and to recombine out deleterious and neutral mutations.
2. The segment recombination frequencies can be controlled by altering the segment input ratios so that shuffling of particular fragments can be enhanced or attenuated as required (*see* **Note 4**). Accordingly, the procedure allows domain-swapping

experiments to be conducted with relative ease, replacing older methods that rely on suitable restriction enzyme sites. It is evident that PCR-induced misincorporation or error-prone mutagenesis can be incorporated as part of the procedure to introduce even more diversity into the products. In this respect, the DOGS method lends itself to random mutagenesis of individual segments to assist in fine-tuning of the encoded gene product. In this case, the CDE primers would be modified so that no degeneracy was available except for the segment to be mutagenized. This would give the investigator control over the extent and nature of mutagenesis of a particular segment by introducing a DNA polymerase without proofreading activity at the appropriate stage in the procedure.

3. The high fidelity Platinum *Pfx* polymerase was used for all overlap extension and PCR to decrease the frequency of PCR misincorporation mutations. *Taq* polymerase should not be used because it catalyzes the nontemplated adenylation of the 3′-ends of PCR products, which can prevent correct segment annealing in subsequent overlap-extension PCR.
4. It is clear from the design of the degenerate primers and the results reported here that altered nucleotide sequences will be generated even using a high-fidelity DNA polymerase, because of the mismatches designed into the degenerate primers. Even greater misincorporation mutagenesis can be generated by using a polymerase without proofreading activity in the amplification and primer extension steps.
5. The formula $r = (gx - x)/(s - x)$ can be used to calculate the exact ratio of parental segments, r, required to be mixed to one part of each of the other genes to achieve the desired levels of chimerization (where g is the total number of genes being shuffled, x is the required average number of parental segments per chimera, and s is the total number of segments per gene). For example, if we wished to shuffle six genes each divided into eight segments, so that, on average, each chimera contained six parental segments, then $g = 6$, $s = 8$, and $x = 6$. Using the above formula, r is calculated to be 15. Therefore, the parental segments should be mixed in a 15:1:1:1:1:1 ratio with segments from the other five genes to obtain chimeras with an average of six parental segments.
6. It is apparent that the procedure lends itself to combination with other gene shuffling and combinatorial mutagenesis techniques for the generation of novel proteins with modified characteristics.

Acknowledgments

This work was supported by grants from the Australian Research Council, the Macquarie University Research Grants scheme, and the Public Good Science Fund (New Zealand).

References

1. Crameri, A., Whitehorn, E. A., Tate, E., and Stemmer, W. P. C. (1996) Improved green fluorescent protein by molecular evolution using DNA shuffling. *Nat. Biotechnol.* **14,** 315–319.

2. Stemmer, W. P. C. (1994) Rapid evolution of a protein in vitro by DNA shuffling. *Nature* **370,** 389–391.
3. Stemmer, W. P. C. (1994) DNA shuffling by random fragmentation and reassembly: in vitro recombination for molecular evolution. *Proc. Natl. Acad. USA* **91,** 10,747–10,751.
4. Schmidt-Dannert, C., Umeno, D., and Arnold, F. H. (2000) Molecular breeding of carotenoid biosynthetic pathways. *Nature Biotechnol.* **18**, 750–753.
5. Kikuchi, M., Ohnishi, K., and Harayama, S. (1999) Novel family shuffling methods for the in vitro evolution of enzymes. *Gene* **236,** 159–167.
6. Morris, D. D., Gibbs, D. D., Chin, C. W., et al. (1998) Cloning of the xynB gene from *Dictyoglomus thermophilum* strain Rt46B.1 and action of the gene-product on kraft pulp. *Appl. Environ. Microbiol.* **64,** 1759–1765.
7. Gibbs, M. D., Nevalainen, K. M. H., and Bergquist, P. L. (2001) Degenerate oligonucleotide gene shuffling (DOGS): a method for enhancing the frequency of recombination with family shuffling. *Gene* **271,** 13–20.
8. Sakka, K., Kojima, Y., Kondo, T., Karita, S., Ohmiya, K., and Shimada, K. (1993) Nucleotide sequence of the *Clostridium stercorarium xynA* gene encoding xylanase A: identification of catalytic and cellulose binding domains. *Biosci. Biotechnol. Biochem.* **57,** 273–277.
9. Yang, V. W., Zhuang, Z., Elegir, G., and Jeffries, T. W. (1995) Alkaline-active xylanase produced by an alkaliphilic *Bacillus* sp isolated from kraft pulp. *J. Ind. Microbiol.* **15,** 434–441.
10. Morris, D. D., Gibbs, M. D., Ford, M., Thomas, J., and Bergquist, P. L. (1999) Family 10 and 11 xylanase genes from Caldicellulosiruptor isolate Rt69B.1. *Extremophiles* **3,** 103–111.
11. Fernandes, A. C., Fontes, C. M., Gilbert, H. J., Hazlewood, G. P., Fernandes, T. H., and Ferreira, L. M. (1999) Homologous xylanases from *Clostridium thermocellum*: evidence for bi-functional activity, synergism between xylanase catalytic modules and the presence of xylan-binding domains in enzyme complexes. *Biochem. J.* **342,** 105–110.
12. Elegir, G., Szakacs, G., and Jeffries, T. W. (1994) Purification, characterization and substrate specificity of multiple xylanases from *Streptomyces* sp strain B-12-2. *Appl. Environ. Microbiol.* **60,** 2609–2615.
13. Rose, T. M., Schultz, E. R., Henikoff, J. G., Pietrokovski, S., McCallum, C. M., and Henikoff, S. (1998) Consensus-degenerate hybrid oligonucleotide primers for amplification of distantly related sequences. *Nucleic Acids Res.* **26,** 1628–1635.
14. Thompson, J. D., Gibson, T. J., Plewniak, F., Jeanmougin, F., and Higgins, D. G. (1997) The ClustalX windows interface: flexible strategies for multiple sequence alignment aided by quality analysis tools. *Nucl. Acids Res.* **24,** 4876–4882.
15. Rice, P., Longden, I., and Bleasby, A. (2000) EMBOSS: The European Molecular Biology Open Software Suite. *Trends Genet.* **16,** 276–277.
16. Ho, S. N., Hunt, H. D., Horton, R. M., Pullen, J. K., and Pease, L. R. (1989) Site-directed mutagenesis by overlap extension using the polymerase chain reaction. *Gene* **77,** 51–59.

17. Teather, R. M. and Wood, P. J. (1982) Use of Congo Red polysaccharide interaction in enumeration and characterisation of cellulolytic bacteria from bovine rumen. *Appl. Environ. Microbiol.* **43,** 777–780.
18. Lever, M. (1973) Colorimetric and fluorometric carbohydrate determination with *p*-hydroxybenzoic acid hydrazide. *Biochem. Med.* **7,** 274–281.
19. Britton, H. T. S. and Robinson, R. A. (1931) Universal buffer solutions and the dissociation constant of veronal. *J. Chem. Soc.* **1,** 1456–1462.

12

M13 Bacteriophage Coat Proteins Engineered for Improved Phage Display

Sachdev S. Sidhu, Birte K. Feld, and Gregory A. Weiss

Summary

This chapter describes a method for increasing levels of protein fusions displayed on the surfaces of M13 bacteriophage particles. By introducing mutations into the anchoring M13 coat protein, protein display levels can be increased by up to two orders of magnitude. Experimental methods are presented for the design, construction, and screening of phage-displayed libraries for improved protein display.

Key Words: Phage display; protein engineering; combinatorial mutagenesis; M13 bacteriophage; major coat protein; viral evolution.

1. Introduction

Phage display is a powerful technology for engineering polypeptides that bind to target molecules of interest *(1–3)*. Proteins fused to phage coat proteins are displayed on phage surfaces and the encoding DNA is packaged within the phage particles *(4)*. The encapsulated DNA simplifies mutagenesis to synthesize libraries of phage-displayed polypeptides and enables ready identification of individual variants. From a library of displayed proteins, high-affinity binding proteins can be selected through in vitro binding to an immobilized target molecule, and the sequences of the selected proteins can be rapidly deduced by DNA sequencing.

Proteins have been displayed on M13 phage in a low-copy format through fusion to either the major coat protein (protein-8 [P8]) or the gene-3 minor coat protein (protein-3 [P3]) encoded by a phagemid vector, with wild-type (wt) P8 and P3 provided *in trans* from a helper phage *(5)*. Polyvalent protein display on P8 has been difficult to achieve, however, because the display levels vary with length and sequence of the fusion protein *(6)*. For example, if every copy of P8

From: *Methods in Molecular Biology, vol. 352: Protein Engineering Protocols*
Edited by: K. M. Arndt and K. M. Müller © Humana Press Inc., Totowa, NJ

is fused to the displayed peptide (a phage system, as opposed to a phagemid system), phages are generally unstable with fusion peptides more than six residues in length *(7)*. Even with a phagemid system, display is highly dependent on the size and nature of the fusion protein *(8)*. Thus, although monovalent phage display on either P3 or P8 has been used to affinity mature many different proteins from moderate to high affinity *(9)*, polyvalent display on P8 for selection of weak binding receptors from naïve libraries has not been practical for large proteins.

Here, we describe a general method for improving the display of proteins by mutating the anchoring P8 moiety. An improvement of almost 100-fold has been reported for two different proteins (streptavidin and human growth hormone [hGH]), and we have found the technique to also work well with other proteins. This dramatic improvement in display is likely because of better accommodation of the fusion protein in the phage coat. In theory, access to a high-copy display format should allow weaker protein functions to be selected initially from naïve libraries with much the same success as it has been used for small peptides. Equally important, the method could extend phage display to investigations of proteins that have previously proven intractable to phage display techniques. Thus, the methods described here should expand the potential for selecting new or modified protein functions by phage display.

2. Materials

2.1. Construction of Mutant P8 Libraries

2.1.1. Preparation of Uracil-Containing Single-Stranded DNA Template

1. 2YT medium: 10 g bacto-yeast extract, 16 g bacto-tryptone, and 5 g NaCl; add water to 1 L, and adjust pH to 7.0 with NaOH; autoclave.
2. 2YT/carb/cmp medium: 2YT, 50 μg/mL carbenicillin, and 5 μg/mL chloramphenicol.
3. 2YT/carb/kan/uridine medium: 2YT, 50 μg/mL carbenicillin, 25 μg/mL kanamycin, and 0.25 μg/mL uridine.
4. Carbenicillin: 5 mg/mL carbenicillin in water, filter sterilize.
5. Chloramphenicol: 50 mg/mL chloramphenicol in ethanol.
6. *Escherichia coli* CJ236 (New England Biolabs, Beverly, MA).
7. Kanamycin: 5 mg/mL kanamycin in water, filter sterilize.
8. Phosphate-buffered saline (PBS): 137 m*M* NaCl, 3 m*M* KCl, 8 m*M* Na_2HPO_4, and 1.5 m*M* KH_2PO_4; adjust pH to 7.2 with HCl; autoclave.
9. Polyethylene glycol (PEG)/NaCl: 20% PEG-8000 w/v and 2.5 *M* NaCl; autoclave.
10. QIAprep Spin M13 Kit (Qiagen, Valencia, CA).
11. M13KO7 helper phage (New England Biolabs).
12. Tris-acetate–EDTA (TAE) buffer: 40 m*M* Tris-acetate and 1 m*M* EDTA; adjust pH to 8.0; autoclave.
13. TAE/agarose gel: TAE buffer, 1% w/v agarose, and 1:5000 v/v 10% ethidium bromide.
14. Uridine: 25 mg/mL uridine in water, filter sterilize.

2.1.2. In Vitro Synthesis of Heteroduplex Covalently Closed Circular Double-Stranded DNA

1. 100 m*M* dithiothreitol (DTT).
2. 25 m*M* deoxyribonucleoside triphosphates: solution with 25 m*M* each of dATP, dCTP, dGTP, and dTTP (Amersham-Pharmacia, Piscataway, NJ).
3. 10 m*M* adenosine-5′-triphosphate (ATP; Amersham-Pharmacia).
4. 10X Tris-Mg (TM) buffer: 0.1 *M* $MgCl_2$ and 0.5 *M* Tris-HCl, pH 7.5.
5. QIAquick Gel Extraction Kit (Qiagen).
6. T4 polynucleotide kinase (New England Biolabs).
7. T4 DNA ligase (Invitrogen, Carlsbad, CA).
8. TAE/agarose gel (*see* **Subheading 2.1.1.**).
9. Ultrapure irrigation United States Pharmacopeia (USP) water (Braun Medical Inc., Irvine, CA).

2.1.3. E. Coli *Electroporation and Phage Propagation*

1. 2YT/carb medium: 2YT and 50 μg/mL carbenicillin.
2. Carbenicillin (*see* **Subheading 2.1.1.**).
3. Electrocompetent *E. coli* SS320 (*see* **Subheading 3.2.4.**).
4. 0.2-cm gap electroporation cuvet (BTX, San Diego, CA).
5. ECM-600 electroporator (BTX).
6. Kanamycin (*see* **Subheading 2.1.1.**).
7. Luria-Bertani (LB)/carb plates: LB agar and 50 μg/mL carbenicillin.
8. PBS (*see* **Subheading 2.1.1.**).
9. PEG/NaCl (*see* **Subheading 2.1.1.**).
10. SOC medium: 5 g bacto-yeast extract, 20 g bacto-tryptone, 0.5 g NaCl, and 0.2 g KCl; add water to bring up to 1.0 L and adjust pH to 7.0 with NaOH; autoclave; add 5.0 mL of autoclaved 2.0 *M* $MgCl_2$ and 20 mL of filter sterilized 1.0 *M* glucose.
11. M13KO7 helper phage (*see* **Subheading 2.1.1.**).

2.1.4. Preparation of Electrocompetent E. Coli *SS320*

1. 1.0 m*M* HEPES, pH 7.4: 4.0 mL of 1.0 *M* HEPES, pH 7.4, in 4.0 L of ultrapure irrigation USP water, filter sterilize.
2. 10% v/v ultrapure glycerol: 100 mL ultrapure glycerol in 900 mL ultrapure irrigation USP water, filter sterilize.
3. 2YT/tet medium: 2YT and 5 μg/mL tetracycline.
4. Magnetic stir bars (2 inch), soaked in ethanol.
5. Superbroth/tet medium: 24 g bacto-yeast extract, 12 g bacto-tryptone, and 5 mL glycerol; add water to 900 mL; autoclave; add 100 mL of autoclaved 0.17 *M* KH_2PO_4, 0.72 *M* K_2HPO_4, and 5 μg/mL tetracycline.
6. Tetracycline: 5 mg/mL tetracycline in water, filter sterilize.
7. Ultrapure glycerol (Invitrogen).
8. Ultrapure irrigation USP water (*see* **Subheading 2.1.2.**).

2.2. Selection and Analysis of P8 Variants That Increase Fusion Protein Display

2.2.1. Selection of Phage From the hGH-P8 Library

1. 0.2% bovine serum albumin (BSA) in PBS.
2. 100 m*M* HCl.
3. 1.0 *M* Tris-base.
4. 96-well maxisorp immunoplates (NUNC, Roskilde, Denmark).
5. *E. coli* XL-1 Blue (Stratagene, La Jolla, CA).
6. PBS-T buffer: PBS and 0.05% Tween-20.
7. PBS-T-BSA buffer: PBS, 0.05% Tween-20, and 0.2% BSA.
8. 2YT/carb/kan medium: 2YT, 50 μg/mL carbenicillin, and 25 μg/mL kanamycin.

2.2.2. Phage Enzyme-Linked Immunosorbent Assay Screen to Assess Levels of hGH Display

1. 1.0 *M* H_3PO_4.
2. 2YT/carb/kan medium (*see* **Subheading 2.2.1.**).
3. 3,3′,5,5′-tetramethylbenzidine/H_2O_2 peroxidase (TMB) substrate (Kirkegaard & Perry Laboratories Inc., Gaithersburg, MD).
4. 96-well maxisorp immunoplates (*see* **Subheading 2.2.1.**).
5. Carbenicillin (*see* **Subheading 2.1.1.**).
6. *E. coli* XL-1 Blue (*see* **Subheading 2.2.1.**).
7. Horseradish peroxidase/anti-M13 antibody conjugate (Amersham-Pharmacia).
8. Kanamycin (*see* **Subheading 2.1.1.**).
9. LB/tet plate: LB agar and 5 μg/mL tetracycline.
10. M13KO7 helper phage (*see* **Subheading 2.1.1.**).
11. PBS (*see* **Subheading 2.1.1.**).
12. PBS-T buffer (*see* **Subheading 2.2.1.**).
13. PBS-T-BSA buffer (*see* **Subheading 2.2.1.**).
14. PEG/NaCl (*see* **Subheading 2.1.1.**).
15. Tetracycline (*see* **Subheading 2.1.4.**).

3. Methods

The methods outlined below describe the design (**Subheading 3.1.**), construction (**Subheading 3.2.**), selection (**Subheading 3.3.1.**), and screening (**Subheading 3.3.2.**) of P8 libraries to identify mutations that improve protein display levels. We describe protocols for the improved display of hGH using the previously described phagemid, pS1607, but the methods are applicable to any protein of interest. The only modifications are the use of a phagemid designed to display the protein of interest in place of pS1607, and the use of a ligand that binds the protein of interest with high affinity in place of hGH binding protein (hGHbp).

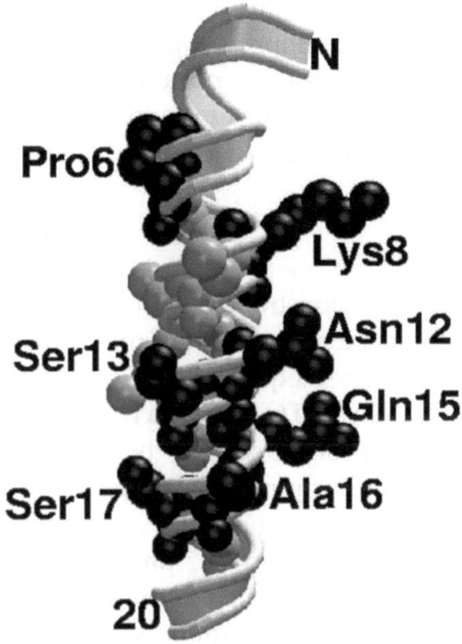

Fig. 1. P8, the major coat protein of filamentous bacteriophage. The first 20 residues of the P8 backbone are shown as a helical ribbon. Residues with wt side chains that are required for efficient incorporation into the phage coat are depicted in light gray (Ala 7, Ala9, Ala10, Phe11, Leu14, and Ala18). Residues targeted for mutagenesis to increase heterologous fusion protein display are colored black and labeled. The P8 shown here is from Ff bacteriophage, but it differs from the M13 P8 in only one position (Asn12 in M13 P8 is an Asp in Ff P8). Atomic coordinates were obtained from the Brookhaven Protein Data Bank (entry 1IFJ) and represented using Visual Molecular Dynamics ***(17)*** and RASTER-3D software ***(18)***.

3.1. Library Design

We have previously shown that, in the context of a phagemid system, the N-terminal half of P8 is extremely tolerant to mutations, and some mutations increase the display of heterologous fusion proteins ***(10,11)***. Subsequently, we showed that only six wt side chains (Ala7, Ala9, Ala10, Phe11, Leu14, and Ala18) within the N-terminal region are required for efficient incorporation of P8 into a wt phage coat ***(10,12)***. These side chains form a compact, hydrophobic epitope (**Fig. 1**) that apparently contributes key interactions required during phage assembly. Furthermore, mutations at sites surrounding this epitope can increase the efficiency of incorporation, and in so doing, increase levels of heterologous protein display ***(10)***. Based on these studies, we identified seven sites (Pro6, Lys8, Asn12, Ser13, Gln15, Ala16, and Ser17), at which mutations are most likely to result in improved protein display (**Fig. 1**). Complete randomization of seven sites within a protein results in approx 10^9 unique amino acid combinations, and this level of diversity can be readily covered by the library diversities (~10^{10}) that can be obtained with the methods described herein.

3.2. Library Construction

Libraries of P8 variants are constructed using an optimized version *(2)* of a previously described oligonucleotide-directed mutagenesis method *(13)*. First, a mutagenic oligonucleotide (sequence: GCCGAGGGTGACGATTAAG-CATAAGCGGCCTTTTAATAACTGTAATAATAAGCGACCGAATATATC) is used to introduce stop codons at the sites to be randomized; oligonucleotide-directed site-specific mutagenesis protocols are provided in **Subheading 3.2.2.** and can be adapted for small-scale mutagenesis to introduce the stop codons (*see* **Note 1**). The resulting "stop template" phagemid can be used as the template for library construction because the presence of stop codons eliminates wt protein display. Uracil-containing (dU) single-stranded (ss)DNA stop template (purified from an *E. coli* dut^-/ung^- host) is then annealed to a mutagenic oligonucleotide (sequence: GCCGAGGGTGACGATNNKGCANNKGCGGC-CTTTNNKNNKCTGNNKNNKNNKGCGACCGAATATATC) designed to replace the stop codons with NNK (N = A/G/C/T, 25% each; K = G/T, 50% each) degenerate codons that encode all 20 natural amino acids. The mutagenic oligonucleotide is used to prime the synthesis of a complementary DNA strand that is ligated to form a covalently closed circular (CCC), double-stranded (ds)DNA heteroduplex. To complete the library construction, the CCC-dsDNA heteroduplex is introduced into an *E. coli* dut^+/ung^+ host by electroporation, and the mismatch is repaired to either the wt or mutant sequence. In an ung^+ strain, the dU template strand is preferentially inactivated and the synthetic, mutant strand is replicated, thus, resulting in efficient mutagenesis (>50%). The use of a template with stop codons at all of the sites to be randomized ensures that only fully mutagenized clones contain open-reading frames that can be displayed on phage. The library members can be packaged into phage particles by coinfection of the *E. coli* host with a helper phage.

3.2.1. Purification of dU-ssDNA Template

Mutagenesis efficiency depends on template purity, and, thus, the use of high-purity dU-ssDNA is critical for successful library construction. We use the Qiagen QIAprep Spin M13 Kit for dU-ssDNA purification, and the following is a modified version of the Qiagen protocol. It yields at least 20 µg of dU-ssDNA for a medium copy number phagemid (e.g., pS1607, which contains a pBR322 backbone), and this is sufficient for the construction of one library (*see* **Note 2**).

1. From a fresh LB/antibiotic plate, pick a single colony of *E. coli* CJ236 (or another *dut⁻/ung⁻* strain) harboring the appropriate phagemid into 1 mL of 2YT medium supplemented with M13KO7 helper phage (10^{10} pfu/mL) and appropriate antibiotics to maintain the host F′ episome and the phagemid. For example, 2YT/carb/cmp medium contains carbenicillin to select for phagemids that carry the β-lactamase gene and chloramphenicol to select for the CJ236 F′ episome. Shake at 200 rpm and 37°C for 2 h and add kanamycin (25 μg/mL) to select for clones that have been coinfected with M13KO7, which carries a kanamycin resistance gene. Shake at 200 rpm and 37°C for 6 h and transfer the culture to 30 mL of 2YT/carb/kan/uridine medium. Shake overnight at 200 rpm and 37°C.
2. Centrifuge for 10 min at 27,000*g* and 4°C (15 krpm in a Sorvall SS-34 rotor). Transfer the supernatant to a new tube containing 1/5 volume of PEG/NaCl and incubate for 5 min at room temperature. Centrifuge 10 min at 12,000*g* and 4°C (10 krpm in an SS-34 rotor). Decant the supernatant. Centrifuge briefly at 2000*g* (4 krpm) and aspirate the remaining supernatant.
3. Resuspend the phage pellet in 0.5 mL of PBS and transfer to a 1.5-mL microcentrifuge tube. Centrifuge for 5 min at 14,000*g* in a microcentrifuge, and transfer the supernatant to a 1.5-mL microcentrifuge tube.
4. Add 7.0 μL of buffer MP (Qiagen) and mix. Incubate at room temperature for at least 2 min.
5. Apply the sample to a QIAprep spin column (Qiagen) in a 2-mL microcentrifuge tube. Centrifuge for 30 s at 6000*g* in a microcentrifuge. Discard the flow-through. The phage particles remain bound to the column matrix.
6. Add 0.7 mL of buffer MLB (Qiagen) to the column. Centrifuge for 30 s at 6000*g* and discard the flow-through.
7. Add another 0.7 mL buffer MLB. Incubate at room temperature for at least 1 min. Centrifuge at 6000*g* for 30 s. Discard the flow-through. The DNA is separated from the protein coat and remains adsorbed to the matrix.
8. Add 0.7 mL buffer PE (Qiagen). Centrifuge at 6000*g* for 30 s and discard the flow-through.
9. Repeat **step 8**. Residual proteins and salt are removed.
10. Centrifuge at 6000*g* for 30 s. Transfer the column to a fresh 1.5-mL microcentrifuge tube.
11. Add 100 μL of buffer EB (Qiagen; 10 m*M* Tris-HCl, pH 8.5) to the center of the column membrane. Incubate at room temperature for 10 min and centrifuge for 30 s at 6000*g*. Save the eluant, which contains the purified dU-ssDNA.
12. Analyze the DNA by electrophoresing 1.0 μL on a TAE/agarose gel. The DNA should appear as a predominant single band, but faint bands with lower electrophoretic mobility are often visible (lane 2 in **Fig. 2**). These are likely caused by secondary structure in the ssDNA.
13. Determine the DNA concentration by measuring absorbance at 260 nm ($A_{260} = 1.0$ for 33 ng/μL of ssDNA). Typical DNA concentrations range from 200 to 500 ng/μL.

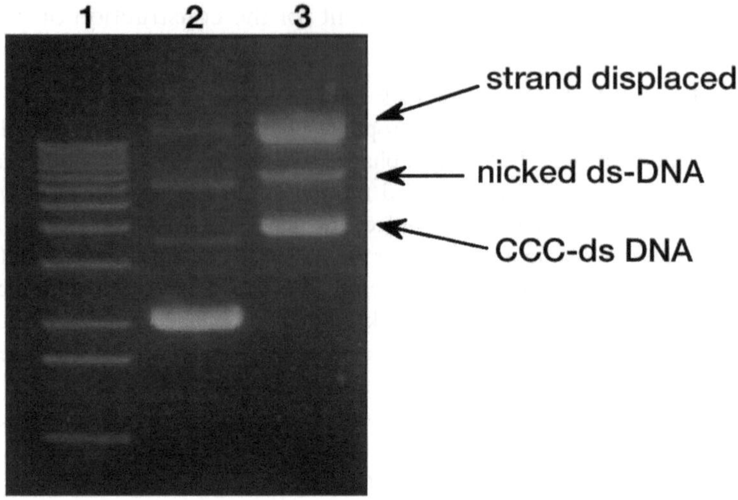

Fig. 2. In vitro synthesis of CCC-dsDNA heteroduplex. The reaction products were electrophoresed on a 1.0% TAE/agarose gel containing ethidium bromide for DNA visualization. Lane 1: 1-kb DNA marker (Gibco BRL); lane 2: the dU-ssDNA template; lane 3: reaction product from **Subheading 3.2.2.** The lower band is correctly extended and ligated CCC-dsDNA, the middle band is nicked dsDNA, and the upper band is strand-displaced dsDNA.

3.2.2. *In Vitro Synthesis of Heteroduplex CCC-dsDNA*

A three-step procedure is used to incorporate the mutagenic oligonucleotide into heteroduplex CCC-dsDNA, using dU-ssDNA as a template. The protocol described here is an optimized, large-scale version of a previously described method ***(13)***. The oligonucleotide is first 5′-phosphorylated (**Subheading 3.2.2.1.**) and then annealed to a dU-ssDNA template (**Subheading 3.2.2.2.**). The oligonucleotide is enzymatically extended and ligated to form heteroduplex CCC-dsDNA (lane 3 in **Fig. 2**), which is then purified and desalted (**Subheading 3.2.2.3.**). This protocol produces approx 20 μg of highly pure, low-conductance CCC-dsDNA. This is sufficient for the construction of a library containing more than 10^{10} unique members (*see* **Note 3**).

3.2.2.1. Oligonucleotide Phosphorylation With T4 Polynucleotide Kinase

1. In a 1.5-mL microcentrifuge tube, combine 0.6 μg of the mutagenic oligonucleotide, 2.0 μL of 10X TM buffer, 2.0 μL of 10 m*M* ATP, and 1.0 μL of 100 m*M* DTT. Add water to bring to a total volume of 20 μL.
2. Add 20 U of T4 polynucleotide kinase. Incubate for 1.0 h at 37°C (*see* **Note 4**).

3.2.2.2. Annealing of the Oligonucleotide to the Template

1. To the 20 µL of phosphorylation reaction mix, add 20 µg of dU-ssDNA template (from **Subheading 3.2.1.**), 25 µL of 10X TM buffer, and water to bring to a final volume of 250 µL. These DNA quantities provide an oligonucleotide to template molar ratio of 3:1, assuming that the oligonucleotide to template length ratio is 1:100.
2. Incubate at 90°C for 3 min, 50°C for 3 min, and 20°C for 5 min (*see* **Note 5**).

3.2.2.3. Enzymatic Synthesis of CCC-dsDNA

1. To the annealed oligonucleotide and template mixture, add 10 µL of 10 m*M* ATP, 10 µL of 25 m*M* dNTPs, 15 µL of 100 m*M* DTT, 30 Weiss U of T4 DNA ligase, and 30 U of T7 DNA polymerase.
2. Incubate overnight at 20°C.
3. Affinity purify, and desalt the DNA using the Qiagen QIAquick DNA Purification Kit. Add 1.0 mL of buffer QG (Qiagen) and mix.
4. Apply the sample to two QIAquick spin columns placed in 2-mL microcentrifuge tubes. Centrifuge at 14,000*g* for 1 min in a microcentrifuge. Discard the flow-through.
5. Add 750 µL of buffer PE (Qiagen) to each column. Centrifuge at 14,000*g* for 1 min. Discard the flow-through and centrifuge at 13 krpm for 1 min. Place the column in a new 1.5-mL microcentrifuge tube.
6. Add 35 µL of ultrapure irrigation USP water to the center of the membrane. Incubate at room temperature for 2 min (*see* **Note 6**).
7. Centrifuge at 14,000*g* for 1 min to elute the DNA. Combine the eluants from the two columns. The DNA can be used immediately for *E. coli* electroporation, or it can be frozen for later use.
8. Electrophorese 1.0 µL of the eluted reaction product alongside the dU-ssDNA template. Use a TAE/agarose gel with ethidium bromide for DNA visualization (**Fig. 2**; *see* **Note 7**).

A successful reaction results in the complete conversion of dU-ssDNA to dsDNA, which has a lower electrophoretic mobility. Usually, at least two product bands are visible and there should be no remaining dU-ssDNA (**Fig. 2**). The product band with the higher electrophoretic mobility represents the desired product: correctly extended and ligated CCC-dsDNA, which transforms *E. coli* efficiently and provides a high mutation frequency (~80%). The product band with the lower electrophoretic mobility is a strand-displaced product resulting from an intrinsic, unwanted activity of T7 DNA polymerase ***(14)***. Although the strand-displaced product provides a low mutation frequency (~20%), it also transforms *E. coli* at least 30-fold less efficiently than CCC-dsDNA. If a significant proportion of the ssDNA template is converted to CCC-dsDNA, a highly diverse library with a high mutation frequency will result. Sometimes a third band is visible, with an electrophoretic mobility between these two product bands (**Fig. 2**). This intermediate band is correctly extended but unligated dsDNA (nicked dsDNA), which results from either insufficient T4 DNA ligase activity or from incomplete oligonucleotide phosphorylation.

3.2.3. E. Coli *Electroporation and Phage Propagation*

To complete the library construction, the heteroduplex CCC-dsDNA must be introduced into an *E. coli* host that contains an F′ episome to enable M13 bacteriophage infection and propagation. Phage-displayed library diversities are limited by methods for introducing DNA into *E. coli*, with the most efficient method being high-voltage electroporation.

We have constructed an *E. coli* strain (SS320) that is ideal for both high efficiency electroporation and phage production ***(2)***. Using a standard bacterial mating protocol ***(15)***, we transferred the F′ episome from *E. coli* XL-1 Blue (Stratagene) to *E. coli* MC1061 (Bio-Rad). The progeny strain was readily selected for double resistance to streptomycin and tetracycline, because *E. coli* MC1061 carries a chromosomal marker for streptomycin resistance and the F′ episome from *E. coli* XL-1 Blue confers tetracycline resistance. *E. coli* SS320 retains the high electroporation efficiency of *E. coli* MC1061, and the presence of an F′ episome enables infection by M13 phage.

1. Chill the purified DNA (20 µg in a minimum volume, from **Subheading 3.2.2.3.**) and a 0.2-cm gap electroporation cuvette on ice. Thaw a 350-µL aliquot of electrocompetent *E. coli* SS320 on ice. Add the cells to the DNA and mix by pipetting several times (avoid introducing bubbles).
2. Transfer the mixture to the cuvet and electroporate. For electroporation, follow the manufacturer's instructions, preferably using a BTX ECM-600 electroporation system with the following settings: 2.5 kV field strength, 129 Ω resistance, and 50 µF capacitance. Alternatively, a BioRad Gene Pulser can be used with the following settings: 2.5 kV field strength, 200 Ω resistance, and 25 µF capacitance.
3. Immediately, rescue the electroporated cells by adding 1 mL SOC medium and transferring to a 250-mL baffled flask. Rinse the cuvette twice with 1 mL SOC medium. Add SOC medium to bring to a final volume of 25 mL and incubate for 20 min at 37°C with shaking at 200 rpm.
4. To determine the library diversity, plate serial dilutions on LB/carb plates to select for the library phagemid (in the case of β-lactamase-encoding phagemids, such as pS1607).
5. Add M13KO7 (4×10^{10} pfu/mL) and incubate for 10 min at 37°C with shaking at 200 rpm.
6. Transfer the culture to a 2-L baffled flask containing 500 mL of 2YT medium, supplemented with antibiotic for phagemid selection (e.g., 2YT/carb medium).
7. Incubate 1 h at 37°C with shaking at 200 rpm and add 25 µg/mL kanamycin. Incubate overnight at 37°C with shaking at 200 rpm.
8. Centrifuge the culture for 10 min at 16,000*g* and 4°C (10 krpm in a Sorvall GSA rotor). Transfer the supernatant to a fresh tube and add 1/5 volume of PEG/NaCl solution to precipitate the phage. Incubate for 5 min at room temperature.

9. Centrifuge for 10 min at 16,000*g* and 4°C in a GSA rotor. Decant the supernatant. Respin briefly and remove the remaining supernatant with a pipet. Resuspend the phage pellet in 1/20 volume of PBS.
10. Pellet insoluble matter by centrifuging for 5 min at 27,000*g* and 4°C (15 krpm in an SS-34 rotor). Transfer the supernatant to a clean tube.
11. Estimate the phage concentration spectrophotometrically (optical density [OD] at 268 nm [OD_{268}] = 1.0 for a solution of 5×10^{12} phage/mL; *see* **Note 8**).

3.2.4. Preparation of Electrocompetent E. coli *SS320*

The following protocol yields approx 12 mL of highly concentrated, electrocompetent *E. coli* SS320 (~3×10^{11} cfu/mL). The cells can be stored indefinitely at –70°C.

1. Inoculate 1 mL 2YT/tet medium with a single colony of *E. coli* SS320 from a fresh LB/tet plate. Incubate 6 to 8 h at 37°C with shaking at 200 rpm.
2. Transfer the culture to 500 mL of 2YT/tet medium in a 2-L baffled flask. Incubate overnight at 37°C with shaking at 200 rpm.
3. Inoculate six 2-L baffled flasks containing 900 mL of superbroth/tet medium with 5 mL of the overnight culture. Incubate at 37°C with shaking at 200 rpm to an OD_{550} of 0.8.
4. Chill three of the flasks on ice for 5 min with occasional swirling. The following steps (**steps 5–12**) should be performed in a cold room, on ice, with prechilled solutions and equipment.
5. Centrifuge at 5000*g* (5.5 krpm) and 4°C for 10 min in a Sorvall GS-3 rotor. Decant the supernatant and add culture from the remaining flasks (these should be chilled while the first set is being centrifuged) to the same tubes. Repeat the centrifugation and decant the supernatant.
6. Fill the tubes with 1.0 m*M* HEPES, pH 7.4, and add sterile magnetic stir bars to facilitate pellet resuspension. Swirl to dislodge the pellet from the tube wall and stir at a moderate rate to resuspend the pellet completely.
7. Centrifuge at 5000*g* (5.5 krpm) and 4°C for 10 min in a Sorvall GS-3 rotor. Decant the supernatant, being careful to retain the stir bar. To avoid disturbing the pellet, maintain the position of the centrifuge tube when removing the tube from the rotor.
8. Fill the tubes with 1.0 m*M* HEPES, pH 7.4. Resuspend the pellet and repeat the centrifugation as in **step 7**. Decant the supernatant.
9. Resuspend each pellet in 150 mL of 10% ultrapure glycerol. Do not combine the pellets.
10. Centrifuge at 5000*g* (5.5 krpm) and 4°C for 15 min in a Sorvall GS-3 rotor. Decant the supernatant and remove the stir bar. Remove remaining traces of supernatant with a pipet.
11. Add 3.0 mL of 10% ultrapure glycerol to one tube and resuspend the pellet by pipetting. Transfer the suspension to the next tube and repeat until all of the pellets are resuspended.
12. Flash-freeze 350-μL aliquots with liquid nitrogen and store at –70°C.

3.3. Selection and Analysis of P8 Variants That Increase Fusion Protein Display

3.3.1. Selection of Phage From the hGH-P8 Library

Phage from the hGH-P8 library described above are cycled through rounds of binding selection with hGHbp *(16)* coated on 96-well Maxisorp immunoplates as the capture target. Phage are propagated in *E. coli* XL-1 Blue with M13KO7 helper phage for further rounds of selection.

3.3.1.1. Coat Wells With Target Protein

1. Coat 8 wells of a 96-well Maxisorp plate with 100 μL of a 5 μg/mL solution of the target protein (e.g., hGHbp) overnight at 4°C. Discard the solution by emptying the plate contents into the sink.
2. To block nonspecific binding to the microtiter wells, add 200 μL of 0.2% BSA in PBS to each well. Shake at room temperature for 1 h.

3.3.1.2. Phage Library Selection

1. Add a solution of the phage library (~10^{12} phage/mL) in PBS-T-BSA buffer to the wells. Shake at room temperature for 2 h. Wash the plate eight times with PBS-T buffer. The stringency of the binding selection can be increased for successive rounds by increasing the number of washes.
2. Elute bound phage, by adding 100 μL of 100 m*M* HCl to each well. Shake vigorously for 5 min at room temperature.
3. Combine the eluants and neutralize with 1/5 volume of 1.0 *M* Tris-base.

3.3.1.3. Propagate Phage for Further Rounds of Selection

1. Add the eluted phage to 10 volumes of XL-1 Blue cells (OD_{550} = 0.5–1.0).
2. Shake at 200 rpm and 37°C for 20 min and remove 10 μL for titers, as described in **Subheading 3.2.3., step 4**.
3. Add M13KO7 helper phage and shake at 200 rpm and 37°C for 45 min.
4. Transfer the culture to 100 mL 2YT/carb/kan medium and shake overnight at 200 rpm and 37°C.
5. Isolate the phage by PEG/NaCl precipitation, as described in **Subheading 3.2.3, steps 8** through **10**.
6. Repeat the selection process five times, using only half of the eluted phage in each round.

3.3.2. Phage Enzyme-Linked Immunosorbent Assay Screen to Assess Levels of hGH Display

After selection for hGH display (**Subheading 3.3.1.**), individual clones are analyzed in a phage enzyme-linked immunosorbent assay *(2,11)* to assess levels of hGH display relative to display with wt P8. Serial dilutions of hGH-P8 phage solutions are incubated in wells containing immobilized hGHbp as the capture target. After washing to remove unbound phage, bound phage are detected spectrophotometrically (at 450 nm), using a reaction catalyzed by a horseradish peroxidase/M13 antibody conjugate. An increase in hGH display is indicated by an increase in the efficiency of hGH-P8 phage capture. From a plot of the phage concentration vs absorbance at 450 nm, the increase in hGH display with a P8 variant relative to that with wt P8 can be estimated by comparing the phage concentrations required to generate a particular absorbance at 450 nm ***(11)***. Clones exhibiting high hGH display are subjected to DNA sequence analysis to determine the sequences of the P8 variants within the hGH-P8 fusions.

1. From a fresh LB/tet plate, pick a single colony of *E. coli* XL-1 Blue harboring the appropriate phagemid into 1 mL of 2YT medium supplemented with M13KO7 helper phage (10^{10} pfu/mL), 50 µg/mL carbenicillin (for phagemid maintenance), and 5 µg/mL tetracycline (for F′ episome maintenance). Shake at 200 rpm and 37°C for 2 h and add 25 µg/mL kanamycin to select for clones that have been coinfected with M13KO7. Shake at 200 rpm and 37°C for 6 h and transfer the culture to 30 mL of 2YT/carb/kan medium. Shake overnight at 200 rpm and 37°C.
2. Centrifuge for 10 min at 27,000*g* and 4°C (15 krpm in a Sorvall SS-34 rotor). Transfer the supernatant to a new tube containing 1/5 volume of PEG/NaCl and incubate for 5 min at room temperature. Centrifuge 10 min at 12,000*g* and 4°C (10 krpm in an SS-34 rotor). Decant the supernatant. Centrifuge briefly at 2000*g* (4 krpm) and aspirate the remaining supernatant.
3. Resuspend the phage pellet in 0.5 mL of PBS-T-BSA buffer and transfer to a 1.5-mL microcentrifuge tube. Centrifuge for 5 min at 14,000*g* in a microcentrifuge, and transfer the supernatant to a 1.5-mL microcentrifuge tube.
4. Determine the phage concentration spectrophotometrically ($\varepsilon_{268} = 1.2 \times 10^8$ /*M*/cm).
5. Prepare fivefold serial dilutions of phage stock, using PBS-T-BSA buffer.
6. Transfer 100 µL of phage solution to Maxisorp immunoplates coated with hGHbp and blocked with BSA (*see* **Subheading 3.3.1.1.**). Incubate for 1 h with gentle shaking.
7. Remove the phage solution and wash eight times with PBS-T buffer.
8. Add 100 µL of horseradish peroxidase/M13 antibody conjugate (diluted 3000-fold in PBS-T-BSA buffer). Incubate for 30 min with gentle shaking.
9. Wash eight times with PBS-T buffer and two times with PBS.
10. Develop the wells with 100 µL of TMB substrate. Stop the reaction with 100 µL of 1.0 *M* H_3PO_4, and read spectrophotometrically at 450 nm in a microtiter plate reader.

4. Notes

1. For mutagenesis to introduce the stop codons, the protocols in **Subheadings 3.2.2.2** and **3.2.2.3** can be scaled down 10-fold. The reaction product can be used directly to transform *E. coli* using any standard procedure.
2. The protocol can be scaled up by inoculating a larger overnight culture and purifying the ssDNA with multiple spin columns.

3. All steps can be scaled-up considerably, with the possible exception of the annealing step. The annealing protocol described here works well with volumes of 250 μL or less. It may also work for larger volumes, but we have not tested this.
4. The phosphorylated oligonucleotide should be used immediately in the subsequent steps.
5. To scale up this reaction, run multiple annealing reactions of 250 μL each.
6. Improved yields of eluted DNA can be obtained by incubating the column at 37°C for 5 min, after the addition of the water.
7. The electrophoretic mobility of circular DNA depends on salt concentrations, pH, and the presence of ethidium bromide. To observe the relative mobilities shown in **Fig. 2**, the DNA must be electrophoresed on a TAE/agarose gel with ethidium bromide added directly to the molten gel rather than to the running buffer.
8. Use the library immediately or add glycerol to a final concentration of 10% and store at –80°C. Some displayed proteins can be denatured by freezing, which could render them unusable for selections. In general, it is best to use libraries immediately.

References

1. Smith, G. P. and Petrenko, V. A. (1997) Phage display. *Chem. Rev.* **97,** 391–410.
2. Sidhu, S. S., Lowman, H. B., Cunningham, B. C., and Wells, J. A. (2000) Phage display for selection of novel binding peptides. *Methods Enzymol.* **328,** 333–363.
3. Sidhu, S. S. (2000) Phage display in pharmaceutical biotechnology. *Curr. Opin. Biotechnol.* **11,** 610–616.
4. Smith, G. P. (1985) Filamentous fusion phage: novel expression vectors that display cloned antigens on the virion surface. *Science* **228,** 1315–1317.
5. Bass, S., Greene, R., and Wells, J. A. (1990) Hormone phage: an enrichment method for variant proteins with altered binding properties. *Proteins* **8,** 309–314.
6. Malik, P., Terry, T. D., Gowda, L. R., et al. (1996) Role of capsid structure and membrane protein processing in determining the size and copy number of peptides displayed on the major coat protein of filamentous bacteriophage. *J. Mol. Biol.* **260,** 9–21.
7. Iannolo, G., Minenkova, O., Petruzzelli, R., and Cesareni, G. (1995) Modifying filamentous phage capsid: limits in the size of the major capsid protein. *J. Mol. Biol.* **248,** 835–844.
8. Kretzschmar, T. and Geiser, M. (1995) Evaluation of antibodies fused to minor coat protein III and major coat protein VIII in bacteriophage M13. *Gene* **155,** 61–65.
9. Clackson, T. and Wells, J. A. (1994) In vitro selection from protein and peptide libraries. *Trends Biotechnol.* **12,** 173–184.
10. Weiss, G. A., Wells, J. A., and Sidhu, S. S. (2000) Mutational analysis of the major coat protein of M13 identifies residues that control protein display. *Protein Sci.* **9,** 647–654.
11. Sidhu, S. S., Weiss, G. A., and Wells, J. A. (2000) High copy display of large proteins on phage for functional selections. *J. Mol. Biol.* **296,** 487–495.
12. Roth, T. A., Weiss, G. A., Eigenbrot, C., and Sidhu, S. S. (2002) A minimized M13 coat protein defines the minimum requirements for assembly into the bacteriophage particle. *J. Mol. Biol.* **322,** 357–367.

13. Kunkel, T. A., Roberts, J. D., and Zakour, R. A. (1987) Rapid and efficient site-specific mutagenesis without phenotypic selection. *Methods Enzymol.* **154,** 367–382.
14. Lechner, R. L., Engler, M. J., and Richardson, C. C. (1983) Characterization of strand displacement synthesis catalyzed by bacteriophage T7 DNA polymerase. *J. Biol. Chem.* **258,** 1174–1184.
15. Miller, J. H. (1972) *Experiments in Molecular Biology* 1st ed., Cold Spring Harbor Laboratory Press, Cold Spring Harbor, NY, p. 190.
16. Fuh, G., Mulkerrin, M. G., Bass, S., et al. (1990) The human growth hormone receptor: secretion from Escherichia coli and disulfide bonding pattern of the extracellular binding domain. *J. Biol. Chem.* **265,** 3111–3115.
17. Humphrey, W., Dalke, A., and Schulten, K. (1996) VMD: visual molecular dynamics. *J. Mol. Graph.* **14,** 27–28, 33–38.
18. Esnouf, R. M. (1997) An extensively modified version of MolScript that includes greatly enhanced coloring capabilities. *J. Mol. Graph. Model.* **15,** 112–113, 132–134.

13

Ribosome-Inactivation Display System

Satoshi Fujita, Jing-Min Zhou, and Kazunari Taira

Summary

We present a novel strategy for the connection of phenotype and genotype in vitro that can be used for the selection of functional proteins. The strategy involves the generation of a stable complex among a ribosome, an messenger RNA and its translated protein, without removal of the termination codon, as a result of the action of the ricin A chain during translation. The technique requires no transfection, no chemical synthesis, no ligation, and no removal of the termination codon. Thus, our novel ribosome-inactivation display system should provide, without loss of the pool population, a reliable, simple, and robust selection system for the in vitro evolution of the properties of proteins in a predictable direction by a combination of randomization and appropriate selection strategies.

Key Words: Ribosome-inactivation display system; in vitro selection; molecular evolution; protein; ribosome display; mRNA display; phage display; ricin A chain; protein–ribosome–mRNA complex.

1. Introduction

During the past decade, several display strategies have provided powerful and efficient techniques for the selection and evolution of peptides and proteins. In these techniques, the coupling of genotype and phenotype is the single most critical determinant. In such selection systems, the specific sequence information (genotype) of members of libraries that encode the selected protein (phenotype) can be determined from the corresponding DNA/RNA that was introduced into the system. The gene encoding the selected protein can then be reamplified for further evolution and analysis. The strategies that have been successfully developed are either cell-dependent, involving, for example, display on the surface of phage *(1)*, other viruses *(2)*, bacteria *(3)* or yeast *(4)*; or they are cell-free, as in the case of ribosome display *(5–7)* and messenger RNA (mRNA) display *(8,9)* systems.

From: *Methods in Molecular Biology, vol. 352: Protein Engineering Protocols*
Edited by: K. M. Arndt and K. M. Müller © Humana Press Inc., Totowa, NJ

The methods that have been developed to date are, however, associated with certain limitations and disadvantages. Because cell-dependent display systems ***(1–4)*** include a necessary in vivo step, the sizes and diversity of sequence libraries are limited by the efficiency of transformation and by the nature of the protein in question. For example, some proteins that are detrimental to cells or that have important regulatory functions within cells cannot be selected. In the ribosome-display system ***(5–7)***, the ribosome forms a stable complex with the translated protein and the mRNA that encodes the protein, because the release of the protein from the complex is slowed by the removal of a termination codon, the addition of $Mg(OAc)_2$ and anti-*ssrA* oligonucleotides ***(6,7)***. However, the yield of the isolated mRNAs after one round of ribosome display is not very high. The alternative, cell-free mRNA display procedure ***(8,9)*** requires careful chemical synthesis and critical purification of puromycin-attached oligonucleotides, which must be ligated to the 3′-end of each mRNA in the sequence libraries. Failure to perform these manipulations appropriately leads to a reduction in the diversity of available libraries.

To solve these problems and to maintain the diversity of sequence libraries, we developed a new strategy that allows us to prepare a protein–ribosome–mRNA ternary complex that is significantly more stable than that obtained by the conventional method (**Fig. 1**). Stabilization was achieved by introducing the gene for a toxin that inactivates eukaryotic ribosomes, the ricin A chain (RTA), downstream of the region of the sequence library that would be translated into a protein library (**Fig. 1**). Ricin, which is composed of A and B chains, is a plant toxin that inhibits protein synthesis by inactivating ribosomes. The B chain is required for the internalization of ricin into cells, and the RTA, which is the active moiety, catalyzes the hydrolysis of a specific *N*-glycosidic bond adjacent to the universally conserved adenosine that is found in a GAGA tetra-loop in 23S–28S ribosomal RNAs (rRNAs) in the large subunit of eukaryotic ribosomes ***(10,11)***. This single depurination, which alters the binding site for elongation factors, inhibits protein synthesis and stalls ribosomes on the translation complex without release of the mRNA or the translated protein ***(12–14)***. Translation is stalled before the termination codon has been reached, and, thus, the recruitment of release factors to the ribosome does not occur. Consequently, the mRNA and the nascent protein remain bound to the ribosome very stably; not just for hours, but for days. In particular, because RTA is able to inactivate approx 1500 ribosomes per minute ***(15)***, extremely stable protein–ribosome–mRNA complexes are rapidly formed under standard translation conditions. Therefore, a key aspect of ribosome-inactivation display system (RIDS) is that the ribosome is stalled on an mRNA via an entirely different mechanism from that of the original ribosome display system. Using this method, we should be able to screen functional proteins without removing the termination codon, under standard translation conditions.

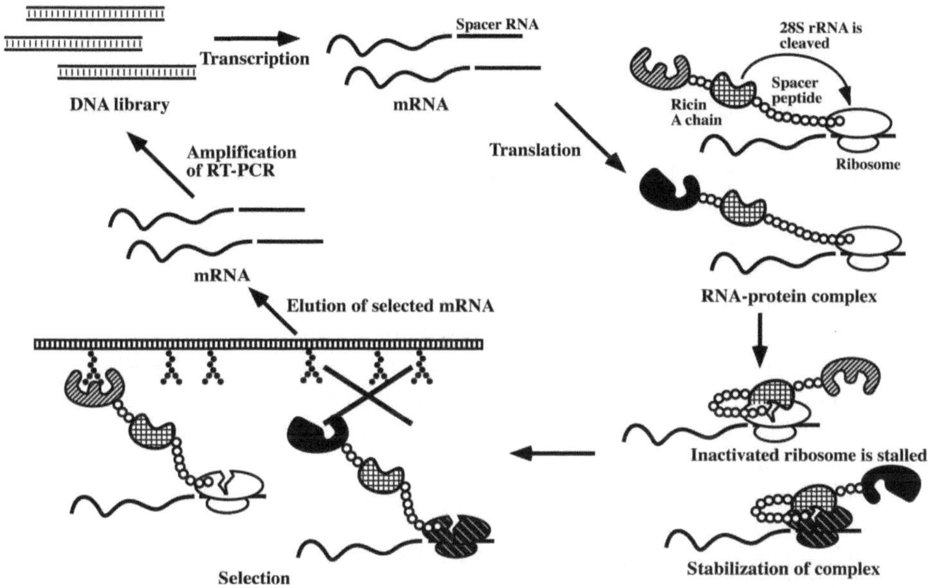

Fig. 1. Schematic representation of the RIDS for screening functional proteins in vitro. Step 1, transcription: the gene for RTA, inserted downstream of a random protein library (or cDNA library) is transcribed by T7 RNA polymerase to yield mRNA. Step 2, translation: mRNA is translated in a rabbit reticulocyte lysate system. Step 3, inactivation: during translation, because the rRNA is inactivated by folded RTA in a *cis* reaction, the ribosome is stalled and a ribosome–mRNA–protein complex is formed. Step 4, selection: the complex of interest is bound to the corresponding affinity matrix. Unwanted complexes are removed by washing. Step 5, elution: the specific complex is dissociated from the matrix by elution with a buffer that contains ethylenediaminetetraacetic acid, and free mRNA is isolated. Step 6, amplification: eluted RNA is amplified by RT-PCR and the resultant cDNA is used for the next cycle or for analysis by cloning and sequencing.

To evaluate the potential usefulness of RIDS, we chose streptavidin and glutathione-*S*-transferase (GST) as target proteins and isolated the respective proteins and their mRNAs using appropriate matrices that contained biotin and glutathione, respectively, as the ligands. We found that it was possible, using RIDS, to screen the functional proteins (streptavidin and GST) without reducing the diversity of the sequence library and without any chemical synthesis. If a complementary DNA (cDNA) library or random DNA library is introduced into the RIDS instead of the model protein (streptavidin or GST), we will be able to select the functional protein.

2. Materials

1. pET30a (Novagen, Darmstadt, Germany).
2. cDNA library or random DNA library.

3. Streptavidin and *GST* genes.
4. *RTA* gene.
5. *Gene III* gene of fd phage derived from pCANTAB 5E (Amersham Biosciences, Piscataway, NJ).
6. Oligonucleotide primers and linkers.
7. T7 Ampliscribe™ kit (Epicentre Technologies, Madison, WI).
8. Cap-structure analog (New England Biolabs, England).
9. DNase I (Epicentre Technologies, Madison, WI).
10. RNeasy™ mini kit (QIAGEN, Hilden, Germany).
11. Flexi® Rabit Reticulocyte Lysate system (Promega, Madison, WI).
12. Recombinant RNasin® Ribonuclease Inhibitor (Promega).
13. Binding (or washing) buffer: 140 m*M* NaCl, 2.7 m*M* KCl, 10.1 m*M* Na_2HPO_4, 1.8 m*M* KH_2PO_4, 0.04% Tween-20, 2% Block Ace™ (Dainippon Pharmaceutical Co.; Japan), and 5 m*M* $MgCl_2$, pH 7.3. Stable at –20°C for up to 1 mo.
14. Ligand beads for the selection.
15. Biotin agarose (Sigma, St. Louis, MO) and glutathione sepharose 4B (Amersham Biosciences).
16. Elution buffer: 1 *M* NaCl and 50 m*M* EDTA.
17. ReverTra Dash® (Toyobo, Japan).
18. KOD Dash® (Toyobo).

3. Methods

The methods described here outline the construction of the expression plasmid and the cycle of RIDS for the selection of functional protein.

3.1. The Construction of Expression Plasmid for RIDS

First, DNA constructs encoding the T7 promoter, protein library, linker, RTA, and spacer should be prepared to construct RIDS (**Fig. 2A**), as described in **Subheadings 3.1.1.** to **3.1.3.** Although it is possible, theoretically, to prepare the double-stranded DNA sequence using a polymerase chain reaction (PCR) method, we recommend making the DNA construct using plasmids, because it may be difficult to connect various motifs (namely, the T7 promoter, protein library, linker, RTA, and spacer) using PCR without nonspecific amplification. In this section, we describe the procedure of plasmid construction for RIDS. DNA manipulations were performed using standard recombinant DNA methods to construct the plasmid, and are not described here in detail because of space limitations.

3.1.1. Construction of the pLRS Expression Plasmid

At first, we prepared the universal plasmid, pLRS, encoding the T7 promoter, linker, RTA, and spacer as the DNA construct for RIDS (**Fig. 2A**). The plasmid is based on the expression system derived from the pET30a plasmid (Novagen). We can construct the protein library for RIDS easily by insertion of a cDNA library or random library at the *Xba*I/*Nhe*I sites in the pLRS plasmid.

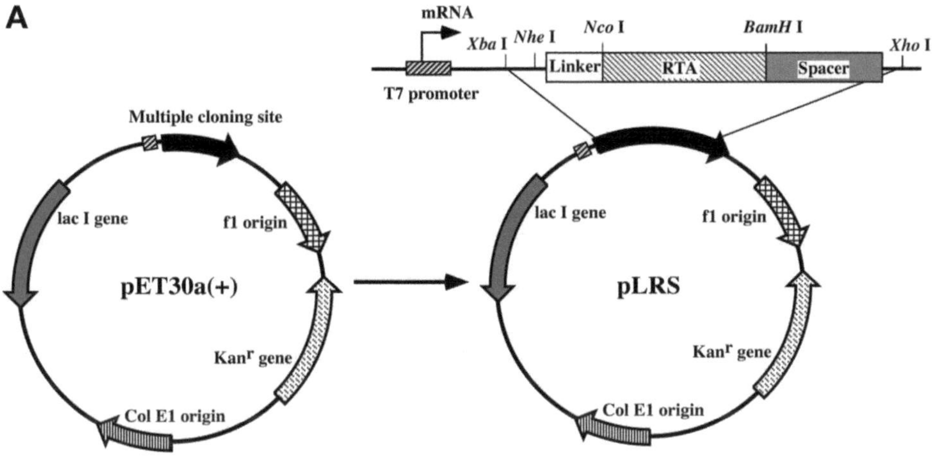

Fig. 2. *(Continued)*

The linker fragment, encoding *Nhe*I, the glycine/serine-rich sequence of *gene III* of M13, *Nco*I, *Bam*HI, *Pst*I, and *Xho*I in sequence from upstream, was synthesized by a DNA synthesizer and was inserted into pET30a plasmid by *Xba*I/*Xho*I downstream of the T7 promoter. DNA coding for RTA was prepared by PCR from pUTA (a gift from Prof. J. Robertus, University of Texas) and inserted downstream of the glycine/serine-rich sequence by *Nco*I/*Bam*HI. DNA coding for the spacer was derived from *gene III* of M13 and inserted downstream of RTA by *Bam*HI/*Pst*I. The sequence of the DNA construct, without inserting the protein library for RIDS, is shown in **Fig. 2B**.

The T7 promoter and the 71-mer sequence between the promoter and the first ATG of the protein library were derived from the sequence of pET30a. A flexible linker upstream of RTA was indispensable to allow nascent proteins to fold into their natural three-dimensional structures to eliminate steric hindrance between the protein library and the downstream RTA. Previous investigations demonstrated that the yield of a selected protein/mRNA is strongly dependent on the length, composition, and sequence of such a linker ***(6,16,17)***. In this study, we used a glycine/serine-rich fragment of 44 amino acids as the linker.

It is important to eliminate the steric hindrance of RTA, because, in the RIDS, stabilization of the mRNA–ribosome–protein complex was achieved by introducing the gene for RTA. In addition to the linker, a spacer at the 3′-terminal end of the open-reading frame (ORF) was also important. The spacer, which was required as an anchor to tether the ribosome, must be of appropriate length so that it can occupy the long tunnel of the ribosome ***(18,19)*** and allow the nascent RTA to fold correctly without any steric hindrance ***(20,21)***. It has been reported that a spacer of at least 20 to 30 amino acids at the carboxyl terminus is required

B

```
       T7 promoter                                   Xba I
GCGAAATTAATACGACTCACTATAGGGGAATTGTGAGCGGATAACAATTCCCCTCTAGAA

                                      Nhe I             <---
ATAATTTTGTTTAACTTTAAGAAGGAGATATACATATGGCTAGCCCAGTAAACGCAGGAG
                                   M  A  S  P  V  N  A  G  G

-----------------------Linker-------------------------------
GAGGAAGCGGCGGAGGTAGCGGAGGTGGAAGCGAAGGCGGAGGGTCAGAGGGGGGTGGTT
  G  S  G  G  G  S  G  G  G  S  E  G  G  G  S  E  G  G  G  S

------------------------------------------------------------
CCGACGGCGGTGGAAGCGAAGGAGGCGGTAGCGAAGGTGGAGGGAGCGGAGGAGGCAGTG
  D  G  G  G  S  E  G  G  G  S  E  G  G  G  S  G  G  G  S  G

------->            Nco I <-------------Ricin A chain-------
GCAGCGGAGACTTCGACTACCCATGGATGATATTTCCCAAACAATACCCAATTATAAACT
  S  G  D  F  D  Y  P  W  M  I  F  P  K  Q  Y  P  I  I  N  F

------------------------------------------------------------
TTACCACAGCGGGTGCCACTGTGCAAAGCTACACAAACTTTATCAGAGCTGTTCGCGGTC
  T  T  A  G  A  T  V  Q  S  Y  T  N  F  I  R  A  V  R  G  R

------------------------------------------------------------
GTTTAACAACTGGAGCTGATGTGAGACATGAAATACCAGTGTTGCCAAACAGAGTTGGTT
  L  T  T  G  A  D  V  R  H  E  I  P  V  L  P  N  R  V  G  L

------------------------------------------------------------
TGCCTATAAACCAACGGTTTATTTTAGTTGAACTCTCAAATCATGCAGAGCTTTCTGTTA
  P  I  N  Q  R  F  I  L  V  E  L  S  N  H  A  E  L  S  V  T

------------------------------------------------------------
CATTAGCGCTGGATGTCACCAATGCATATGTGGTAGGCTACCGTGCTGGAAATAGCGCAT
  L  A  L  D  V  T  N  A  Y  V  V  G  Y  R  A  G  N  S  A  Y

------------------------------------------------------------
ATTTCTTTCATCCTGACAATCAGGAAGATGCAGAAGCAATCACTTATCTTTTCACTGATG
  F  F  H  P  D  N  Q  E  D  A  E  A  I  T  Y  L  F  T  D  V

------------------------------------------------------------
TTCAAAATCGATATACATTCGCCTTTGGTGGTAATTATGATAGACTTGAACAACTTGCTG
  Q  N  R  Y  T  F  A  F  G  G  N  Y  D  R  L  E  Q  L  A  G

------------------------------------------------------------
GTAATCTGAGAGAAAATATCGAGTTGGGAAATGGTCCACTAGAGGAGGCTATCTCAGCGC
  N  L  R  E  N  I  E  L  G  N  G  P  L  E  E  A  I  S  A  L

------------------------------------------------------------
TTTATTATTACAGTACTGGTGGCACTCAGCTTCCAACTCTGGCTCGTTCCTTTATAATTT
  Y  Y  Y  S  T  G  G  T  Q  L  P  T  L  A  R  S  F  I  I  C
```

Fig. 2. *(Continued)*

```
------------------------------------------------------------
GCATCCAAATGATTTCAGAAGCAGCAAGATTCCAATATATTGAGGGAGAAATGCGCACGA
 I  Q  M  I  S  E  A  A  R  F  Q  Y  I  E  G  E  M  R  T  R

------------------------------------------------------------
GAATTAGGTACAACCGGAGATCTGCACCAGATCCTAGCGTAATTACACTTGAGAATAGTT
 I  R  Y  N  R  R  S  A  P  D  P  S  V  I  T  L  E  N  S  W

------------------------------------------------------------
GGGGGAGACTTTCCACTGCAATTCAAGAGTCTAACCAAGGAGCCTTTGCTAGTCCAATTC
 G  R  L  S  T  A  I  Q  E  S  N  Q  G  A  F  A  S  P  I  Q

------------------------------------------------------------
AACTGCAAAGACGTAATGGTTCCAAATTCAGTGTGTACGATGTGAGTATATTAATCCCTA
 L  Q  R  R  N  G  S  K  F  S  V  Y  D  V  S  I  L  I  P  I

--------------------Ricin A chain---------------->
TCATAGCTCTCATGGTGTATAGATGCGCGCCTCCACCGAGCTCACAGTTTGGGATATCGT
 I  A  L  M  V  Y  R  C  A  P  P  P  S  S  Q  F  G  I  S  S

                                    <-----------------------
                                   BamH I
CGACCGGAGGAGGAGGTGGCGGCGGTGGCGCTGCATCGGATCCATTCGTTTGTGAATATC
 T  G  G  G  G  G  G  G  G  A  A  S  D  P  F  V  C  E  Y  Q

-----Spacer (gene III)--------------------------------------
AAGGCCAATCGTCTGACCTGCCTCAACCTCCTGTCAATGCTGGCGGCGGCTCTGGTGGTG
 G  Q  S  S  D  L  P  Q  P  P  V  N  A  G  G  G  S  G  G  G

------------------------------------------------------------
GTTCTGGTGGCGGCTCTGAGGGTGGCGGCTCTGAGGGTGGCGGTTCTGAGGGTGGCGGCT
 S  G  G  G  S  E  G  G  G  S  E  G  G  G  S  E  G  G  G  S

------------------------------------------------------------
CTGAGGGTGGCGGTTCCGGTGGCGGCTCCGGTTCCGGTGATTTTGATTATGAAAAAATGG
 E  G  G  G  S  G  G  G  S  G  S  G  D  F  D  Y  E  K  M  A

------------------------------------------------------------
CAAACGCTAATAAGGGGGCTATGACCGAAAATGCCGATGAAAACGCGCTACAGTCTGACG
 N  A  N  K  G  A  M  T  E  N  A  D  E  N  A  L  Q  S  D  A

------------------------------------------------------------
CTAAAGGCAAACTTGATTCTGTCGCTACTGATTACGGTGCTGCTATCGATGGTTTCATTG
 K  G  K  L  D  S  V  A  T  D  Y  G  A  A  I  D  G  F  I  G

----------------Spacer (gene III)-------->                Xho
GTGACGTTTCCGGCCTTGCTAATGGTAATGGTGCTACTGGTCTGCAGGTTAAGAATTCTC
 D  V  S  G  L  A  N  G  N  G  A  T  G  L  Q  V  K  N  S  R

 I
GAGCACCACCACCACCACCAC
```

Fig. 2. (**A**) Schematic drawing of the pET30a(+) plasmid and the pLRS plasmid. We can construct a protein library for RIDS easily by insertion of a cDNA library or random library at the *Xba*I/*Nhe*I sites in the pLRS plasmid. (**B**) DNA sequence of pLRS used for the RIDS constructs. The sequence shows the T7 promoter, linker, RTA, and spacer.

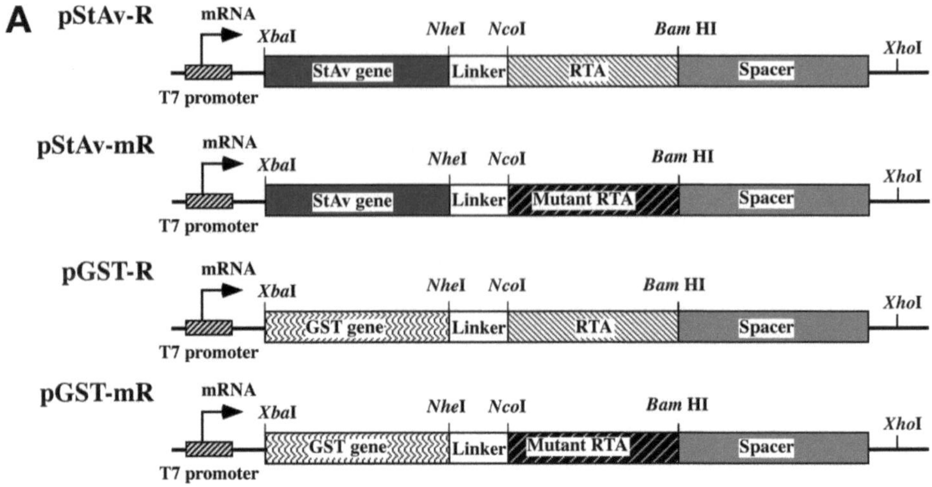

Fig. 3. *(Continued)*

to retain the activity of an enzyme displayed on a ribosome ***(22,23)***. Thus, the amount of mRNA isolated by the ribosome display system is influenced by the length of the spacer and the secondary structure of its 3′-end ***(6,18,24)***. We introduced a long and a short spacer sequence, separately, at the 3′-terminal end of the ORF and compared the effects of the two spacers on translation and selection. We found that the long spacer (of 404 amino acids, encoded by full-length *gene III*, in its entirety, from the filamentous phage M13) was not suitable for translation and selection in our system. Therefore, in all of our experiments, we used a fragment of 121 amino acids as the spacer.

3.1.2. Construction of pStAv-R and pGST-R Plasmids

To confirm the validity of the proposed method and the potential usefulness of RIDS, we introduced the gene for streptavidin or *GST* into a DNA sequence library as the model study (**Fig. 3A**). These proteins have frequently been fused to newly discovered proteins and/or the molecules with which they interact for the successful functional characterization of such newly discovered, fused proteins ***(25–28)***. Streptavidin and GST bind to biotin and glutathione, respectively, with high affinity and specificity. Thus, it should be possible to isolate and characterize the protein–ribosome–mRNA ternary complex with ease.

DNA encoding for residues of streptavidin and GST were excised from the plasmids, pSTA (a gift from Prof. M. Sisido, Okayama University) and pGEX-4T-3 (Amersham Biosciences), respectively. These fragments were amplified by PCR and ligated separately into the *Xba*I/*Nhe*I sites in the pLRS plasmid. The sequences of the DNA construct encoding streptavidin or GST are shown in **Fig. 3B,C**.

B

```
        T7 promoter                                       Xba I
GCGAAATTAATACGACTCACTATAGGGGAATTGTGAGCGGATAACAATTCCCCTCTAGAA

ATAATTTTGTTTAACTTTAAGAAGGAGATATACATATGGCGAGCATGACTGGTGGACAGC
                                       M  A  S  M  T  G  G  Q  Q

                           <--------Streptavidin gene-------------
AAATGGGTACCGAATTCCATATGGACCCGTCCAAGGACTCCAAAGCTCAGGTTTCTGCAG
  M  G  T  E  F  H  M  D  P  S  K  D  S  K  A  Q  V  S  A  A

------------------------------------------------------------
CCGAAGCTGGTATCACTGGCACCTGGTATAACCAACTGGGGTCGACTTTCATTGTGACCG
  E  A  G  I  T  G  T  W  Y  N  Q  L  G  S  T  F  I  V  T  A

------------------------------------------------------------
CTGGTGCGGACGGAGCTCTGACTGGCACCTACGAATCTGCGGTTGGTAACGCAGAATCCC
  G  A  D  G  A  L  T  G  T  Y  E  S  A  V  G  N  A  E  S  R

------------------------------------------------------------
GCTACGTACTGACTGGCCGTTATGACTCTGCACCTGCCACCGATGGCTCTGGTACCGCTC
  Y  V  L  T  G  R  Y  D  S  A  P  A  T  D  G  S  G  T  A  L

------------------------------------------------------------
TGGGCTGGACTGTGGCTTGGAAAAACAACTATCGTAATGCGCACAGCGCCACTACGTGGT
  G  W  T  V  A  W  K  N  N  Y  R  N  A  H  S  A  T  T  W  S

------------------------------------------------------------
CTGGCCAATACGTTGGCGGTGCTGAGGCTCGTATCAACACTCAGTGGCTGTTAACATCCG
  G  Q  Y  V  G  G  A  E  A  R  I  N  T  Q  W  L  L  T  S  G

------------------------------------------------------------
GCACTACCGAAGCGAATGCATGGAAATCGACACTAGTAGGTCATGACACCTTTACCAAAG
  T  T  E  A  N  A  W  K  S  T  L  V  G  H  D  T  F  T  K  V

-------------------------------------------Streptavidin gene-
TTAAGCCTTCTGCTGCgAGCATTGATGCTGCCAAGAAAGCAGGCGTAAACAACGGTAACC
  K  P  S  A  A  S  I  D  A  A  K  K  A  G  V  N  N  G  N  P

------------------->Nhe I
CTCTTGACGCTGTTCAGCAAGCTAGC
  L  D  A  V  Q  Q
```

Fig. 3. *(Continued)*

3.1.3. Construction of pStAv-mR and pGST-mR Plasmids

To confirm that RTA helps to maintain a stable ribosome complex, in a control experiment, we made a gene for an inactive, mutant RTA by site-specific mutagenesis, changing functional amino acids as follows: glutamic acid 177 to glutamine; arginine 180 to histidine; and glutamic acid 208 to aspartic acid. Mutation of these three amino acids in RTA completely abolished its activity

C

```
     T7 promoter                                       Xba I
GCGAAATTAATACGACTCACTATAGGGGAATTGTGAGCGGATAACAATTCCCCTCTAGAA

                                   <---GST gene-------------
ATAATTTTGTTTAACTTTAAGAAGGAGATATACATATGTCCCCTATACTAGGTTATTGGA
                                    M  S  P  I  L  G  Y  W  K

------------------------------------------------------------
AAATTAAGGGCCTTGTGCAACCCACTCGACTTCTTTTGGAATATCTTGAAGAAAAATATG
 I  K  G  L  V  Q  P  T  R  L  L  L  E  Y  L  E  E  K  Y  E

------------------------------------------------------------
AAGAGCATTTGTATGAGCGCGATGAAGGTGATAAATGGCGAAACAAAAAGTTTGAATTGG
 E  H  L  Y  E  R  D  E  G  D  K  W  R  N  K  K  F  E  L  G

------------------------------------------------------------
GTTTGGAGTTTCCCAATCTTCCTTATTATATTGATGGTGATGTTAAATTAACACAGTCTA
 L  E  F  P  N  L  P  Y  Y  I  D  G  D  V  K  L  T  Q  S  M

------------------------------------------------------------
TGGCCATCATACGTTATATAGCTGACAAGCACAACATGTTGGGTGGTTGTCCAAAAGAGC
 A  I  I  R  Y  I  A  D  K  H  N  M  L  G  G  C  P  K  E  R

------------------------------------------------------------
GTGCAGAGATTTCAATGCTTGAAGGAGCGGTTTTGGATATTAGATACGGTGTTTCGAGAA
 A  E  I  S  M  L  E  G  A  V  L  D  I  R  Y  G  V  S  R  I

------------------------------------------------------------
TTGCATATAGTAAAGACTTTGAAACTCTCAAAGTTGATTTTCTTAGCAAGCTACCTGAAA
 A  Y  S  K  D  F  E  T  L  K  V  D  F  L  S  K  L  P  E  M

------------------------------------------------------------
TGCTGAAAATGTTCGAAGATCGTTTATGTCATAAAACATATTTAAATGGTGATCATGTAA
 L  K  M  F  E  D  R  L  C  H  K  T  Y  L  N  G  D  H  V  T

------------------------------------------------------------
CCCATCCTGACTTCATGTTGTATGACGCTCTTGATGTTGTTTTATACATGGACCCAATGT
 H  P  D  F  M  L  Y  D  A  L  D  V  V  L  Y  M  D  P  M  C

------------------------------------------------------------
GCCTGGATGCGTTCCCAAAATTAGTTTGTTTTAAAAAACGTATTGAAGCTATCCCACAAA
 L  D  A  F  P  K  L  V  C  F  K  K  R  I  E  A  I  P  Q  I

------------------------------------------------------------
TTGATAAGTACTTGAAATCCAGCAAGTATATAGCATGGCCTTTGCAGGGCTGGCAAGCCA
 D  K  Y  L  K  S  S  K  Y  I  A  W  P  L  Q  G  W  Q  A  T

------------------GST gene----------->Nhe I
CGTTTGGTGGTGGCGACCATCCTCCAAAATCGGATCTGGCTAGC
 F  G  G  G  D  H  P  P  K  S  D  L
```

Fig. 3. **(A)** Schematic drawing of the DNA constructs used for the model study of RIDS. Streptavidin and *GST* genes were inserted between *Xba*I and *Nhe*I in the pLRS

(29,30). To construct pStAv-mR and pGST-mR, the gene for RTA in each plasmid (pStAV-R or pGST-R) was replaced by a mutant gene for RTA by standard procedures ***(31)*** with the following primers: 5′-TTG CAT CCA AAT GAT TTC ACA AGC AGC ACA CTT CCA ATA TAT TAG GGA GAA ATG-3′ (underlining indicates mutations E177A and R180H) and 5′-GAT CCT AGC GTA ATT ACA CTT GAC GAT CCT AGC GTA ATT ACA CTT GA-3′ (underlining indicates mutation E208D).

3.2. The Cycle of RIDS for the Selection of Functional Protein

Digest the modified pLRS containing the sequence of the random library or cDNA library (in our model study, pStAv-R, pGST-R, pStAv-mR, and pGST-mR) in the 3′-terminal region of the ORF by *Xho*I to yield linear DNA by standard recombinant DNA methods. The digestion is indispensable for the termination of the transcription.

3.2.1. Transcription of the DNA Library to mRNA

DNAs should be transcribed by T7 RNA polymerase with a cap-structure analog. In our model study, DNAs were transcribed by a T7 Ampliscribe™ kit, according to the supplier's recommendation (Epicentre Technologies Co.).

1. Add 1 μg of linear DNA to 20 μL of the transcription mixture including a 3 m*M* cap-structure analog, 7.5 m*M* rATP, rCTP, and rUTP; 0.75 m*M* rGTP; and 10 m*M* dithiothreitol, reaction buffer, and AmpliScribe T7 Enzyme Solution, according to the supplier's recommendation.
2. Incubate the reaction mixture for 2 to 4 h at 37°C to produce each mRNA (we call respective mRNAs: StAv-R, pGST-R, pStAv-mR, and GST-mR).
3. After the transcriptional reaction, digest the template DNA by DNase I completely, because residual template DNA influences the selection step.
4. Purify mRNA with an RNeasy™ mini kit, according to the supplier's recommendation (QIAGEN).

3.2.2. Affinity Selection of the Ribosome–mRNA–Protein Complex and Isolation of mRNA

Prepare the RNA-encoding library (in our model study, StAv-R, StAv-mR, GST-R, and GST-mR) and Flexi® Rabbit Reticulocyte Lysate system (Promega; *see* **Note 1**).

Fig. 3. *(Continued)* plasmid to construct pStAv-R and pGST-R. To confirm that RTA helps to maintain a stable ribosome complex, in a control experiment, we made pStAv-mR and pGST-mR encoding mutant (inactive) RTA by site-specific mutagenesis. **(B)** DNA sequence of pStAv-R used for the RIDS constructs. **(C)** DNA sequence of pGST-R used for the RIDS constructs.

1. Add 2 μg mRNA to 40 μL of the translation mixture including 33 μL of Flexi® Rabbit Reticulocyte Lysate, 40 m*M* KCl, 40 μ*M* total amino acid mixture, and 40 U Recombinant RNasin® Ribonuclease Inhibitor (Promega), and adjust the volume of the solution to 50 μL with distilled water. Do not add dithiothreitol and Mg^{2+} ion (*see* **Note 1**).
2. Incubate for 20 min at 30°C. In our study, three sets of mRNA were translated (2 μg of each set: a mixture of mRNAs in the ratio of 1:1 or each mRNA encoding streptavidin or GST [StAv-R and GST-R] as a control; *see* **Note 2**).
3. After translation, add 1 mL of the appropriate binding buffer (*see* **Heading 2., item 13**) and ligand-immobilized beads. In our study, we add 10 μL of a suspension of biotin–agarose or glutathione beads (*see* **Note 3**).
4. Incubate for 1 h at 4°C with gentle rotation for the binding reaction. Although the incubation can be performed at room temperature, we recommend performing the reaction at 4°C, because nonspecific binding was sometimes detected in our study at reverse transcriptase (RT)-PCR (*see* **Note 4**).
5. After the binding reaction, centrifuge the mixture for 1 min at 500*g* to precipitate the ligand-immobilized beads.
6. Remove supernatant containing unbound mRNA and protein.
7. Add 200 μL of washing buffer (*see* **Subheading 2., item 13**) and mix gently for 30 s.
8. Centrifuge the mixture for 1 min at 500*g* to precipitate ligand-immobilized beads.
9. Remove supernatant containing unbound mRNA and protein.
10. Repeat the wash **steps 7** to **9** two to three times.
11. Add 100 μL of elution buffer (*see* **Subheading 2, item 16**) and shake vigorously at room temperature for 30 min to isolate the bound mRNA from ligand-immobilized beads.
12. Purify eluted mRNA with the RNeasy™ kit, according to the supplier's recommendation, and concentrate purified mRNA by vacuum pump.

3.2.3. Reverse Transcription

Perform the reverse transcription reaction using RT. In our model study, the reverse transcription reaction was performed using ReverTra Ace® (Toyobo), according to the supplier's recommendation.

1. Mix the purified mRNA, 4 μL of 10 μ*M* RT primer, 2 μL of each 10 m*M* dNTP, and 4 μL of 5X RT buffer. Adjust the volume of the solution to 18 μL with distilled water, without adding RNase inhibitor and ReverTra Ace, according to the supplier's recommendation. An RT primer (5′-GTG TAG CTT TGC ACA GTG GC-3′) that recognizes the upstream portion of the RTA sequence was used.
2. Denature the mixture at 65°C for 5 min and chill on ice immediately.
3. Add 1 μL of RNase inhibitor and 1 μL of ReverTra Ace to the mixture.
4. Incubate at 42°C for 1 h for the reverse transcription reaction. To inactivate the enzyme, incubate at 99°C for 5 min.

3.2.4. Polymerase Chain Reaction

Perform PCR using thermostable DNA polymerase. In our model study, PCR was performed by KOD Dash® (Toyobo) according to the supplier's recommendation.

1. Mix 20 μL cDNA, 10 μL of 10X PCR buffer, 10 μL of each 2 m*M* dNTP, 1 μL of 10 μ*M* RT primer (as the down primer), and 2 μL of 10 μ*M* (as the up primer). Adjust the volume of the solution to 99 μL with distilled water, without adding KOD Dash. The up primer (5′-AAT TTT GTT TAA CTT TAA GAA GGA G-3′) that recognizes the downstream portion of the T7 promoter sequence was used.
2. Denature the mixture at 98°C for 2 min and chill on ice immediately.
3. Add 0.5 μL of KOD Dash.
4. Perform PCR cycles (1 min at 98°C, followed by 15 to 30 cycles of 10 s at 98°C, 2 s at 55°C, and 30 s at 72°C; and finished by 10 min at 72°C). Incubate at 98°C for 1 min.

4. Notes

1. The Flexi® Rabbit Reticulocyte Lysate system provides flexibility of reaction conditions in comparison with standard rabbit reticulocyte lysate systems. The reducing agents, including dithiothreitol, should be omitted for the display of proteins containing disulfide bridges. The concentrations of Mg^{2+} and K^+ ions were optimized by referring to the ribosome display method ***(32)***.
2. We occasionally detected a high background in the affinity selection of GST by glutathione sepharose. However, we did not detect background in the affinity selection of GST by antibody against GST instead of glutathione sepharose. The result may show that several GST displayed on ribosomes cannot dimerize because of steric hindrance, because only GST dimers recognize glutathione.
3. The ribosomal complexes are stabilized by adding 5 to 50 m*M* of the $MgCl_2$.
4. Although we succeeded in the selection of GST or streptavidin from the pool of the mixture of GST and streptavidin at room temperature, we occasionally detected nonspecific binding. It seems that the ribosomal complexes are stabilized by chilling the translation mixture. Surprisingly, by adding $MgCl_2$ to the binding buffer, and chilling the translation mixture, a small amount of StAv-mR mRNA can bind to biotin agarose despite the presence of a stop codon. It seems that an mRNA–ribosome–protein complex is formed in the ribosome display system with a stop codon.

Acknowledgments

The authors thank Prof. J. D. Robertus of the University of Texas for the kind gift of plasmid pUTA, and Prof. M. Sisido and Dr. T. Hosaka of the University of Okayama for plasmid pGSH. The authors also thank members of the Taira laboratory, especially Dr. Loura Nelson, for helpful comments and discussions. This research was supported by grants from Promotion of Basic Research Activities for

Innovative Biosciences and by grants from the Ministry of Economy, Trade and Industry of Japan. S. F. is the recipient of a Japan Society for the Promotion of Science research fellowship. We recently developed a seemingly more effective selection system by exploiting the interaction between a tandemly fused MS2 coat-protein (MSp) dimer and the RNA sequence of the corresponding specific binding motif, C-variant (Cv), which increase the stability of the mRNA–ribosome–protein complex allowing selection at room temperature. We call this system an Advanced Ribosome-display with Strengthened Association (ARiSA) system ***(33–35)***.

References

1. Winter, G., Griffiths, A. D., Hawkins, R. E., and Hoogenboom, H. R. (1994) Making antibodies by phage display technology. *Annu. Rev. Immunol.* **12,** 433–455.
2. Kasahara, N., Dozy, A. M., and Kan, Y. M. (1994) Tissue-specific targeting of retroviral vectors through ligand-receptor interactions. *Science* **266,** 1373–1376.
3. Georgiou, G., Poetschke, H. L., Stathopoulos, C., and Francisco, J. A. (1993) Practical applications of engineering Gram-negative bacterial cell surfaces. *Trends Biotechnol.* **11,** 6–10.
4. Kieke, M. C., Cho, B. K., Boder, E. T., Kranz, D. M., and Wittrup, K. D. (1997) Isolation of anti-T cell receptor scFv mutants by yeast surface display. *Protein Eng.* **10,** 1303–1310.
5 Mattheakis, L. C., Bhatt, R. R., and Dower, W. J. (1994) An in vitro polysome display system for identifying ligands from very large peptide libraries. *Proc. Natl. Acad. Sci. USA* **91,** 9022–9026.
6. Hanes, J. and Plückthun, A. (1997) In vitro selection and evolution of functional proteins by using ribosome display. *Proc. Natl. Acad. Sci. USA* **94,** 4937–4942.
7. He, M. Y., Menges, M., Groves, M. A. T., et al. (1999) Selection of a human anti-progesterone antibody fragment from a transgenic mouse library by ARM ribosome display. *J. Immunol. Methods* **231,** 105–117.
8. Roberts, R. W. and Szostak, J. W. (1997) RNA-peptide fusions for the in vitro selection of peptides and proteins. *Proc. Natl. Acad. Sci. USA* **94,** 12,297–12,302.
9. Nemoto, N., Miyamoto-Sato, E., Husimi, Y., and Yanagawa, H. (1997) In vitro virus: Bonding of mRNA bearing puromycin at the 3′-terminal end to the C-terminal end of its encoded protein on the ribosome in vitro. *FEBS Lett.* **414,** 405–408.
10. Endo, Y. and Tsurugi, K. (1987) RNA *N*-glycosidase activity of ricin a chain. Mechanism of action of the toxic lectin ricin on eukaryotic ribosomes. *J. Biol. Chem.* **262,** 8128–8130.
11. Endo, Y., Mitsui, K., Motizuki, M., and Tsuruki, K. (1987) The mechanism of action of ricin and related toxic lectins on eukaryotic ribosomes. The site and the characteristics of the modification in 28S ribosomal RNA caused by the toxins. *J. Biol. Chem.* **262,** 5908–5912.
12. Moazed, D., Robertson, J., and Noller, H. (1989) Interaction of elongation factors EF-G and EF-Tu with a conserved loop in 23S rRNA. *Nature* **334,** 362–364.

13. Kudlicki, W., Kitaoka, Y., Odom, O. W., Kramer, G., and Hardesty, B. (1995) Elongation and folding of nascent ricin chains as peptidyl-tRNA on ribosomes: the effect of amino acid deletions on these processes. *J. Mol. Biol.* **252,** 203–212.
14. Munishkin, A. and Wool, I. G. (1997) The ribosome-in-pieces: binding of elongation factor EF-G to oligoribonucleotides that mimic the sarcin/ricin and thiostrepton domains of 23S ribosomal RNA. *Proc. Natl. Acad. Sci. USA* **94,** 12,280–12,284.
15. Eiklid, K., Olsnes, S., and Pihl, A. (1980) Entry of lethal doses of abrin, ricin, and modeccin into the cytosol of HeLa cells. *Exp. Cell Res.* **126,** 321–326.
16. Mössner, E., Koch, H., and Plückthun, A. (2001) Fast selection of antibodies without antigen purification: adaptation of the protein fragment complementation assay to select antigen-antibody pairs. *J. Mol. Biol.* **308,** 115–122.
17. Liu, R., Barrick, J. E., Szostak, J. W., and Roberts, R. W. (2000) Optimized synthesis of RNA-protein fusions for in vitro protein selection. *Methods Enzymol.* **318,** 268–293.
18. Malkin, L. I. and Rich, A. (1967) Partial resistance of nascent polypeptide chains to proteolytic digestion due to ribosomal shielding. *J. Mol. Biol.* **26,** 329–346.
19. Smith, W. P., Tai, P. C., and Davis, B. D. (1978) Interaction of secreted nascent chains with surrounding membrane in bacillus subtilis. *Proc. Natl. Acad. Sci. USA* **75,** 5922–5925.
20. Komar, A. A., Kommer, A., Krasheninnikov, I. A., and Spirin, A. S. (1997) Cotranslational folding of globin. *J. Biol. Chem.* **272,** 10,646–10,651.
21. Fedorov, A. N. and Baldwin, T. O. (1997) Cotranslational protein folding. *J. Biol. Chem.* **272,** 32,715–32,718.
22. Kudlicki, W., Chirgwin, J., Kramer, G., and Hardesty, B. (1995) Folding of an enzyme into an active confirmation while bound as a peptidyl-tRNA to the ribosome. *Biochemistry* **34,** 14,284–14,287.
23. Makeyev, E. V., Kolb, V. A., and Spirin, A. S. (1996) Enzymatic activity of the ribosome-bound nascent polypeptide. *FEBS Lett.* **378,** 166–170.
24. Schaffitzel, C., Hanes, J., Jermutus, L., and Plückthun, A. (1999) Ribosome display: an in vitro method for selection and evolution of antibodies from libraries. *J. Immunol. Methods* **231,** 119–135.
25. Wilson, D. S., Keefe, A. D., and Szostak, J. W. (2001) The use of mRNA display to select high-affinity protein-binding peptides. *Proc. Natl. Acad. Sci. USA* **98,** 3750–3755.
26. Doi, N. and Yanagawa, H. (1999) STABLE: protein-DNA fusion system for screening of combinatorial protein libraries in vitro. *FEBS Lett.* **457,** 227–230.
27. Cheadle, C., Ivashchenko, Y., South, V., et al. (1994) Identification of a Src SH3 domain binding motif by screening a random phage display library. *J. Biol. Chem.* **269,** 24,034–24,039.
28. Gram, H., Schmitz, R., Zuber, J. F., and Baumann, G. (1997) Identification of phosphopeptide ligands for the Src-homology 2 (SH2) domain of Grb2 by phage display. *Eur. J. Biochem.* **246,** 633–637.
29. Frankel, A., Welsh, P., Richardson, J., and Robertus, J. D. (1990) Role of arginine 180 and glutamic acid 177 of ricin toxin a chain in enzymatic inactivation of ribosomes. *Mol. Cell Biol.* **10,** 6257–6263.

30. Kim, Y., Mlsa, D., Monzingo, A. F., Ready, M. P., Frankel, A., and Robertus, J. D. (1992) Structure of a ricin mutant showing rescue of activity by a noncatalytic residue. *Biochemistry* **31,** 3294–3296.
31. Sambrook, J., Fritsch, E. F., and Maniatis, T. (1989) *Molecular Cloning: A Laboratory Manual.* Cold Spring Harbor Laboratory Press, Cold Spring Harbor, NY.
32. Hanes, J., Jermutus, L., and Plückthun, A. (2000) Selecting and evolving functional proteins in vitro by ribosome display. *Methods Enzymol.* **328,** 404–430.
33. Sawata, Y. S., Wada, A., and Taira, K. (2003) An advanced ribosome-display with strengthened association (*ARiSA*) for *in vitro* selection of a peptide aptamer with strong affinity. Manuscript in preparation.
34. Sawata, S. Y., and Taira, K. (2003) Modified peptide selection *in vitro* by introduction of a protein-RNA interaction. *Protein Eng.* **16,** 1115–1124.
35. Sawata, S. Y., Suyama, E., and Taira, K. (2004) A system based on specific protein-RNA interactions for analysis of target protein-protein interactions *in vitro*: successful selection of membrane-bound Bak-Bcl-xL proteins *in vitro*. *Protein Eng. Des. Sel.* **17,** 501–508.

14

Compartmentalized Self-Replication

A Novel Method for the Directed Evolution of Polymerases and Other Enzymes

Farid J. Ghadessy and Philipp Holliger

Summary

Compartmentalized self-replication (CSR) is a novel method for the directed evolution of enzymes and, in particular, polymerases. In its simplest form, CSR consists of a simple feedback loop involving a polymerase that replicates only its own encoding gene (self-replication). Self-replication occurs in discrete, spatially separate, noncommunicating compartments formed by a heat-stable water-in-oil emulsion. Compartmentalization ensures the linkage of phenotype and genotype (i.e., it ensures that each polymerase replicates only its own encoding gene to the exclusion of those in the other compartments). As a result, adaptive gains by the polymerase directly (and proportionally) translate into genetic amplification of the encoding polymerase gene. CSR has proven to be a useful strategy for the directed evolution of polymerases directly from diverse repertoires of polymerase genes. In this chapter, we describe some of the CSR protocols used successfully to evolve variants of *T. aquaticus* Pol I (*Taq*) polymerase with novel and useful properties, such as increased thermostability or resistance to the potent inhibitor, heparin, from a repertoire of randomly mutated *Taq* polymerase genes.

Key Words: Compartmentalized self-replication; directed evolution; in vitro selection; emulsion; *Taq* polymerase.

1. Introduction

The ability to self-replicate, to faithfully copy the genome, is a defining characteristic of all life. In present day organisms, this fundamental process is carried out by members of a diverse class of enzymes: the polynucleotide polymerases *(1)*. Apart from genome replication, polynucleotide polymerases perform a range of core functions within the cell, including transcription, DNA repair, and telomere maintenance. Recently, it was found that specialized polymerases are also involved in a range of diverse processes, ranging from adaptive mutation and antibody

From: *Methods in Molecular Biology, vol. 352: Protein Engineering Protocols*
Edited by: K. M. Arndt and K. M. Müller © Humana Press Inc., Totowa, NJ

affinity maturation to protection against DNA damage by ultraviolet radiation *(2)*, and, possibly, RNA interference *(3)*. Aberrant polymerase function has been implicated in the pathogenesis of cancer *(4)* and many polymerases, in particular, viral polymerases, are important drug targets. Finally, polymerases have been central to the development of modern biology, enabling DNA sequencing, polymerase chain reaction (PCR), site-directed mutagenesis, and complementary DNA cloning, and are also crucial for emerging technologies, such as molecular computing and nanobiotechnology. We reasoned that a better understanding of polymerase function may, therefore, not only provide insights into fundamental cellular processes, but may also enable novel applications in biotechnology and potentially accelerate the design of antiviral drugs.

Although great progress has been made in the understanding of polymerase function through the pioneering structural studies of T. Steitz and others, and through careful biochemical and kinetic analysis of wild-type and mutant polymerases *(5)*, the ability to alter polymerase properties in a predictable manner or to tailor polymerases for existing or novel applications has lagged behind.

1.1. Polymerase Engineering by Design

Attempts have been made to alter polymerase function by the use of protein engineering. For example, variants of *Taq* polymerase (Stoffel fragment and Klentaq) have been generated by full or partial deletion of its 5′ to 3′ exonuclease domain; these show improved thermostability and fidelity, although at the cost of reduced processivity *(6,7)*. In addition, the availability of high-resolution structures has allowed the rational design of mutants with improved properties (for example, *Taq* mutants with improved properties of dideoxynucleotide incorporation for cycle sequencing; **ref.** ***8***). Site-directed mutagenesis has also yielded polymerase variants with an increased capability to incorporate ribonucleotides *(9,10)*, reduced pausing *(11)*, as well as numerous polymerases with altered fidelity *(12)*. Grafting of the thioredoxin-binding loop of T7 DNA polymerase onto the *Escherichia coli* Pol I Klenow fragment lead to an impressive increase in processivity *(13)*.

1.2. Polymerase Engineering by Repertoire Selection

Genetic approaches have also been used for polymerase design. For example, Loeb and co-workers have selected active mutant polymerases by complementation of a *polA12*, *recA718* bacterial strain with *Taq* polymerase *(14)*, but also with human immunodeficiency virus reverse transcriptase *(15)* and human pol β *(16)*. Complementation selection was used to probe the mutability of the polymerase active site, and screening of the selected mutants has yielded polymerase variants with a range of properties, including reduced fidelity or an increased capability to incorporate ribonucleotides. Recently, the same approach was used to select for pol η function in a *RAD30, RAD52* yeast strain, which yielded a mutant of pol η with increased activity ***(17)***.

Phage display technology has been a highly successful method for repertoire selection, in particular, for the directed evolution of molecular interactions, allowing, for example, the isolation of peptide hormone mimics or specific antibodies directly from repertoires of human V-genes. Recently, phage display has been adapted for the selection of catalytic activity by proximal display of both substrate and enzyme on the phage particle. This concept has been used successfully by Jestin and Winter to enrich for active over inactive polymerases, relying on the *in cis* incorporation of a tagged nucleotide into a template–primer duplex substrate tethered to the phage particle ***(18)***. More recently, Romesberg and coworkers ***(19)*** used a similar approach to select for a variant of the Stoffel fragment that incorporates rNTPs with efficiencies approaching those of the wild-type enzyme for dNTP substrates.

Although the genetic complementation approach is severely limited in the properties that can be selected for, the phage methods seem much more versatile. However, selection conditions have to be compatible with phage viability, and the intramolecular tethering of the substrate may favor the selection of polymerases with a low affinity for the template–primer duplex and/or a poor processivity. However, both methods should potentially be able to detect extremely weak polymerase activities, requiring, potentially, only a single dNTP incorporation for selection.

1.3. Polymerase Engineering by Compartmentalized Self-Replication

We have pursued another new strategy for the evolution of enzymes and, in particular, polymerases, called compartmentalized self-replication (CSR; **ref.** ***20***). In CSR, individual polymerase variants are isolated in separate compartments. Provided with appropriate reagents, each polymerase replicates only its own encoding gene, to the exclusion of those in other compartments (i.e., it self-replicates; **Fig. 1**). Consequently, only genes encoding active polymerases are replicated, whereas inactive variants disappear from the gene pool. Among differentially active variants, the more active variants will make proportionally more copies of their own encoding gene (i.e., will produce more "offspring"). As a result, the copy number of a polymerase gene after replication reflects the catalytic activity of the polymerase it encodes under the given selection conditions. Thus, polymerase genes encoding the most active polymerases that are best adapted to the selection conditions are going to increase in number and will come to dominate the gene population.

Segregation of self-replication into discrete, physically separate compartments is critical to ensure linkage of phenotype and genotype during CSR (i.e., to ensure that adaptive gains of each polymerase only benefit the replication of its own

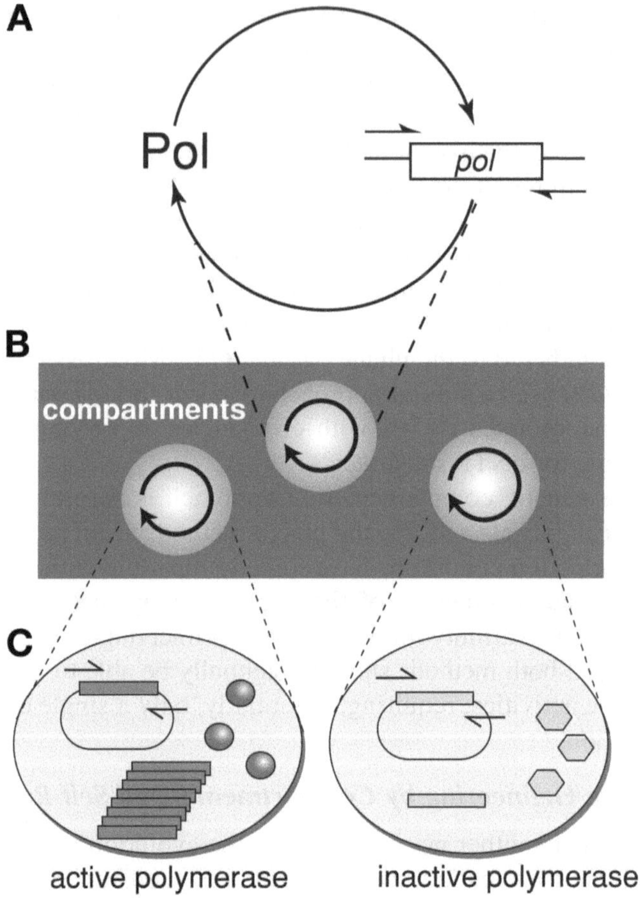

Fig. 1. **(A)** CSR is based on a simple feedback loop consisting of a polymerase that replicates only its own encoding gene. **(B)** Compartmentalization serves to isolate individual self-replication reactions from each other. **(C)** In such a system, adaptive gains directly (and proportionally) translate into genetic amplification of the encoding gene, i.e., only genes encoding active polymerases (dark gray spheres) are replicated, while inactive variants (light gray hexagons) fail to amplify their own gene and disappear from the gene pool.

encoding gene). Compartment integrity must not be compromised during the selection process, which, in the case of polymerases from thermophilic organisms, such as *Taq* polymerase, includes prolonged exposure to high temperatures. Furthermore, compartments must not allow exchange of macromolecules, such as DNA or protein (although exchange of small molecules is permissible and, in certain circumstances, desirable). Many different approaches have been described for the encapsulation of enzymatic reactions, including lipid vesicles ***(21)***. We chose water-in-oil emulsions ***(22)***. We modified the original surfactant composition and water-to-oil phase ratio to increase heat stability and product yield of CSR. The resulting emulsions are stable for prolonged periods at temperatures exceeding 90°C and allow emulsion–PCR reactions with yields approaching those of solution–PCR ***(20)***.

The high stability of the CSR emulsion allows selection for activity under a wide range of experimental conditions. CSR also exerts a stringent selection for polymerase activity: in a trial selection, active *Taq* wild-type polymerase was recovered from dilution into a 10^6 excess of an inactive *Taq* mutant (D785H/E786V) in a single round of CSR. Together, these make CSR a powerful method for the directed evolution of polymerases. Indeed, starting from two modestly sized libraries of randomly mutated *Taq* genes and using only three rounds of CSR each, we isolated variants of *Taq* polymerase with 11-fold increased thermostability or more than 130-fold increased resistance to heparin, a potent inhibitor of all known polymerases and a widely used anticoagulant ***(20)***.

We anticipate a plethora of applications for CSR and other methods for polymerase selection, including the directed evolution of "designer" polymerases that are tailor-made for existing applications, such as dye-terminator sequencing, clinical or forensic PCR, mutagenesis, and so on. Polymerase stability, tolerance to inhibitors, processivity, fidelity, affinity for template–primer duplex, substrate specificity, and so on, may be altered, and such polymerases will considerably expand the scope of existing applications, such as PCR, and enable completely novel ones. For example, polymerases with altered substrate specificity may allow an expansion of the genetic alphabet, allow the creation and in vitro evolution of ribozymes or deoxyribozymes with an expanded chemical repertoire, or facilitate the realization of ultrafast single-molecule sequencing strategies. Ultimately, it may become possible to expand not just substrate specificity but also polymerase chemistry, which may allow the creation of DNA-like polymers of defined sequence and an expanded chemical alphabet, which could be replicated at will. Such a strategy would allow the extension of molecular evolution to material science.

Self-replication and compartmentalization have also been studied with an aim to define plausible scenarios for molecular self-organization and prebiotic evolution. Specifically, ingenious selection strategies have been used in an attempt to recreate an RNA-based RNA replicase capable of self-replication. Already, the laboratories of Bartel and Joyce have succeeded in creating different ribozymes capable of template-directed addition of multiple nucleotide triphosphates ***(23,24)***. Once a RNA replicase is created, compartmentalization, as used in CSR (or by another strategy), is likely to be needed to ensure that it replicates only its own kind ***(25)*** and to prevent takeover by parasitic RNA species that notoriously plague in vitro replication systems.

Finally, the CSR strategy is not limited to polymerases but may be applied to other enzymes via interdependent catalytic networks. We have already demonstrated an example of this, using nucleoside diphosphate kinase (NDK) and *Taq* polymerase, which cooperate through reciprocal catalysis, whereby NDK produces the nucleotide triphosphate substrates required for the replication of its own *ndk* gene by the polymerase ***(20)***. Thus, only *ndk* genes encoding an active enzyme are replicated. Additional stages may be added to such a cooperative CSR cycle for the evolution of both single enzymes, or of reaction pathways. A more generic selection system for catalysis built around the "polymerase engine" of CSR is envisaged, whereby coupled catalytic reactions either produce substrates for the polymerase or consume inhibitors, thus, allowing replication of the genes encoding the enzymes to proceed. Therefore, CSR can potentially be extended to any enzyme, provided that suitable substrate chemistry can be found to feed into the basic CSR loop.

In conclusion, CSR is a powerful new activity-based selection strategy, and we anticipate many applications of CSR in the directed evolution of enzymes and, in particular, polymerases, as well as for generic studies of in vitro evolution and molecular cooperation. Here, we provide some detailed protocols for the implementation of CSR selection using *E. coli* as the expression host and *Taq* polymerase as the target gene.

2. Materials

1. *E. coli* suppressor strain TG1 (K12, Δ[*lac-pro*], *sup*E, *thi*, *hsd*D5/F′ *tra*D36, *pro*A$^+$B$^+$, *lac*Iq, *lac*ZΔM15) is used for propagation of plasmids and polymerase expression.
2. 2X TY medium (per liter; **ref. *26***): 16 g bacto-tryptone, 10 g bacto-yeast extract, and 5 g NaCl; supplemented with 0.1 mg/mL ampicillin (Amp; 2X TYA).
3. TYE agar plates (per liter; **ref. *26***): 10 g bacto-tryptone, 5 g bacto-yeast extract, 8 g NaCl, and 15 g bacto-agar; supplemented with 0.1 mg/mL Amp.
4. Anhydrotetracycline (ACROS).
5. 10X SuperTaq polymerase buffer (HT Biotech, Cambridge, UK).
6. Tetramethyl ammonium chloride (TMAC; Sigma).
7. DNase-free pancreatic RNase (Roche).
8. CSR oil phase: 4.5% (v/v) Span 80 (Fluka), 0.4% (v/v) Tween-80 (Sigma), and 0.05% (v/v) Triton X-100 (Sigma) in light mineral oil (Sigma; *see* **Subheading 3.3.** for preparation).
9. 2-mL round-bottom Biofreeze vials (Costar, Cambridge MA).
10. Magnetic stirrer bar.
11. Thin-walled PCR tubes (Roche).
12. Diethyl ether (Sigma).
13. Polyethylene glycol (PEG) 800/$MgCl_2$ solution: 30% (v/v) PEG 800 (BDH Biochemical) and 30 m*M* $MgCl_2$.
14. Tris-HCl–EDTA (TE) buffer: 10 m*M* Tris-HCl, pH 8.0, and 1 m*M* EDTA.

15. 20 U/μL *Dpn*I (New England Biolabs).
16. PCR purification kit (Qiagen).
17. EB (Qiagen): 10 m*M* Tris-HCl, pH 8.0.
18. ExoSap (USB).

3. Methods

The following protocol describes the setting up of a core CSR selection using *Taq* polymerase variants expressed in *E. coli*. This protocol will select for variants present in a library that are active under standard PCR conditions. Methods to generate combinatorial libraries using nucleotide analogs ***(27)*** and error-prone PCR ***(28)*** are well-documented and will not be described here. Primer 1 (5′-CAG GAA ACA GCT ATG ACA AAA ATC TAG ATA ACG AGG GCA A-3′) is specific to expression vector pASK75 ***(29)*** and is 5′ to the *Taq* gene that is cloned into the *Xba*I and *Sal*I sites of the vector pASK75. Primer 2 (5′-GTA AAA CGA CGG CCA GTA CCA CCG AAC TGC GGG TGA CGC CAA GCG-3′) is specific to expression vector pASK75 and is 3′ to the *Taq* gene. Both primers contain 5′ overhangs that enable out-nested PCR using primers 3 (5′-CAG GAA ACA GCT ATG AC-3′) and 4 (5′-GTA AAA CGA CGG CCA GT-3′) to enrich for CSR selectants.

3.1. Bacterial Expression of Taq Polymerase Using the Tet Promotor

1. Inoculate a culture with a single colony. Grow in 2 mL of 2X TY (*see* **Heading 2., item 2**) and 0.1 mg/mL Amp (2X TYA) at 37°C overnight.
2. Dilute the overnight culture 1:100 into fresh 2X TYA. Grow to optical density (OD) at 600 nm (OD_{600}) = 0.6 to 0.9 at 37°C (this should take no longer than 2–3 h).
3. Induce expression from the Tet promotor of pASK75 by addition of ACROS (a non-toxic tetracycline derivative), to a final concentration of 0.2 μg/mL.
4. Continue shaking for 2 to 4 h at 37°C.

3.2. Harvesting Polymerase Expressor Cells for CSR

1. Centrifuge cells in a benchtop centrifuge (1300 rcf [3 krpm], 15 min), discard the supernatant, and resuspend the cells gently in an equal volume of 1X *Taq* buffer.
2. Repeat the centrifugation and resuspension step, but this time resuspend cells in 1/10 culture volume of 1X *Taq* buffer. The cell pellet should resuspend easily.
3. Measure OD_{600} of an aliquot of the resuspension to determine cell number (OD_{600} = 1 is equivalent to 8×10^8 cells/mL).

3.3. Set Up CSR

1. Prepare aqueous phase CSR mix on ice in 1X *Taq* buffer, comprising 1 μ*M* primers 1 and 2, 0.25 m*M* dNTPs, 50 μ*M* TMAC (optional; *see* **Note 1**), 0.05% (v/v) deoxyribonuclease-free pancreatic ribonuclease (optional), and induced expressor cells (final concentration, 1×10^9 cells/mL).

2. Prepare oil-phase CSR by mixing light mineral oil with 4.5% (v/v) Span 80, 0.4% (v/v) Tween-80, and 0.05% (v/v) Triton X-100, under constant stirring at room temperature. Because of the high viscosity of the surfactants, it is advisable to prepare a large volume of oil-phase CSR (>50 mL) and use cut-off pipet tips to dispense the surfactants. The oil-phase CSR, once prepared, can be stored in the dark at room temperature for 1 mo.
3. Add 200 µL of aqueous-phase CSR mix drop-wise (~5–10 µL per drop, one drop every 5 s) to 400 µL of oil-phase CSR in a 2-mL round-bottom Biofreeze vial under constant stirring (1000 rpm) with a magnetic stir bar.
4. After addition of the last drop, continue stirring for 5 min. The emulsion should be creamy white and viscous (*see* **Note 2**).

3.4. CSR

1. Transfer the emulsion mix to 0.5-mL thin-walled PCR tubes (100 µL/tube) and add two drops of mineral oil to prevent evaporation (*see* **Note 3**).
2. Carry out PCR thermocycling, typically using 20 cycles with the profile 94°C (1 min), 60°C (1 min), 72°C (5 min), preceded by an initial 5-min incubation at 94°C to lyse bacterial cells and destroy background enzymatic activities.

After thermocycling, the emulsion phase should remain creamy white and should be overlaid with a clear oil phase. The volume may have reduced but this does not indicate coalescence of the emulsion compartments. Coalescence and breaking up of the emulsion manifests itself by the appearance of a clear layer of aqueous phase just beneath the white emulsion phase. This may indicate the presence of a variety of factors that interfered with the creation of a stable emulsion, including, for example, insufficient mixing.

3.5. Work-Up

1. Recover the aqueous-phase CSR by extraction of the emulsion with a double volume (200 µL) of diethyl ether. Mix by vortexing at maximum speed for 20 s, and spin at 20,000 rcf (13 krpm) for 2 min in a benchtop centrifuge.
2. Two liquid layers should be apparent. Carefully remove the lower, often cloudy, aqueous phase from underneath the clear organic phase and transfer into a fresh 1.5-mL Eppendorf tube.

There are two different methods for the workup of CSR reactions that we have found to work.

3.5.1. PEG Extraction Method

1. Extract the aqueous phase once with phenol–chloroform (1/1 v/v), and then again with chloroform–isoamyl alcohol (24/1 v/v) alone.
2. To the aqueous phase, add 0.5 volumes of PEG 800/$MgCl_2$ solution (*see* **Subheading 2., item 13**) and mix well by pipetting up and down carefully several times.

3. Centrifuge the sample at 20,000 rcf (13 krpm) for 10 min at room temperature in a benchtop centrifuge. Discard the supernatant (containing unincorporated primers and dNTPs) and resuspend the pellet in TE (*see* **Subheading 2., item 14**).
4. Unacceptable carryover of unselected polymerases can be further reduced by adding 20 to 50 U *Dpn*I to the aqueous phase extracted using ether, and incubating for 1 h at 37°C (*see* **Note 4**).
5. To ensure complete removal of primers, further purify CSR products from the aqueous phase using a PCR purification kit (Qiagen). Include two washes with 35% (w/v) guanidinium hydrochloride during purification to ensure complete removal of any residual primers. Elute purification products in 50 μL of provided buffer EB (Qiagen; *see* **Subheading 2., item 16**).

3.5.2. Quick Work-Up

1. As in **Subheading 3.5.1.**, **step 5** purify CSR products from the aqueous phase using a PCR purification kit (Qiagen). Include one wash with 35% (w/v) guanidinium hydrochloride during purification to ensure complete removal of any residual primers. Elute purification products in 50 μL of provided buffer EB.
2. Take 7 μL of the elution product and add 1 μL of 10X *Taq* buffer, 1 μL of 20 U/μL *Dpn*I, and 1 μL ExoSAP, and incubate for 1 h at 37°C and 15 min at 85°C.

3.6. Pull Through

1. Rescue purified selection products by reamplifying using out-nested primers 3 and 4. Typically, 20% of purified selection products can be carried over into 50 μL of a rescue PCR reaction (i.e., 10 μL out of the 50 μL of purified products). However, it may be advisable to use more or less depending on the initial result (*see* **Notes**). The cycling conditions for reamplification should be optimized according to the yield of the selection products, typically 25 cycles with the profile 94°C (30 s), 50°C (30 s), and 72°C (5 min) will yield a strong reamplification band. The number of cycles may need to be adjusted depending on the result (*see* **Notes**).
2. Visualize rescued selectants using standard agarose gel electrophoresis. A successful selection is indicated by the presence of a band of the correct size (2.5 kbp, in the case of *Taq* polymerase) in the selection, and the absence of such (usually the absence of any band) in the negative control (*see* **Notes 5** and **6**).

The CSR band can be further gel purified, restricted, and recloned into the vector pASK75 using standard procedures. After transformation into TG1 (*see* **Heading 2., item 1**), transformants can be either screened as in **Subheading 3.7.** or pooled, grown, and induced for a further round of selection, as described in **Subheadings 3.2.–3.6.**

3.7. Quick Screening of Taq Polymerase Variants

1. To screen rescued selectants, transformed colonies are replica-spotted on TYE-Amp plates (*see* **Subheading 2., item 3**) and inoculated into a 96-well tissue culture plate containing 100 μL of 2X TYA (*see* **Subheading 2., item 2**) per well.

2. After 6 to 10 h, add 20 µL of 2X TYA containing 1.2 µg/mL ACROS and leave for a further 2 to 4 h for expression.
3. Using a multichannel pipet, transfer 2 µL of induced cells directly from each well into 30 µL of PCR mix used in selection (*see* **Subheadings 3.3.** and **3.4.**) in a 96-well PCR plate.
4. Overlay the reactions with a drop of oil, and carry out PCR with conditions similar to those used in selection (*see* **Subheading 3.4., step 2**), although the number of cycles can be increased to 30 to identify weaker selectants. It is important to include a wild-type control as a benchmark. Identify positive selectants on a standard agarose gel.

4. Notes

1. TMAC is used to increase primer specificity and may not be necessary for certain selection conditions.
2. To test a newly prepared oil phase, a standard PCR reaction can be emulsified using the protocol in **Subheadings 3.3.–3.4.** The yield of PCR product generated in emulsion should be comparable to the nonemulsified reaction. There should be no significantly visible deterioration of the emulsion after thermal cycling, usually manifested by a clear aqueous layer at bottom of PCR tube.
3. CSR yields seem to be reduced if reactions are not overlaid with oil or if a hot lid is used.
4. This removes methylated plasmid DNA that can serve as nonspecific template during workup of selectants. Unmethylated amplification products are not cut.
5. It is important to include a negative control in all CSR selections. We find omission of dNTPs to a CSR reaction that is identical in all other respects to the selection reaction to be a useful negative control. After reamplification, there should be no visible amplification products in the negative control. The presence of selection products can indicate incomplete removal of primers and parent plasmid, which can reduce selection factors by carryover of unselected polymerase genes to the next round of selection.
6. If reamplification provides only a weak signal, then either one of the out-nested primers 3 or 4 may be used with an appropriate gene specific primer. This sometimes produces a more-specific signal. In any case, it is important to always include the negative (–dNTP) control.

References

1. Steitz, T. A. (1999) DNA polymerases: structural diversity and common mechanisms. *J. Biol. Chem.* **274,** 17,395–17,398.
2. Goodman, M. F. (2002) Error-prone repair DNA polymerases in prokaryotes and eukaryotes. *Annu. Rev. Biochem.* **71,** 17–50.
3. Nishikura, K. (2001) A short primer on RNAi: RNA-directed RNA polymerase acts as a key catalyst. *Cell* **107,** 415–418.
4. Goldsby, R. E., Lawrence, N. A., Hays, L. E., et al. (2001) Defective DNA polymerase-delta proofreading causes cancer susceptibility in mice. *Nat. Med.* **7,** 638, 639.

5. Joyce, C. M. and Steitz, T. A. (1994) Function and structure relationships in DNA polymerases. *Annu. Rev. Biochem.* **63,** 777–822.
6. Barnes, W. M. (1992) The fidelity of Taq polymerase catalyzing PCR is improved by an N-terminal deletion. *Gene* **112,** 29–35.
7. Lawyer, F. C., Stoffel, S., Saiki, R. K., et al. (1993) High-level expression, purification, and enzymatic characterization of full-length Thermus aquaticus DNA polymerase and a truncated form deficient in 5′ to 3′ exonuclease activity. *PCR Meth. Appl.* **2,** 275–287.
8. Li, Y., Mitaxov, V., and Waksman, G. (1999) Structure-based design of Taq DNA polymerases with improved properties of dideoxynucleotide incorporation. *Proc. Natl. Acad. Sci. USA* **96,** 9491–9496.
9. Patel, P. H. and Loeb, L. A. (2000) Multiple amino acid substitutions allow DNA polymerases to synthesize RNA. *J. Biol. Chem.* **275,** 40,266–40,272.
10. Astatke, M., Ng, K., Grindley, N. D., and Joyce, C. M. (1998) A single side chain prevents Escherichia coli DNA polymerase I (Klenow fragment) from incorporating ribonucleotides. *Proc. Natl. Acad. Sci. USA* **95,** 3402–3407.
11. Ignatov, K. B., Bashirova, A. A., Miroshnikov, A. I., and Kramarov, V. M. (1999) Mutation S543N in the thumb subdomain of the Taq DNA polymerase large fragment suppresses pausing associated with the template structure. *FEBS Lett.* **448,** 145–148.
12. Kunkel, T. A. and Bebenek, K. (2000) DNA replication fidelity. *Annu. Rev. Biochem.* **69,** 497–529.
13. Bedford, E., Tabor, S., and Richardson, C. C. (1997) The thioredoxin binding domain of bacteriophage T7 DNA polymerase confers processivity on Escherichia coli DNA polymerase I. *Proc. Natl. Acad. Sci. USA* **94,** 479–484.
14. Suzuki, M., Baskin, D., Hood, L., and Loeb, L. A. (1996) Random mutagenesis of Thermus aquaticus DNA polymerase I: concordance of immutable sites in vivo with the crystal structure. *Proc. Natl. Acad. Sci. USA* **93,** 9670–9675.
15. Kim, B., Hathaway, T. R., and Loeb, L. A. (1996) Human immunodeficiency virus reverse transcriptase. Functional mutants obtained by random mutagenesis coupled with genetic selection in Escherichia coli. *J. Biol. Chem.* **271,** 4872–4878.
16. Sweasy, J. B. and Loeb, L. A. (1993) Detection and characterization of mammalian DNA polymerase beta mutants by functional complementation in Escherichia coli. *Proc. Natl. Acad. Sci. USA* **90,** 4626–4630.
17. Glick, E., Vigna, K. L., and Loeb, L. A. (2001) Mutations in human DNA polymerase eta motif II alter bypass of DNA lesions. *EMBO J.* **20,** 7303–7312.
18. Jestin, J. L., Kristensen, P., and Winter, G. (1999) A method for the selection of catalytic activity using phage display and proximity coupling. *Angew. Chem. Int. Ed.* **38,** 1124–1127.
19. Xia, G., Chen, L., Sera, T., Fa, M., Schultz, P. G., and Romesberg, F. E. (2002) Directed evolution of novel polymerase activities: mutation of a DNA polymerase into an efficient RNA polymerase. *Proc. Natl. Acad. Sci. USA* **99,** 6597–6602.
20. Ghadessy, F. J., Ong, J. L., and Holliger, P. (2001) Directed evolution of polymerase function by compartmentalized self-replication. *Proc. Natl. Acad. Sci. USA* **98,** 4552–4557.

21. Oberholzer, T., Albrizio, M., and Luisi, P. L. (1995) Polymerase chain reaction in liposomes. *Chem. Biol.* **2,** 677–682.
22. Tawfik, D. S. and Griffiths, A. D. (1998) Man-made cell-like compartments for molecular evolution. *Nature Biotechnol.* **16,** 652–656.
23. Johnston, W. K., Unrau, P. J., Lawrence, M. S., Glasner, M. E., and Bartel, D. P. (2001) RNA-catalyzed RNA polymerization: accurate and general RNA-templated primer extension. *Science* **292,** 1319–1325.
24. McGinness, K. E., Wright, M. C., and Joyce, G. F. (2002) Continuous in vitro evolution of a ribozyme that catalyzes three successive nucleotidyl addition reactions. *Chem. Biol.* **9,** 585–596.
25. Szostak, J. W., Bartel, D. P., and Luisi, P. L. (2001) Synthesizing life. *Nature* **409,** 387–390.
26. Sambrook, J., Fritsch, E. F., and Maniatis, T. (1990) *Molecular Cloning: A Laboratory Manual.* Cold Spring Harbor Laboratory, Cold Spring Harbor, NY.
27. Zaccolo, M. and Gherardi, E. (1999) The effect of high-frequency random mutagenesis on in vitro protein evolution: a study on TEM-1 beta-lactamase. *J. Mol. Biol.* **285,** 775–783.
28. Vartanian, J. P., Henry, M., and Wain-Hobson, S. (1996) Hypermutagenic PCR involving all four transitions and a sizeable proportion of transversions. *Nucleic Acids Res.* **24,** 2627–2631.
29. Skerra, A. (1994) Use of the tetracycline promoter for the tightly regulated production of a murine antibody fragment in Escherichia coli. *Gene* **151,** 131–135.

15

Synthesis of Degenerated Libraries of the *Ras*-Binding Domain of *Raf* and Rapid Selection of Fast-Folding and Stable Clones With the Dihydrofolate Reductase Protein Fragment Complementation Assay

François-Xavier Campbell-Valois and Stephen W. Michnick

Summary

The protein-engineering field is mainly concerned with the design of novel enzyme activities or folds and with understanding the fundamental sequence determinants of protein folding and stability. Much effort has been put into the design of methods to generate and screen libraries of polypeptides. Screening for the ability of proteins to bind with high affinity and/or specificity is most often approached with phage display technologies. In this chapter, we present an alternative to phage display, performed totally in vivo, based on the dihydrofolate reductase (DHFR) protein-fragment complementation assay (PCA). We describe the application of the DHFR PCA to the selection of degenerated sequences of the *ras*-binding domain (RBD) of *raf* for correct folding and binding to *ras*. Our screening system allows for enrichment of the libraries for the best-behaving sequences through iterative competition experiments, without the discrete library screening and expansion steps that are necessary in in vitro approaches. Moreover, the selected clones can be processed rapidly to purification by Ni–nitrilotriacetic acid (NTA) affinity chromatography in 96-well plates. Our methods are particularly suitable for the designing and screening of libraries aimed at studying sequence folding and binding determinants. Finally, it can be adapted for library-against-library screening, thus, allowing for coevolution of interacting proteins simultaneously.

Key Words: Protein-fragment complementation assays; dihydrofolate reductase; bacterial survival assay; phage display; protein–protein interactions; protein engineering; protein folding; degenerated libraries; polymerase chain reaction; binding assays; 6xHis-tag affinity purification.

1. Introduction

Since the development of recombinant DNA technologies in the 1970s and 1980s, numerous ingenious approaches have been exploited to synthesize and

From: *Methods in Molecular Biology, vol. 352: Protein Engineering Protocols*
Edited by: K. M. Arndt and K. M. Müller © Humana Press Inc., Totowa, NJ

screen oligonucleotide libraries to discover those that code for novel protein sequences displaying a desired characteristic, be it enzyme activity, binding, or stability of a protein under selected conditions *(1–8)*. In any given case, to tackle such protein engineering efforts, one must have two methods in place: a strategy to generate a diverse library of sequences, and an efficient way to screen for the desired characteristics of the products of the library. The choice of library synthesis method is crucial to providing a sufficiently large sequence search space, such that a maximum number of choices are available from which sequences coding for desired characteristics can be found. Thus, it is not surprising that the development of such strategies has been, and still is, a focus of research in the field. There are a few examples reported in the literature of studies in which a region of a protein is completely randomized or highly degenerated to answer questions regarding protein folding *(9,10)*. Nevertheless, examples of truly and highly degenerated libraries to explore sequence space in search of a novel fold, binding capability, or enzyme activity are rare *(6,11)*. The inherent limitations of generating highly randomized libraries and subsequently searching for the few sequences that display the desired characteristics in a huge sequence space have prompted efforts toward the design of methods that explore more-limited library sets. These include DNA shuffling strategies or completely alternative approaches that allow for recombination between genes devoid of any sequence homology *(12–18)*.

Even the most cleverly designed libraries will not yield useful products without an adequate screening and selection strategy. For example, an ideal way to screen a library coding for an enzyme activity, stability, or ability to bind to a target protein is to express the library in a cell or organisms in which expression of library members with the desired characteristics confers specific growth capabilities on selective medium, in harsh conditions or in a specific genetic background *(1,19–22)*. Such examples are, unfortunately, rare and thus protein engineers have sought approaches that are more general to screen libraries *(23–26)*. More specifically, binding assays can often serve the general purpose of selecting expressed polypeptides from a DNA library that are properly folded; stable; and whose binding to some molecule, whether it be another protein, nucleic acid, organic substrate, or transition state analog, imply the specific function desired. The most well-established method of choice to do this is the phage display strategy (**refs.** ***4***, ***5***, and ***27***; reviewed in **refs.** ***28–30***). In this strategy, the expansion and the screening of libraries are performed in discrete steps, taking place, respectively, in vivo and in vitro. The method is well-suited to proteins that bind to small molecules or peptides that can be easily crosslinked to a solid phase support, but it is not straightforward to adapt for studies of protein–protein interactions or for library-against-library screening. More recently, protein-fragment complementation assay (PCA) has emerged as an alternative technology (**refs.** ***31*** and ***32***; reviewed in **refs.** ***33*** and ***34***). The PCA strategy relies on the association and folding of a reporter protein or enzyme from fragments, driven by the interaction of two proteins to which the fragments are fused. The reconstitution of the reporter protein fold and, thus, detectable catalytic activity, depends on the interaction of the fused proteins. In particular, a simple survival-selection assay has been developed for screening libraries in *Escherichia coli*, based on the murine dihydrofolate reductase (mDHFR) as the reporter PCA ***(32)***. In *E. coli*, as in all prokaryotes and eukaryotes, the DHFR product, tetrahydrofolate, is necessary for the synthesis of thymine, glycine, serine, and adenine, whereas, in prokaryotes, it is also required for synthesis of pantothenate. DHFR activity is, thus, absolutely required for cell growth and division in the absence of a source of DHFR end products. *E. coli* can be made dependant on expression of recombinant mDHFR by treatment of the cells with trimethoprim, a folate analog that is 12,000 times more potent an inhibitor of *E. coli* than mammalian DHFRs ***(35)***. Thus, the principle of the mDHFR PCA is that two proteins fused to complementary fragments of mDHFR must be coexpressed and interact together in *E. coli* grown in minimal (M9) medium supplemented with trimethoprim for cells to grow and divide ***(31)***. In a first demonstration of a library-against-library screen, the DHFR PCA was used to identify optimally heterodimerizing pairs of leucine zipper-forming sequences from individual libraries containing 6×10^{10} possible combinations of sequences. Competition experiments and "library shuffling" strategies were devised to improve library screening coverage, to further optimize dimerizing pairs, and, finally, to identify a "winner pair" ***(32,36)***. More recently, DHFR PCA was adapted for screening and selection of single-chain antibodies in vivo ***(37)***. The all-in-one genetic screening approach of the DHFR PCA strategy is the key feature allowing for simple performance of library-against-library screening, because selective pressure is applied concomitantly on both library populations during several cycles, without the tedious alternation between discrete expansion and screening steps associated with phage display. Thus, PCA truly allows for the study of sequences covariation of oligomeric partners.

The results obtained with leucine zippers convinced us that the assay could be useful for tackling problems that are more challenging. In the zipper studies, only a handful of key amino acid positions were varied, and only between two and four amino acid substitutions were allowed. Based on previous theoretical work ***(38)***, we have attempted to rigorously and exhaustively determine the sequence determinants for folding of the *ras*-binding domain (RBD) of the serine/threonine protein kinase, *raf* ***(39)***. The premise of our approach is similar to a previously published strategy aimed at identifying sequences that support rapid folding and stability of proteins selected by phage display ***(40)***. The principle, as applied to the *raf* RBD, is as follows: if the sequence of a given RBD variant folds rapidly to the correct structure and is sufficiently stable, it should interact with its natural binding partner, the small GTPase, *ras*. Fusing the RBD library to one complementary fragment of DHFR and *ras* to the other, and then co-expressing these in *E. coli*, grown under selective pressure as described earlier in this section, it can be reasoned that fast-folding and stable members of the RBD library will interact with *ras* and allow for the reconstitution of DHFR activity and the rescue of cell growth. We chose an efficient and realistic way to design libraries that allowed for exploring the maximum sequence diversity in a meaningful way, while creating libraries of reasonable size. Based on the questions we chose to address in these studies, a meaningful way to explore sequence space is to generate libraries in which only a small stretch of contiguous residues are varied at a time ***(10)***. Examination of the RBD structure allows one to dissect it into 13 regions corresponding to individual β-turns or loops, β-strands, and one α-helix (two libraries were generated for this region corresponding to amino and carboxyl termini of the helix), ranging in length from four to eight amino acids ***(41,42)***. On this basis, we created 13 degenerated libraries, in which each wild-type codon is replaced by a NNK codon (where N is any nucleotide, and K is G or T) that allows the insertion of the 20 amino acids at each varied position in the sequence. These libraries were screened for binding to *ras* by the DHFR PCA in *E. coli*. In addition to the screening being performed entirely in vivo, a key advantage of this approach is that expressed RBD library members that interact well with *ras* can be purified for physical analysis without having to switch to another expression system.

Herein, we present the protocols and proposed trouble-shooting strategies, based on the technical challenges that we have encountered in the design and synthesis of the degenerated libraries and in their screening with the DHFR PCA. Hopefully, these protocols are general enough to be useful not only to those interested in folding, but more generally to problems requiring the optimization of protein–protein interactions.

2. Materials

2.1. Library Synthesis

1. Oligonucleotide primers (IDT). The primers with positions at which multiple bases are allowed are hand mixed to assure that the desired ratio of each base is respected. These are sodium dodecyl sulfate (SDS)-polyacrylamide gel electrophoresis (PAGE) purified.
2. *Taq* polymerase (Fermentas).
3. Agarose gel: agarose (Bioshop) Dark Reader™ (Clare Chemical Research) and Gelstar™ (Biowittaker Molecular Applications).
4. Gel purification, QIAEX™ II or, preferably, QIAquick™ gel extraction kit (Qiagen).

2.2. Library Cloning and Recovery

1. Plasmid pQE-32 Δ F [1,2] (derived from plasmid pQE-32 distributed by Qiagen; F [1,2] indicates DHFR fragment 1).
2. Plasmid pREP4 (harbors *lac* repressor and kanamycin as selectable marker. Cells in which protein is expressed from the pQE-32 plasmid, such as those used in these studies, must contain this vector to limit expression from the otherwise very leaky *tac* promoter contained in the pQE-32 plasmid. Distributed by Qiagen).
3. Ligation, T4 DNA ligase (Fermentas), and adenosine triphosphate (Pharmacia).
4. SS320 electrocompetent cells (*see* **Subheading 2.7.**).
5. Genepulser™ II electroporation apparatus (Bio-Rad).
6. Electroporation cuvet with 2-mm-wide slot (Invitrogen).
7. 10 mL SOC medium transformation: 1% tryptone, 0.5% yeast extract, and 0.5% NaCl, supplemented with 0.4% glucose, 2.5 m*M* KCl, and 10 m*M* $MgCl_2$.
8. LB-agar supplemented with 10 μg/mL tetracycline, 10 μg/mL spectinomycin, and 100 μg/mL ampicillin in 100-mm petri dishes.
9. 100 to 250 mL of LB medium per library. LB medium is supplemented with 0.2% glucose, 0.25X M9 salts solution (*see* **Subheading 2.3., item 6** for recipe), 10 μg/mL tetracycline, 10 μg/mL spectinomycin, and 100 μg/mL ampicillin (Bioshop).
10. Plasmid Midi Kit (12143) (Qiagen).

2.3. Library Screening

1. BL21 electrocompetent cells transformed with pREP4 (*see* **Subheading 2.2., item 2**) and then transformed with pQE-32 *ras*-F [3] (F [3] indicates DHFR fragment 2).
2. Genepulser™ II electroporator system (BioRad) or Electroporator 2510 (Eppendorf).
3. Electroporation cuvet with 1-mm-wide slot (Invitrogen).
4. SOC medium (*see* **Subheading 2.2., item 7**).
5. Phosphate-buffered saline (PBS). For 1 L final volume in water, combine 8 g NaCl, 0.2 g KCl, 1.44 g Na_2HPO_4, and 0.24 g KH_2PO_4. The pH is adjusted to 7.4 with HCl, and the solution is autoclaved.
6. M9 minimal medium supplemented with the appropriate antibiotics as described **Heading 1.** (hence dubbed "selective medium"). For 1 L of complete medium, combine: 740 mL of 2.5% noble agar (Difco), 200 mL of 5X M9 salts (for 1 L, 64 g Na_2HPO_4, 15 g KH_2PO_4, 2.5 g NaCl, and 5 g NH_4Cl; composition given in **ref. *43***), 2 mL of 1 *M* $MgSO_4$, 1 mL of 100 m*M* $CaCl_2$, and 20 mL of 20% glucose solution. All salts and glucose are cell-culture grade, from any source, such as Sigma, Fisher, or ICN except: 100 μg/mL ampicillin, 25 μg/mL kanamycin, and 1m*M* isopropyl-β-D-thio-galacto-pyranoside (IPTG) (Bioshop), 10 μg/mL trimethoprim (ICN), 800 μg/mL casamino acids (Difco),and 10 μg/mL thiamine (Fisher). All solutions must be prepared with deionized water and sterilized by filtration (for antibiotics, casamino acids, IPTG, and thiamine; store at –20°C) or by autoclave (for salts; store at room temperature). Note that Mg salts, $CaCl_2$, $MgSO_4$, and glucose solutions must be autoclaved separately. The reconstituted medium is poured into 100- or 150-mm petri dishes. The reconstituted medium can be kept at 4°C for up to 2 mo.

7. Plasmid Midi Kit (Qiagen) or alkaline lysis maxiprep.
8. Restriction enzymes: *Hpa*I, *Xma*I, *Eco*NI, and *Xba*I (NEB or Fermentas).
9. XL-1 Blue chemiocompetent cells (*see* **Subheading 2.8.**).

2.4. Clones Competition Experiment

1. Glass culture or 15-mL conical tubes (Corning).
2. Solid and liquid selective medium (same protocol as in **Subheading 2.3., item 6**, except that agar is not added for the liquid medium).
3. Plasmid Midi Kit (Qiagen) or alkaline lysis maxiprep.
4. LB medium supplemented with 100 μg/mL ampicillin and 25 μg/mL kanamycin.

2.5. Isolation of Clones and Sequencing

1. Restriction enzymes: *Hpa*I, *Xma*I, *Eco*NI, and *Xba*I (NEB or Fermentas).
2. XL-1 Blue competent cells.
3. 24-well plates (Corning), LB-agar with 100 μg/mL ampicillin.
4. 2-mL V-shaped 96-well culture block (VWR).
5. Montage™ Plasmid Miniprep 96 kit (Millipore, LSKP 096 01) or smaller scale prep kit, such as QIAprep™ Spin Miniprep Kit (27104) (Qiagen), depending on the number of samples to be processed.
6. Vacuum manifold Multiscreen Resist™ (Millipore, MAVM 096 OR).
7. Oligonucleotide primer for sequencing specific to the plasmid harboring the library (IDT).

2.6. Protein Purification

1. Appropriate restriction endonucleases (*Sal*I and *Xho*I, in this case) and reagents necessary for ligation (*see* **Subheading 2.2.**).
2. BL21 pREP4 competent cells and LB with 100 μg/mL ampicillin and 25 μg/mL kanamycin petri dishes.
3. Terrific broth (TB) medium (12 g tryptone, 24 g yeast extract, 4 mL glycerol, 2.31 g KH_2PO_4, and 12.54 g K_2HPO_4; add water to a final volume of 1 L) supplemented with 100 μg/mL ampicillin and 25 μg/mL kanamycin.
4. 50-mL conical tubes (Corning).
5. A centrifuge and rotor that accommodate 96-well plate, such as Eppendorf 5810 or 5810 R and A-4-62, respectively.
6. Ni-NTA Spin Kit or Ni-NTA Superflow™ 96 Biorobot Kit (Qiagen), depending on the number of samples to be processed. An affordable alternative to the Superflow 96 Biorobot Kit is the following: we use Ni-NTA Superflow resin (Qiagen), 0.25-mm glass fiber filter 96-well plates (3510), 0.2-μm polyvinylidene fluoride (PVDF) membrane 96-well plates (3504), 96-well volume extender (3584), and fraction collector (3958) from Corning.
7. Vacuum manifold Multiscreen Resist™ and a large collection and sealing block (Millipore; respectively, MAVM 096 OR and OT).

8. Buffer A: 6 *M* guanidinium-HCl (Gdn-HCl), 0.1 *M* NaH_2PO_4, and 0.01 *M* Tris-HCl, pH 8.0; supplemented with 10 µ*M* phenylmethyl sulfonyl fluoride, 7.2 m*M* β-mercaptoethanol, 15–25 m*M* imidazole, and 300 m*M* NaCl.
9. Buffer B: same as buffer A, but pH 6.3 and supplemented with 7.2 m*M* β-mercaptoethanol and, sometimes, 15 m*M* imidazole.
10. Buffer E: 4 *M* Gdn-HCl and 0.025 *M* NaOAc, pH 4.5.
11. Dithiothreitol (DTT).
12. 6 *M* KOH.

2.7. Preparation of SS320 and BL21 pREP4 pQE-32 ras-F [3] Electrocompetent Cells

1. Overnight (O/N) preculture of SS320 or BL21 pREP4 pQE-32 *ras*-F [3].
2. 500 mL SOB medium: 2% tryptone, 0.5% yeast extract, 1 mL 10 m*M* NaCl, 2.5 m*M* KCl, 10 m*M* $MgCl_2$, and 10 m*M* $MgSO_4$; supplemented with 0.2% glucose for SS320 strain.
3. 500 mL LB medium supplemented with 0.2% glucose for BL21 pREP4 strain.
4. 2 L of ice-cold autoclaved deionized water.
5. 10% autoclaved glycerol solution (Bioshop).

2.8. Preparation of XL-1 Blue and BL21 pREP4 Chemiocompetent Cells

1. O/N preculture of XL-1 Blue or BL21 pREP4.
2. 500 mL SOB medium supplemented with 0.2% glucose for SS320 strain.
3. 500 mL LB medium supplemented with 0.2% glucose for BL21 pREP4 strain.
4. Transformation buffer: 10 m*M* Pipes, 15 m*M* $CaCl_2$ and 250 m*M* KCl, pH 6.7, with KOH. After the pH is set, $MnCl_2$ is added to a concentration of 55 m*M*.
5. Dimethylsulfoxide.

3. Methods

3.1. General Considerations

3.1.1. Steric Requirements in PCA

The spatial orientation of the PCA fragments is crucial to whether the PCA reporter protein can fold from its cognate fragments, and is determined by the orientations of the amino or carboxyl termini of the interacting proteins in the complex formed (*see* **Fig. 1** for schematization of spatial considerations encountered in designing linkers). In the design of the protein–PCA fragment fusions, it is, therefore, important to determine, *a priori*, whether the fragments would be brought together by a given combination of fusion constructs in such a way that the topology of the native structure could be achieved from a given configuration of the fusions. The two main factors that will determine whether correct folding can be attained are the orientation of the fusion (carboxyl- and/or amino-terminal) and second, the length of polypeptide linkers between the individual fragments and the proteins to which they are fused. Our experience has shown that

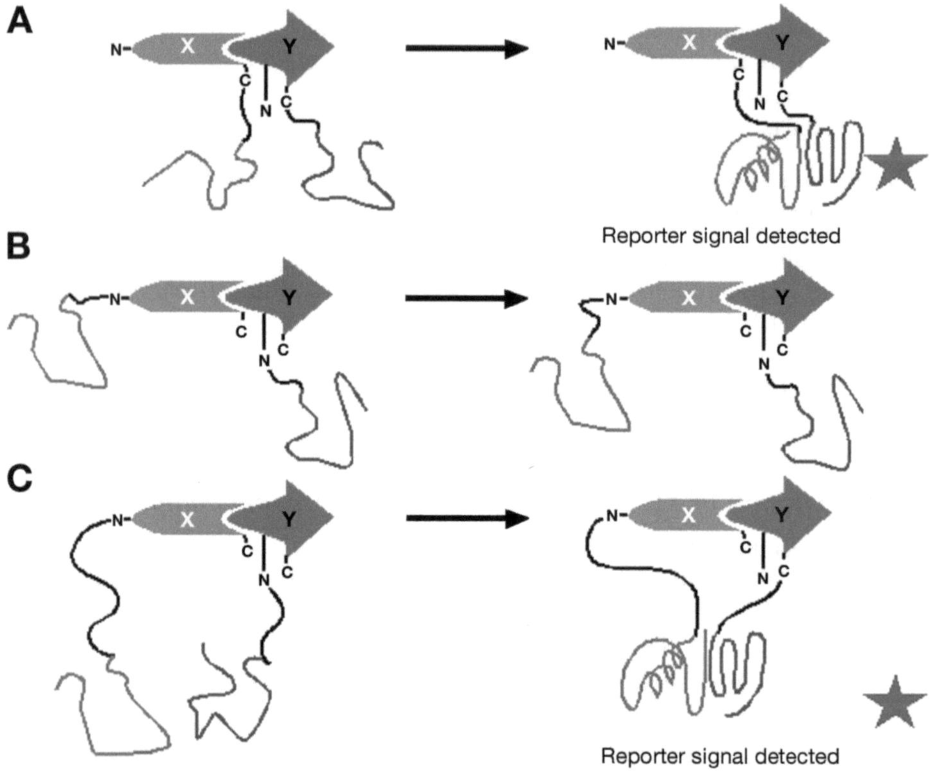

Fig. 1. Schematic structure of a heterodimeric complex formed between proteins X and Y in which their respective amino termini (N) are far apart whereas their carboxyl termini (C) are proximal in space and do not directly participate in the binding interface. In this case, the fragments should be attached as depicted in **(A)**. The number of amino acids in each linker can then be determined by assuming that a peptide bond is approx 3.75 Å long and by considering structural and geometrical constraints of the complex X-Y of interest and of the PCA reporter protein in its native and engineered format. If the crystal structure of the X–Y complex is not available, linker length has to be determined empirically. In some cases, it may not be possible to make the fusions in an optimal orientation. For example, in the (A) case, the carboxyl terminus of X has to be free for binding to Y. One could design constructs as pictured in **(B)**. In this case, it is obvious that the short linkers designed for (A) would not allow for reconstitution of native topology of the PCA reporter. Nevertheless, if longer linkers are used instead, as depicted in **(C)**, the folding of the reporter from its cognate fragment is possible, thus, making this protein fusion orientation adequate for the PCA-based protein-engineering screening strategy.

linkers constituted of repeats of GGGGS behave better in *E. coli*, yeast, and mammalian cells based on tests of a number of different interacting proteins. We rationalize that this type of linker improves flexibility and solubility of the fusions, thus, easing their reassembly. Moreover, they ensure metabolic stability because of their lack of susceptibility to naturally occurring proteolytic activities. Although these types of linkers are preferable, they are not essential for productive fragment complementation ***(31,32,39)***. However, bulky hydrophobic and rigid amino acids, such as proline and β-branched amino acids, should be avoided. In most protein-engineering problems, the structure of a protein complex of interest is already known and the orientation of the complex and requirement of linkers of a given length can be deduced. For the DHFR PCA, the spatial requirements for fusion of proteins at C- and N-termini are clear ***(31,44)***. For example, if the proteins of interest are fused, respectively, to the carboxyl terminus of F [1,2] and the amino terminus of F [3], the inserted linker can be quite short or may even not need to be included in the constructs, because it respects the normal topology of the enzyme. However, if the oligomers are fused to the amino termini of both fragments, the topology of the enzyme is permuted, thus, requiring a minimum of two amino acids (each peptide bond is approximated to 3.75 Å) in each linker to permit productive fragment reassembly, because the distance between the two amino termini is approx 10 Å. This orientation was chosen in the case of *ras* and the RBD; however, examination of the complex between *raf* RBD and the highly similar *ras* homolog, *rap1A* ***(45)***, reveals that the carboxyl terminus of each monomer is located 40 Å apart, thus, requiring that a minimum of six amino acids have to be added to the linker of each construct. For the library screening, we have fixed the length of each linker to 14 amino acids total, including the restriction sites, to make sure that sufficient flexibility is allowed.

3.1.2. Controls and Stringency

Before beginning any protein-engineering study and library screening with the DHFR PCA, one should perform rigorous controls to assess the sensitivity and stringency of the assay for the specific test system. Ideally, this means that, before beginning library screens, the investigator should know, roughly, what is the dissociation constant (K_d) limit of detection of the PCA for a given interaction. The sensitivity limit (maximum K_d for which a PCA response is detected) varies among different interacting pairs of proteins, but, in addition to the K_d, it is modified by factors such as the level of expression of each fusion, the amount of soluble vs the total expression of protein fusions, and the intrinsic properties of the proteins, such as their stability, solubility, and their kinetics of folding and binding. If the PCA is very sensitive and can detect very weak interactions between two specific proteins, it could prove impossible to distinguish clones that have the best-desired properties from any other; that is, the assay can be too sensitive, resulting in a loss of stringency. Thus, it is important to maintain a balance between these two factors. To assess these issues, general controls should be performed before PCA studies, although not all of these controls are necessarily relevant to a specific protein-engineering study, however, we will come back to this issue later.

1. *Spurious reassembly*: for the PCA to work, assembly of fragments from weak or nonspecific interactions cannot be allowed. The effective sensitivity is intrinsic to a given interacting protein pair test system and PCA and can be assessed by **controls 4** and **6**. If the sensitivity is too high, for example, if growth occurs for proteins that should not interact together (*see* **Fig. 2A**), sensitivity can be reduced by decreasing expression levels and/or by using stringency mutants as described in **control 2** (*see* **Fig. 2B**).
2. *Stringency mutants*: the effects on DHFR PCA of side-chain truncation mutant at fragment interface, such as Ile114 of F [3] have been reported ***(31,32)***. The problem encountered in the latter study was that when clones expressing leucine zipper-forming pairs that formed complexes with varying efficiencies were compared, it was impossible to distinguish them based on growth rates or numbers of colonies formed. In contrast, by inserting the mutation Ile114Ala on F [3], we changed the sensitivity of the PCA to an appropriate value for the leucine zipper system. A measure of this change in sensitivity was the "selection factor" in single-step selection, defined as the number of cotransformed cells plated divided by the number of colonies surviving under selective conditions. The result was an increase in stringency, such that it allowed us to distinguish the best from the more poorly behaving heterodimerizing pairs of leucine zippers in a reasonable number of iterative competition steps ***(32)***.
3. *Fragment swapping*: an observed interaction between binding proteins should occur regardless of the PCA fragments to which either of the proteins are attached. Therefore, an interaction observed with one protein-fragment configuration should give comparable result if proteins and fragments are swapped.
4. *Noninteracting proteins*: a PCA response should not be observed if a protein that is not known to interact with either of two interacting proteins being tested is used as a PCA partner (*see* **Fig. 2A**), nor should overexpression of this protein alone compete for the known interaction.
5. *Ability to titrate and to decrease the observed reporter activity by competition*: the observed reporter activity should vary with the relative expression ratio of each of the protein-fragment constructs. In addition, the PCA response should be diminished by simultaneous overexpression of one or the other interacting partners alone. However, one should bear in mind that the respective solubility and stability of each construct, the affinity of the interacting proteins for each other vs their cellular concentration, and the sensitivity of the PCA affect the ability to titrate the reporter activity. Modulation of reporter activity could then be achieved only by lowering expression level of the fusion constructs or by combining this type of control with **control 2** and/or **control 6** to reduce the complementation efficiency.
6. *Disrupting the interaction*: insertion of specific point or deletion mutation in one of the complex monomer's that is known to disrupt or diminish the interaction should also affect the PCA response in a predictable way.

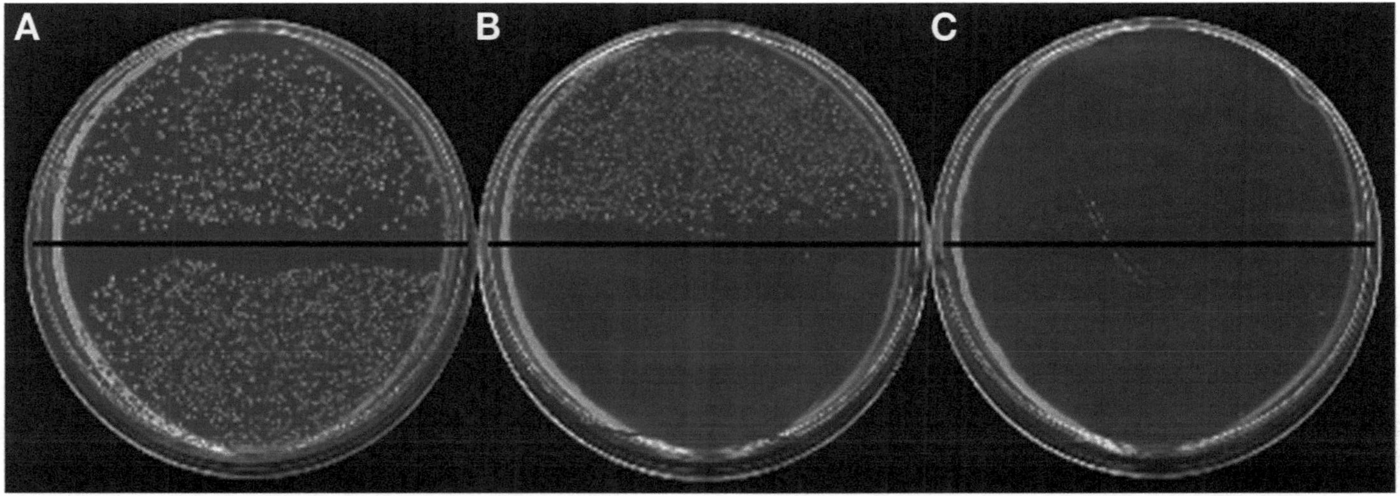

Fig. 2. The caspase-activated deoxyribonuclease (CAD) domains of inhibitor of CAD (*icad*) and *cad* were fused respectively to F [1,2] and F [3], F [3] Ile114Val, or Ile114Ala. These *cad* constructs were cotransformed into BL21 pREP4 along with either pQE-32 Δ *icad*-F [1,2] or RBD of *raf*-F [1,2] and plated on selective medium on the upper and lower parts of the petri dishes, respectively, and incubated for 24 h at 30°C. **(A)** pQE-32 *cad*-F [3] I114 (wild-type DHFR fragment F[3]); **(B)** pQE-32 *cad*-F [3] I114V; and **(C)** pQE-32 *cad*-F [3] I114A. The cotransformation of *cad* fusions with the RBD of *raf* fusions served as an internal control, as described in **Subheading 3.1.2.,** particularly as described under **controls 1** and **2**. These tests allow for determination of the sensitivity limit of the PCA. Because there should not be any significant interaction between *cad* and the RBD of *raf*, no interaction should be detected by the PCA, and, thus, no colony formation observed on selective medium. The use of Ile114Val mutant is, thus, ideal for assuring sufficient stringency, because it allows for growth of cells cotransformed with the relevant *cad* and *icad* constructs, whereas, in contrast to the wild-type F[3] fusion construct in (A), it does not lead to colony formation for cells cotransformed with the constructs of the RBD of *raf* and *cad*. The F[3] Ile114Ala mutant does not allow growth of either the positive *icad–cad* or the negative control *icad–raf* RBD pairs of fusions.

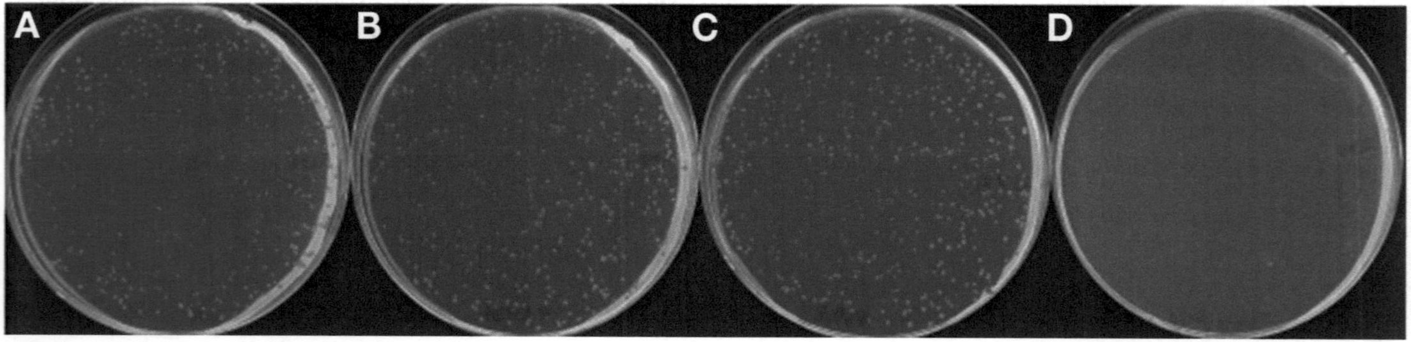

Fig. 3. BL21 pREP4 cells were cotransformed with plasmids expressing *ras*-F [3] and either the RBD of wild-type *raf* or mutants fused to F [1,2]. The cells were allowed to grow for 48 h on selective media and the petri dishes were scanned. RBD of *raf* **(A)** wild type (K_d = 0.13 μ*M*); **(B)** K65M (K_d = 0.40 μ*M*); **(C)** V69A (K_d = 0.95 μ*M*); and **(D)** R89L (K_d > 100 μ*M*). These mutants and others not shown allowed us to situate the sensitivity limit of our detection assay in the 10 to 100 μ*M* range.

In a protein-engineering project in which the model system under study is very well-characterized, only **controls 1**, **2**, **4**, and **6** are essential for establishing the specificity and stringency of the assay. In the case of the RBD–*ras* interaction, a comprehensive mutagenesis study and its effect on the K_d for binding of the RBD had already been published ***(46)***. These data permitted us to engineer several mutants that reduce the K_d for association of RBD–*ras* more than three orders of magnitude (*see*, e.g., **Fig. 3**). These mutants and others were tested in the DHFR PCA, allowing us to establish that the assay is able to detect binding for the RBD to *ras* for mutants with a K_d on the order of 1 μ*M*. In addition, published mutants that destabilize the protein fold, such as core hydrophobic residues (valine, leucine, or isoleucine) side-chain truncation to alanine, could be used as a stringency test.

3.2. Library Synthesis

1. To have a nonbiased library, we first generated a template in which the region to be varied was deleted and replaced by a stop codon, inserting also a frame shift and a unique restriction site allowing for its unequivocal identification (*see* **Note 1**).
2. To generate each library, we synthesized two PCR products that partially overlap (typically a 18–20 basepair [bp] hybridization region). For example, for PCR 1, we used one primer hybridizing in the promoter region of our vector (120 bp upstream of the start codon; *see* **Fig. 4**) and one primer hybridizing in the region immediately 5′ of the section targeted for degeneracy. For PCR 2, we used one two-arms oligonucleotide and a primer that hybridizes to the F [1,2] (120 bp downstream of the 3′-end of the open-reading frame; *see* **Fig. 4**). Typically, the PCR program was set as following: 1 min hot start at 94°C, 25 cycles of 20 s at 94°C, 30 s at 52°C, and 30 s at 72°C (*see* **Note 2**). Finally, the reactions were run for 10 min at 72°C to ensure completion of the elongation.
3. The PCR products are analyzed on an agarose gel. If the desired product is obtained, the remainder of the PCR product is loaded on a gel. We have advantageously used Gelstar™ and the Dark Reader™ (*see* **Note 3**) to visualize PCR products on agarose gels. It permits observation of bands under blue light (400–500 nm), wavelengths that do not damage DNA, in contrast to ultraviolet light (this allows one to cut the bands out and to proceed easily, in parallel, to the generation of several libraries).
4. Next, the bands are gel purified with Qiaex™ II (*see* **Note 4**).
5. Approximately 300 ng of the PCR product from PCR 1 and 2 are combined (*see* **Fig. 4** and **Note 5**) with 0.2 μ*M* of the terminal primers (hybridizing in the promoter and in F [1,2]) that anneal in regions 5′ and 3′, respectively, to the product of PCR 1 and 2. The PCR 3 program is the following: 1 min hot-start at 94°C, 10 cycles of 20 s at 94°C, 30 s at 52°C, and 30 s at 72°C. Finally, 10 min at 72°C to ensure completion of the elongation (*see* **Note 6**).
6. The entry vector pQE-32 Δ F [1,2] (*see* **Note 7**) and the resultant PCR products are digested with the appropriate restriction enzyme (*Sph*I and *Xho*I, in this case).
7. Bands are purified according to **Subheading 3.2., step 4**.

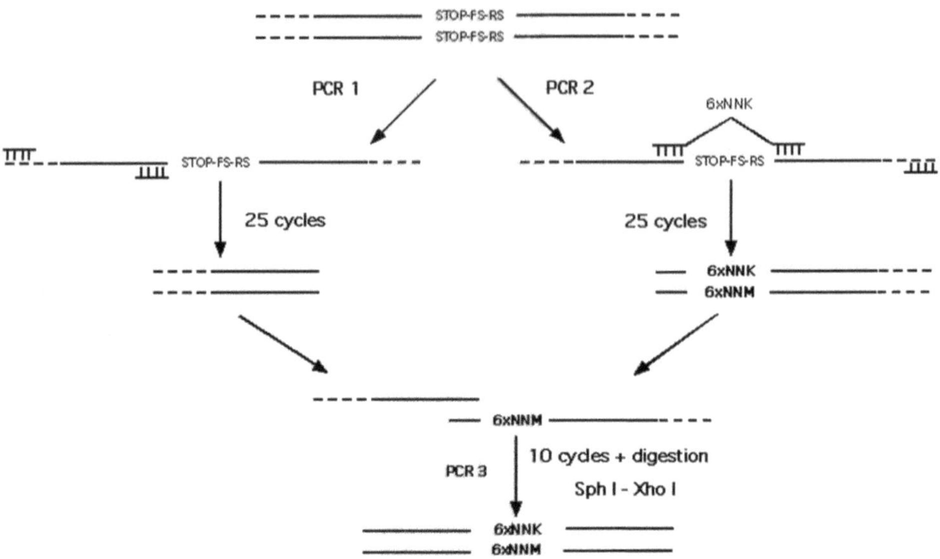

Fig. 4. Schematic representation of the strategy for the synthesis of degenerated libraries (*see* **Subheading 3.2.** for detail). The strategy is divided into three distinct steps: first, the template is obtained by replacing the wild-type sequence of the region to be varied by an in-frame stop codon, a frame shift (FS), and a unique restriction site (RS) to allow for identification of the mutant (the solid line and dashed line correspond, respectively, to the DNA of the gene of interest and of the vector). Next, two PCR reactions are performed on this template. PCR 1 product corresponds to the 5′-end of the gene, whereas PCR 2 corresponds to the 3′-end of the gene. In the latter case, the STOP-FS-RS sequence in the template is replaced by the appropriate number of NNK degenerated codons. Note that the products from the first round of PCR partially hybridize through their 3′- and 5′-ends. Thus, in PCR 3, the products from the first round PCRs act as the template and primers for the reaction. After a low number of cycles, usually 10, the full-length degenerated gene is recovered by digestion with the restriction enzymes *Sph* I and *Xho*I. The library is then ready to be cloned.

3.3. Library Cloning and Recovery

1. The ideal insert vs vector ratio for ligation is 2:1 to 3:1. We try to limit the concentration of DNA ligated to 10 ng/μL and we use 1 m*M* adenosine triphosphate. We allow the ligation to proceed O/N at 16°C (*see* **Note 8**).
2. The enzyme is heat inactivated at 65°C, chloroform extracted, and precipitated with ethanol. The DNA pellet is then air-dried for several minutes, and resuspended in 30 μL of deionized water before electroporation.
3. SS320 *E. coli* strain (*see* **Note 9**) electrocompetent cells are prepared the same day (*see* **Subheading 3.8.**). The ligation reaction from **step 2** is mixed in 300 μL of resuspended SS320 cells. The mix is transferred to 2-mm-wide electroporating cuvets and electroporated on the Genepulser™ II. The apparatus parameters are

adjusted to the following settings for the pulse: 2.5 kV, 25 μF, and 200 to 400 Ω. For optimal results, the time constant should be between 3.8 and 4.5 for 200 Ω and 7.6 and 9.0 for 400 Ω. Immediately after the pulse, 1 mL of ice-cold SOC medium is added to the cuvets. Cells are transferred to a 15-mL conical tube, the cuvets are washed two times with SOC medium to maximally recuperate the electroporated cells, and the cells are allowed to recover for 30 min in 5 mL SOC medium (*see* **Subheading 2.2., item 7** for description) at 37°C, with moderate shaking.

4. The efficiency of the ligation and cloning is evaluated by counting the colonies formed for plating of 1×10^{-4} of the electroporated bacteria (*see* **Note 10**). The remainder is directly inoculated into 250 mL of properly supplemented LB medium in a 500-mL flask (**Subheading 2.2., item 9**).
5. The DNA is isolated with Qiagen Midiprep Kit or similar kits, or, alternatively, by alkaline lysis maxiprep *(47)*.

3.4. Library Screening

1. 100 ng of the pooled library clones recovered from **Subheading 3.3., step 5** is ethanol precipitated and electroporated in 65 μL of BL21 pREP4 cells already harboring a plasmid expressing pQE-32 *ras*-F [3] (*see* **Note 11**) with 1-mm-wide electroporating cuvets. The apparatus parameters are adjusted to the following settings for the pulse: 1.25 to 1.6 kV, 25 μF, and 200 Ω. The time constant is varied from 3.7 to 4.2 on the Genepulser™ II or from 4.0 to 4.6 on Electroporator 2510. The cells are allowed to recover during 30 min in SOC medium at 37°C with moderate shaking.
2. The cells are washed twice with cold PBS to remove traces of SOC medium.
3. The cells are plated on selective medium, as described in **Subheading 2.3., item 6** and allowed to grow for 24 to 72 h at 30°C (*see* **Note 12**). Again, to be able to assess the efficiency of transformation and the ratio of clones in the library that rescue cell growth, a fraction of the electroporated cells, on the order of 1×10^{-3}, should be plated separately to allow colony counting and comparison with a positive control. For example, in our procedure, we transform the same mass of a vector expressing the wild-type RBD fusion to F [1,2] and we plate a dilution of 10^{-4} to 10^{-5} of the electroporated cells. All measures should be taken to avoid cross-contamination of the library pool mix with wild-type positive controls at every step of these manipulations.

3.5. Clonal Competition Experiment

This procedure is adapted from **ref.** ***32***.

1. After the appropriate incubation period, the cells plated at **Subheading 3.4., step 3** are harvested with a small volume of selective medium and incubated in 25 mL of selective medium at 30°C in a shaker at 250 rpm (*see* **Note 13**).
2. After 24 h of incubation, an aliquot of 1 μL of the saturated culture is diluted in 2 mL of fresh selective medium.
3. **Step 2** can be repeated until the targeted enrichment of the library is reached. Normally, we observe that the pool is greatly enriched for one to a few clones after 12 passages (12 d). However, it could vary from system to system, depending on

various factors, such as the level of degeneracy of the libraries and the use of stringency mutants (*see* **Subheading 3.1.2.**).

4. At any step, a 10^{-4} to 10^{-5} diluted aliquot of the saturated culture can be plated on solid selective medium to qualitatively check the efficiency of the competition. The heterogeneity in colony size should decrease and the average size of colonies should increase with each successive competition step.
5. At any passage, the clones represented in a pool mix can be recovered by inoculating 2 mL LB (100 μg/mL ampicillin and 25 μg/mL kanamycin) with 10 μL of the pool mixture and incubated O/N. Then, DNA is prepared with QIAprep™ (*see* **Note 14**).

3.6. Isolation of Clones and Sequencing

This step consists of the isolation of the library-bearing plasmid from independent clones or from the mixed pools obtained by the manipulations described in **Subheadings 3.4.** and **3.5**.

1. 300 ng of DNA from the pool of clones is digested with a mixture of restriction enzymes that recognize sites present in the pREP4 and pQE-32 *ras*-F [3] plasmids, but absent in the library plasmid. For this purpose, we used *Hpa*I, *Xma*I, *Eco*NI, and *Xba*I (*see* **Note 15**).
2. One-tenth of the digested DNA is transformed in XL-1 Blue chemiocompetent cells, and 20 μL is plated on 24-well plates containing LB-agar with 100 μg/mL of ampicillin (*see* **Note 15**).
3. Colonies are picked and incubated in LB, and supplemented with 100 μg/mL of ampicillin, in the appropriate culture vessels.
4. High-quality DNA minipreps are prepared for sequencing. For processing 96 samples in plates, we use the Montage™ kit. We have used QIAprep™ column kit for smaller-scale preps.
5. For sequencing, we have used a primer that anneals only to the library plasmid, i.e., inside F [1,2] (*see* **Note 16**).
6. Sequencing.

3.7. Protein Purification and Characterization

After analysis of the obtained sequences, clones of interest are selected and rearrayed on the appropriate number of 96-well plates. At this moment, clones can be retransformed in XL-1 Blue cells and frozen to serve as backup stock.

1. The selected clones at this step are recloned to express them as fusions with the 6xHis-tag only, i.e., without the DHFR fragment (*see* **Note 17**). The expression of the 6xHis clones is verified by an induction test (*see* **Note 18**).
2. The preps of the clones that express correctly are performed with a Montage™ kit and rearrayed again.
3. These clones are then transformed into BL21 pREP4 cells and plated on LB-agar medium containing 100 μg/mL of ampicillin and 25 μg/mL kanamycin. The plates are incubated at 37°C O/N. When processing several clones in parallel, we use 24-well plates. In this case, no more than 20 μL of competent cells should be used

per transformation, and this should be the maximum volume to be plated per well. Plating more than the maximum volume will not allow the plated cells to absorb completely into the medium.

4. The following day, one colony for each selected clone is picked and incubated O/N in 2.5 mL of LB supplemented with the appropriate antibiotics at 37°C with moderate shaking (*see* **Note 19**).
5. The saturated cultures are diluted 1:10 in 25 mL of TB supplemented with the appropriate antibiotics (*see* **Note 20**).
6. The cultures are then incubated for 90 to 120 min at 37°C in the shaker and IPTG is added at 1 m*M*. After 2 to 4 h of induction, the cells are harvested, and the protein can be purified immediately or stored at –80°C (*see* **Note 21**).
7. The cell pellets are resuspended in 1 mL of Buffer A (*see* **Subheading 2.6., item 8**) by agitation at room temperature until the solution becomes translucent, and then arrayed in a 96-well block (*see* **Note 22**). Most of the insoluble material is then removed by centrifugation at 3200*g* for 40 min on Eppendorf A-4-62 rotor.
8. The 0.2-µm PVDF 96-well plate is filled with 200 µL of 50% Ni-NTA Superflow™ resin. A volume extender is assembled on top of the 0.25-mm glass-fiber 96-well plate. This assembly is placed on the sealing block on top of the vacuum manifold (*see* **Note 23**). The samples (900 µL) are then applied into the first filter plate and 100 µL of ethanol is added to reduce the risk of cross-contamination. Approximately 500 mbar of pressure is applied until all of the samples are completely drawn through the plate (*see* **Note 24**).
9. The PVDF plate assembly should now contain the resin and the samples filtrate. The pressure is interrupted, and the PVDF plate assembly is moved from the collection chamber to be fitted on the sealing block on top of the vacuum manifold. Then, approx 100 mbar of pressure is applied until all of the samples are completely drawn through the resin (*see* **Note 25**).
10. Wash twice with 800 µL of Buffer B (*see* **Subheading 2.6., item 9**). For all wash steps, the vacuum pressure is set at 500 mbar (*see* **Note 26**).
11. Place a fraction collector in the collection chamber. The samples are eluted four times with 100 µL of Buffer E (*see* **Subheading 2.6., item 10**, and **Note 26**) at 100 mbar of pressure. The eluate is supplemented with 1 m*M* DTT and the pH is adjusted to 5 (*see* **Note 27**). The samples are now ready for immediate characterization (*see* **Note 28**).

3.8. Preparation of SS320 and BL21 pREP4 pQE-32 ras-F [3] Electrocompetent Cells

Cells were prepared according to **ref. *48***:

1. Inoculate a single colony of SS320 or BL21 pREP4 pQE-32 *ras*-F [3] into 5 mL of LB medium (for BL21) or 5 mL of SOB medium (for SS320). Grow 5 h to O/N at 37°C with moderate shaking.
2. Inoculate 2.5 mL of the culture into 500 mL of LB medium (for BL21) or 500 mL of SOB medium (for SS320) in a sterile 2-L flask. Grow at 37°C, shaking at 300 rpm, to an optical density (OD) at 600 nm (OD_{600}) of approx 0.5 to 0.7.

3. Chill cells in an ice-water bath 10 to 15 min and transfer to a prechilled 1-L centrifuge bottle.
4. Centrifuge cells for 20 min at 5000*g*.
5. Pour off supernatant and resuspend the pellet in 5 mL of ice-cold water. Add 500 mL of ice-cold water and mix well. Centrifuge cells as in **step 4**.
6. Pour off supernatant immediately and resuspend the pellet by swirling in the remaining liquid.
7. Add another 500 mL of ice-cold water, mix well, and centrifuge again as in **step 4**.
8. Pour off supernatant immediately and resuspend the pellet by swirling in the remaining liquid.
9. If fresh cells are to be used for electroporation, place suspension in a prechilled, 50-mL polypropylene tube, and centrifuge for 10 min at 5000*g* and 2°C. Estimate the pellet volume (usually ~500 µL from a 500-mL culture) and add an equal volume of ice-cold water to resuspend cells (on ice). Aliquot 50 to 300 µL cells into prechilled microcentrifuge tubes. The cell density is approx 2×10^{11} cells/mL.
10. If frozen cells are to be used for electroporation, add 40 mL of ice-cold 10% glycerol to the cells from **step 8** and mix well. Centrifuge cells as described in **step 9**. Estimate the pellet volume and add an equal volume of ice-cold 10% glycerol to resuspend cells (on ice). Place 50 to 300 µL cells into prechilled microcentrifuge tubes and freeze on dry ice. Store at –80°C.

3.9. Preparation of XL-1 Blue and BL21 pREP4 Chemiocompetent Cells

Cells were prepared according to H. Inoue et al. ***(49)***, with one slight modification: the cells are washed only once after the first centrifugation step. The bottles containing the cell pellets are placed, inverted, on a piece of paper at 4°C to remove most traces of medium (*see* **Note 29**).

1. Inoculate 200 mL of SOB or LB medium in a 2-L flask from an O/N culture of the respective cell strain, and grow to an OD_{600} of 0.6 at 18°C with vigorous shaking (200–250 rpm).
2. Chill cells on ice for 10 min and transfer the culture to a 500-mL centrifuge bottle.
3. Centrifuge cells at 2500*g* for 10 min at 4°C.
4. Resuspend the pellet in 80 mL of ice-cold transformation buffer, incubate in an ice bath for 10 min, and centrifuge as in **step 3**.
5. Gently resuspend the cell pellet in 20 mL of transformation buffer, and add DMSO with gentle swirling to a final concentration of 7%.
6. Incubate in an ice bath for 10 min.
7. Aliquot the cell suspension, and freeze in liquid nitrogen.

4. Notes

1. Before PCR, the primers are phosphorylated with T4 polynucleotide kinase to allow ligation. The primers are designed to anneal respectively to each strand, adjacent to the region to be deleted. We used high proofreading polymerases, such as Pfu or Pfu Turbo for this type of PCR. The number of PCR cycles is kept low

(10–16 cycles). The product of the PCR is a linear version of the plasmid without the deleted region. After the PCR, the resultant reaction mixture is digested with *Dpn*I to digest the template DNA. After ethanol precipitation, approx 1/10 of the PCR product is ligated (25 ng PCR product/50 µL ligation reaction) and transformed. The positive clones are screened for the insertion of the appropriate restriction site. A frame shift is included to reduce the possibility of stop codon read-through. The PCR protocol is derived from the ExSite™ PCR-based site-directed mutagenesis kit (Stratagene).

2. We used *Taq* polymerase because the sequences we wanted to amplify were relatively short. For longer genes, we suggest Pfu or Pfu Turbo polymerase, which give more reliable results than Vent polymerase in our hands with this protocol. If the difference in size between the product of PCR 1 and PCR 2 is too large, thus particularly for large genes, we would recommend a mega-primer protocol ***(50,51)***. Both approaches with slight modification could be useful to generate combinatorial libraries in which regions far apart in the sequence are varied simultaneously.
3. The Gelstar™ is diluted 1×10^{-4} to 5×10^{-5} in TAE agarose gel. Gelstar is much more labile than ethidium bromide and, thus, the gels should be prepared on the day that they will be used. In addition, it is difficult to easily quantify DNA with Gelstar™ because the signal becomes saturated at lower quantities of DNA, and, thus, it is difficult to distinguish different amounts of DNA above a certain threshold. In addition, one must be careful not to load large amounts of DNA per well because this could dramatically affect the migration of the samples (typically, we do not load more than 300 ng of plasmid DNA per 25-µL well). Gel staining with ethidium bromide also permits visualization of DNA with the blue light of the Dark Reader, although more DNA must be loaded per well (>1.5 µg for digested plasmid DNA) to allow for visualization of small fragments or PCR products. In the worst case, ultraviolet ethidium bromide visualization can, of course, be used, but care should be taken to process the bands as quickly as possible.
4. The QIAquick™ gel purification protocol is preferable for generating several libraries at the same time, but is more expensive. As an alternative to the procedure described in **Subheading 3.2., step 3**, the electrophoresis step can be replaced by a 2-h, 37°C incubation of the PCR product with the restriction enzyme *Dpn*I to remove the plasmid template DNA. Then the PCR products are purified according to the QIAquick™ PCR purification protocol.
5. This amount could be increased in proportion to the size of the gene under study. The relative quantities of product from PCR 1 and PCR 2 added to PCR 3 are adjusted according to their relative molecular weights.
6. The number of amplification steps in PCR 3 has to be minimized because failure to do so could be detrimental to library representation. For this reason, the number of cycles is minimized and the concentration of the two terminal primers is reduced (*see* **Fig. 3**). It is worth optimizing this step to obtain the maximum quantity of product in a minimum number of cycles. We recommend using *Taq* polymerase, as described in **Note 2**.

7. This vector was obtained by intramolecular religation of pQE-32 digested with *Nhe*I and *Xba*I (compatible cohesive restriction sites). This procedure removed approx 850 bp between the terminator and the origin of replication. The stability and the expression level of the open-reading frame harbored in the new vector was exactly the same as the original, as far as we could judge, and allowed the number of cells transformed per microgram of ligated plasmid DNA to be increased by an order of magnitude. Furthermore, the *Xho*I site present in the original vector was removed to allow use of this restriction site for library cloning.
8. We obtain best results with these ratios, and when we use a vector that is not dephosphorylated.
9. The SS320 cells are obtained by mating MC1061 with XL-1 Blue ***(30)***. An equal volume of both strains at an OD_{600} of 0.6 are mixed and incubated at 37°C for 1 h with smooth shaking (50 rpm). The conjugation is stopped by increasing the shaking to 250 rpm for 5 min. The new strains are isolated by plating on LB petri dishes supplemented with 10 µg/mL tetracycline and spectinomycin. The strains obtained combine the elevated electrocompetence of MC1061 (up to 5×10^{10} colonies/µg of supercoiled plasmid in our laboratory) with the episome overexpressing $LacI_q$ of XL-1 Blue necessary for cells transformed with pQE vectors. This strain could be replaced by any appropriate one for a given expression system.
10. The colonies are counted and the number multiplied by the dilution factor. With 300 ng of vectors resuspended in 30 µL water and electroporated in 300 µL of SS320, we usually obtained between 10^6 to 10^7 colonies.
11. We originally chose to use pQE vector because we wanted to be able to perform the screening and Ni-NTA affinity purification without having to change vectors. The *ras*-F [3] protein fusion is expressed from a pQE-32-derived vector ***(31)***. This plasmid contains the same origin of replication and antibiotic resistance as the plasmid harboring the RBD of *raf* libraries. The selective pressure of the trimethoprim forces the cells to keep both plasmids (i.e., reconstitution of mDHFR from complementary fragments requires that both fusions are expressed). We had engineered a pQE-32-derived plasmid with an alternative origin of replication and antibiotic resistance to express the *ras* fusion, but the stringency of the screening was greatly diminished if we used this vector. In the present case, we mean by a decrease of stringency that the DHFR PCA selects clones that, based on sequence data, should not have the ability to bind to *ras*. These conclusions were drawn from experiments in which sequences of clones of the RBD of *raf* selected by the screening assay revealed that they contained stop codons and aberrant sequences in a region important for binding to *ras,* suggesting that these interactions were nonspecific. The problems we have encountered in these conditions could probably be corrected by inserting the destabilizing mutants in F [3], Ile114Val, or Ile114Ala. We have not tested this yet. We think the stringency is high in the configuration we have chosen, because the BL21 pREP4 cells harboring pQE-32 *ras*-F [3] have the maximum number of copies for the ColE1 origin (harbored by pQE vectors) before the electroporation of the library, thus, allowing conservation of only a minimum number of copies of the library plasmid, as is necessary for

growth in the selective conditions. A good alternative for future screens could be to use vector expressing the bait fusion construct at a much lower level or to use DHFR destabilizing mutants ***(31,32)***.

12. Alternatively, directly inoculate 25 mL of liquid M9 minimal medium with the electroporated cells. This is particularly useful if one wants to directly start a competition experiment.
13. We have observed that the stringency, particularly for the first passages, is improved if cells are plated on solid medium before the competition experiment, meaning that growth on solid phase allows for faster selection of the most-efficient clones.
14. We recommend using QIAprep™ spin kit at this stage, because the DNA extracted by alkaline lysis methods from BL21 pREP4 is not of good quality. In addition, as described in the manufacturer's protocol, the columns are washed once with Qiagen PB buffer when preparing plasmid DNA.
15. **Steps 2** and **3** obliterate the necessity to check whether the cells are transformed with the appropriate plasmid before sequencing. Of the 700 colonies treated according to these procedures, we have never encountered a single colony harboring the pQE-32 *ras*-F [3] or pREP4 plasmid.
16. We used the same primer that anneals to the 3′ region of PCR 2 and 3 (*see* **Subheading 3.2., step 5**).
17. We recommend removing the DHFR fragment before further in vitro characterization, unless it is required for specific experiments, because it diminishes yields, complicates purification procedures, and might modify protein characteristics. We recommend engineering the plasmid in such a way that the DHFR fragments could be removed and the plasmid religated intramolecularly. To do so, we have used the compatible restriction sites, *Xho*I and *Sal*I, to clone, respectively, the library and F [1,2] fragment.
18. O/N saturated cultures are diluted 1:10 in TB with 100 μg/mL ampicillin and incubated at 37°C at 300 rpm in a 96-well culture block. After 90 min, the cultures are induced with 1 m*M* IPTG. After 4 h, 60 μL aliquots of each clone are pipetted, and 1 volume of 2X SDS-PAGE loading buffer is added. Sample, markers, and protein induction test samples are loaded on 15% acrylamide SDS-PAGE, and the protein bands are visualized by Coomassie brilliant blue staining. Alternatively, expression could be check by Western blot with an antibody directed against the 6xHis-tag (Qiagen).
19. Before proceeding to the large-scale purification, we recommend making test purifications of several isolated clones to check yields obtained with different culture volumes. We recommend purification under denatured conditions, because it is easier to perform, for obvious reasons, if several purifications are processed in parallel. Nevertheless, if the protein is very soluble, one could design a simple native-state purification protocol amenable to such a scale. In our case, 25-mL cultures of BL21 pREP4 cells expressing *raf* RBD mutants, processed as described in **Subheading 3.7.**, yielded 400 μL of protein solution at concentrations of 200 to 600 μg/mL.
20. Depending on the quantity of protein needed or the expression level, the volume of culture can vary from 5 to 50 mL. Read **Note 19** for more details.

21. Alternatively, protein expression could be induced O/N at 30°C.
22. Otherwise, if the quantity of clones to analyze was lower, we used Ni-NTA spin columns. If more protein from fewer independent clones is needed, regular Ni-NTA agarose column purification is the method of choice. In this case, it may be useful and reasonable to perform the purification under native conditions.
23. The described system only works with Millipore large collector and sealing block. It is also possible to replace vacuum steps by centrifugation.
24. A gauge is included with the Millipore manifold.
25. If required, the flow-through and the filtrate from the diverse washing steps are collected by placing a fraction collector in the collection chamber, in **Subheading 3.7., steps 9** and **10**.
26. In our case, the protein is eluted under denaturing conditions and the sample diluted sufficiently to reduce the denaturant to a concentration that does not interfere with protein function. The concentration of the denaturant used in the elution buffer should be fixed according to the stability of the protein studied. Nevertheless, if it is desired to elute protein in its native condition, Qiagen recommends using a gradient of decreasing concentration of Gdn-HCl. The protein is then eluted with the appropriate buffer, without denaturant. One could also simply purify the protein under native conditions (*see* the Qiagen Expressionist and Ni-NTA spin column handbook). We have improved elution by using diluted glacial acetic acid, in quantities sufficient to buffer 25 or 50 m*M* NaOAc to pH 5.0 (*see* **Note 27** concerning pH adjustment after elution). Urea could also be used as denaturant, provided that it will denature the protein under study at reasonable concentrations. In addition, care must be taken because urea in solution equilibrates with cyanate. However, urea does not precipitate SDS even at high concentrations, thus, facilitating the characterization of the protein at individual purification steps by SDS-PAGE electrophoresis.
27. DTT is added to the sample after elution because it is not recommended for use with Ni-NTA. The pH of the eluate can be adjusted to a relevant value by the addition of NaOH or of an appropriate weak base, such as Tris or NaOAc. The quantity of base to be added is determined empirically on larger volumes with a pH meter.
28. If the samples are not going to be characterized immediately, we recommend adding 40% glycerol (v/v) and 1 m*M* NaN_3 for storing of samples at –20°C. The glycerol can then be removed by dialysis or ultrafiltration before characterization. Alternatively, samples without glycerol can be flash frozen in liquid nitrogen and preserved at –80°C.
29. BL21 pREP4 chemiocompetent cells prepared according to QIAEXpressionnist Handbook with TFB buffer (RbCl) are usually more competent, but because we needed to transform only super-coiled DNA in our protocols, the Inoue method was satisfactory ***(48)***.

Acknowledgments

We are grateful to Emil Pai, David Waugh, and Shigekazu Nagata for the cDNAs of h-*ras*, *raf* RBD, and the CAD domain's of *cad* and *icad*, respectively. We also want to thank Dimitri Sans for carefully reading this manuscript, Jérôme

Dupras for help with the CAD constructs, and Joelle Pelletier and other members of the laboratory who contributed to the development of the DHFR PCA. F.-X. C.-V. is a Canadian Institutes of Health Research and Fonds pour la formation de cherocheors et l'aide à la recherhe scholar.

References

1. Chen, K. Q. and Arnold, F. H. (1991) Enzyme engineering for nonaqueous solvents: random mutagenesis to enhance activity of subtilisin E in polar organic media. *Biotechnology (NY)* **9,** 1073–1077.
2. Chen, K. Q., Robinson, A. C., Van Dam, M. E., Martinez, P., Economou, C., and Arnold, F. H. (1991) Enzyme engineering for nonaqueous solvents. II. Additive effects of mutations on the stability and activity of subtilisin E in polar organic media. *Biotechnol. Prog.* **7,** 125–129.
3. Iffland, A., Tafelmeyer, P., Saudan, C., and Johnsson, K. (2000) Directed molecular evolution of cytochrome c peroxidase. *Biochemistry* **39,** 10,790–10,798.
4. Scott, J. K. and Smith, G. P. (1990) Searching for peptide ligands with an epitope library. *Science* **249,** 386–390.
5. Lowman, H. B., Bass, S. H., Simpson, N., and Wells, J. A. (1991) Selecting high-affinity binding proteins by monovalent phage display. *Biochemistry* **30,** 10,832–10,838.
6. Keefe, A. D. and Szostak, J. W. (2001) Functional proteins from a random-sequence library. *Nature* **410,** 715–718.
7. Imanaka, T., Shibazaki, M., and Takagi, M. (1986) A new way of enhancing the thermostability of proteases. *Nature* **324,** 695–697.
8. Shih, P. and Kirsch, J. F. (1995) Design and structural analysis of an engineered thermostable chicken lysozyme. *Protein. Sci.* **4,** 2063–2072.
9. Riddle, D. S., Santiago, J. V., Bray-Hall, S. T., et al. (1997) Functional rapidly folding proteins from simplified amino acid sequences. *Nat. Struct. Biol.* **4,** 805–809.
10. Kim, D. E., Gu, H., and Baker, D. (1998) The sequences of small proteins are not extensively optimized for rapid folding by natural selection. *Proc. Natl. Acad. Sci. USA* **95,** 4982–4986.
11. Kamtekar, S., Schiffer, J. M., Xiong, H., Babik, J. M., and Hecht, M. H. (1993) Protein design by binary patterning of polar and nonpolar amino acids. *Science* **262,** 1680–1685.
12. Stemmer, W. P. (1994) Rapid evolution of a protein in vitro by DNA shuffling. *Nature* **370,** 389–391.
13. Crameri, A., Raillard, S. A., Bermudez, E., and Stemmer, W. P. (1998) DNA shuffling of a family of genes from diverse species accelerates directed evolution. *Nature* **391,** 288–291.
14. Zhao, H., Giver, L., Shao, Z., Affholter, J. A., and Arnold, F. H. (1998) Molecular evolution by staggered extension process (StEP) in vitro recombination. *Nat. Biotechnol.* **16,** 258–261.
15. Ostermeier, M., Shim, J. H., and Benkovic, S. J. (1999) A combinatorial approach to hybrid enzymes independent of DNA homology. *Nat. Biotechnol.* **17,** 1205–1209.

16. Lutz, S. and Benkovic, S. J. (2000) Homology-independent protein engineering. *Curr. Opin. Biotechnol.* **11,** 319–324.
17. Coco, W. M., Levinson, W. E., Crist, M. J., et al. (2001) DNA shuffling method for generating highly recombined genes and evolved enzymes. *Nat. Biotechnol.* **19,** 354–359.
18. Murakami, H., Hohsaka, T., and Sisido, M. (2002) Random insertion and deletion of arbitrary number of bases for codon-based random mutation of DNAs. *Nat. Biotechnol.* **20,** 76–81.
19. Crameri, A., Dawes, G., Rodriguez, E., Jr., Silver, S., and Stemmer, W. P. (1997) Molecular evolution of an arsenate detoxification pathway by DNA shuffling. *Nat. Biotechnol.* **15,** 436–438.
20. Christians, F. C., Scapozza, L., Crameri, A., Folkers, G., and Stemmer, W. P. (1999) Directed evolution of thymidine kinase for AZT phosphorylation using DNA family shuffling. *Nat. Biotechnol.* **17,** 259–264.
21. MacBeath, G., Kast, P., and Hilvert, D. (1998) Redesigning enzyme topology by directed evolution. *Science* **279,** 1958–1961.
22. Ostermeier, M., Nixon, A. E., Shim, J. H., and Benkovic, S. J. (1999) Combinatorial protein engineering by incremental truncation. *Proc. Natl. Acad. Sci. USA* **96,** 3562–3567.
23. O'Neil, K. T. and Hoess, R. H. (1995) Phage display: protein engineering by directed evolution. *Curr. Opin. Struct. Biol.* **5,** 443–449.
24. Roberts, R. W. and Szostak, J. W. (1997) RNA-peptide fusions for the in vitro selection of peptides and proteins. *Proc. Natl. Acad. Sci. USA* **94,** 12,297–12,302.
25. Firestine, S. M., Salinas, F., Nixon, A. E., Baker, S. J., and Benkovic, S. J. (2000) Using an AraC-based three-hybrid system to detect biocatalysts in vivo. *Nat. Biotechnol.* **18,** 544–547.
26. Hanes, J., Jermutus, L., and Plückthun, A. (2000) Selecting and evolving functional proteins in vitro by ribosome display. *Methods Enzymol.* **328,** 404–430.
27. Smith, G. P. (1985) Filamentous fusion phage: novel expression vectors that display cloned antigens on the virion surface. *Science* **228,** 1315–1317.
28. Dunn, I. S. (1996) Phage display of proteins. *Curr. Opin. Biotechnol.* **7,** 547–553.
29. Forrer, P., Jung, S., and Plückthun, A. (1999) Beyond binding: using phage display to select for structure, folding and enzymatic activity in proteins. *Curr. Opin. Struct. Biol.* **9,** 514–520.
30. Sidhu, S. S., Lowman, H. B., Cunningham, B. C., and Wells, J. A. (2000) Phage display for selection of novel binding peptides. *Methods Enzymol.* **328,** 333–363.
31. Pelletier, J. N., Campbell-Valois, F. -X., and Michnick, S. W. (1998) Oligomerization domain-directed reassembly of active dihydrofolate reductase from rationally-design fragments. *Proc. Natl. Acad. Sci. USA* **95,** 12,141–12,146.
32. Pelletier, J. N., Arndt, K. M., Plückthun, A., and Michnick, S. W. (1999) An in vivo library-versus-library selection of optimized protein-protein interactions. *Nat. Biotechnol.* **17,** 683–690.

33. Michnick, S. W., Remy, I., Campbell-Valois, F. X., Vallee-Belisle, A., and Pelletier, J. N. (2000) Detection of protein-protein interactions by protein fragment complementation strategies. *Methods Enzymol.* **328,** 208–230.
34. Michnick, S. W. (2001) Exploring protein interactions by interaction-induced folding of proteins from complementary peptide fragments. *Curr. Opin. Struct. Biol.* **11,** 472–477.
35. Appleman, J. R., Prendergast, N., Delcamp, T. J., Freisheim, J. H., and Blakley, R. L. (1988) Kinetics of the formation and isomerization of methotrexate complexes of recombinant human dihydrofolate reductase. *J. Biol. Chem.* **263,** 10,304–10,313.
36. Arndt, K. M., Pelletier, J. N., Muller, K. M., Alber, T., Michnick, S. W., and Plückthun, A. (2000) A heterodimeric coiled-coil peptide pair selected in vivo from a designed library-versus-library ensemble. *J. Mol. Biol.* **295,** 627–639.
37. Mossner, E., Koch, H., and Plückthun, A. (2001) Fast selection of antibodies without antigen purification: adaptation of the protein fragment complementation assay to select antigen-antibody pairs. *J. Mol. Biol.* **308,** 115–122.
38. Michnick, S. W. and Shakhnovich, E. (1998) A strategy for detecting the conservation of folding-nucleus residues in protein superfamilies. *Fold. Des.* **3,** 239–251.
39. Campbell-Valois, F. X., Tarassov, K., and Michnick, S. W. (2005) Massive sequence perturbation of a small protein. *Proc. Natl. Acad. Sci. USA* **102,** 14,988–14,993.
40. Gu, H., Yi, Q., Bray, S. T., Riddle, D. S., Shiau, A. K., and Baker, D. (1995) A phage display system for studying the sequence determinants of protein folding. *Protein Sci.* **4,** 1108–1117.
41. Emerson, S. D., Waugh, D. S., Scheffler, J. E., Tsao, K. L., Prinzo, K. M., and Fry, D. C. (1994) Chemical shift assignments and folding topology of the Ras-binding domain of human Raf-1 as determined by heteronuclear three-dimensional NMR spectroscopy. *Biochemistry* **33,** 7745–7752.
42. Emerson, S. D., Madison, V. S., Palermo, R. E., et al. (1995) Solution structure of the Ras-binding domain of c-Raf-1 and identification of its Ras interaction surface. *Biochemistry* **34,** 6911–6918.
43. Sambrook, J., Fritsch, E. F., and Maniatis, T. (1989) Bacterial media, antibiotics and bacterial strains, in *Molecular Cloning* (Nolan, C., ed.), vol. 3. Cold Spring Harbor Laboratory Press, Cold Spring Harbor, NY, p. A.3.
44. Remy, I., Wilson, I. A., and Michnick, S. W. (1999) Erythropoietin receptor activation by a ligand-induced conformation change. *Science* **283,** 990–993.
45. Nassar, N., Horn, G., Herrmann, C., Scherer, A., McCormick, F., and Wittinghofer, A. (1995) The 2.2 A crystal structure of the Ras-binding domain of the serine/threonine kinase c-Raf1 in complex with Rap1A and a GTP analogue. *Nature* **375,** 554–560.
46. Block, C., Janknecht, R., Herrmann, C., Nassar, N., and Wittinghofer, A. (1996) Quantitative structure-activity analysis correlating Ras/Raf interaction in vitro to Raf activation in vivo. *Nat. Struct. Biol.* **3,** 244–251.
47. Sambrook, J., Fritsch, E. F., and Maniatis, T. (1989) Alkalyne lysis and PEG preparation for large scale DNA preparation, in *Molecular Cloning*, vol. 1 (Nolan, C., ed.) Cold Spring Harbor Laboratory Press, Cold Spring Harbor, NY, pp. 1.38–1.41.

48. Seidman, C. E., Struhl, K., Sheen, J., and Jessen, T. (1997) Introduction of plasmid DNA into cells, basic protocol 2, in *Current Protocols in Molecular Biology*, vol. 1, Suppl 37 (Ausubel, F. M., Brent, R., Kingston, R. E., Moore, D. D., Seidman, J. G., Smith, J. A., and Struhl, K., eds.) John Wiley & Sons, Inc., New York, NY, pp. 1.8.4–1.8.5.
49. Inoue, H., Nojima, H., and Okayama, H. (1990) High efficiency transformation of Escherichia coli with plasmids. *Gene* **96,** 23–28.
50. Sarkar, G. and Sommer, S. S. (1990) The "megaprimer" method of site-directed mutagenesis. *Biotechniques* **8,** 404–407.
51. Brons-Poulsen, J., Petersen, N. E., Horder, M., and Kristiansen, K. (1998) An improved PCR-based method for site directed mutagenesis using megaprimers. *Mol. Cell Probes* **12,** 345–348.

16

A General Method of Terminal Truncation, Evolution, and Re-Elongation to Generate Enzymes of Enhanced Stability

Jochen Hecky, Jody M. Mason, Katja M. Arndt, and Kristian M. Müller

Summary

Improving enzyme stability is a highly desirable design step in generating enzymes able to function under extreme conditions, such as elevated temperatures, while having the additional benefit of being less susceptible to cleavage by proteases. For these reasons, many different approaches and techniques have been devised in constructing such proteins, but the results to date have been of mixed success. Here, we present a robust method involving the terminal truncation, random mutagenesis and fragmentation, recombination, elongation, and finally, selection at physiological temperatures, to generate an enzyme with improved stability. Three cycles of directed evolution comprising of random mutagenesis, DNA shuffling, and selection at 37°C were used, using the bacterial enzyme TEM-1 β-lactamase as a model protein to yield deletion mutants with in vivo ampicillin resistance levels comparable to wild-type (wt) enzyme. Kinetic studies demonstrate the selected mutant to have a significantly improved thermostability relative to its wt counterpart. Elongation of this mutant to the full-length gene resulted in a β-lactamase variant with dramatically increased thermostability. This technique was so fruitful that the evolved enzyme retained its maximum catalytic activity even 20°C above its wt parent protein optimum. Thus, structural perturbation by terminal truncation and subsequent compensation by directed evolution at physiological temperatures is a fast, efficient, and highly effective way to improve the thermostability of proteins without the need for selecting at elevated temperatures.

Key Words: Protein stabilization; protein stability; terminal truncation; thermostability; enzyme activity; DNA shuffling; random mutagenesis; elongation.

1. Introduction

If an enzyme were able to be modified for increased stability, a number of new options could become open for exploitation. For example, the protein would be active over a greater range of temperatures, and would be likely to have an

From: *Methods in Molecular Biology, vol. 352: Protein Engineering Protocols*
Edited by: K. M. Arndt and K. M. Müller © Humana Press Inc., Totowa, NJ

improved half-life resulting from a lower sensitivity to proteases. Stability, together with increased specificity and activity, is one of the most sought-after properties of a protein to be improved and applied either industrially or pharmacologically.

A large number of factors enhancing thermostability has been identified so far ***(1–4)***. Examples include increased rigidity and compactness, more core hydrophobic residues, and increased van der Waals interactions. Improved thermostability can also be achieved by the introduction of metal binding sites, additional disulfide bridges ***(5,6)***, backbone cyclization, and by the shortening or deletion of loop regions ***(7)***. Two of the main factors of high protein thermostability seem to be increased hydrogen bonding and the formation of ionic interaction networks ***(8–10)***.

However, because a complete understanding of thermostability at the molecular level remains elusive ***(11)***, the generation of thermostable proteins continues to be a challenging task. Computational and comparative approaches have been established to aid in the identification of stabilizing interactions and both have been applied successfully for these purposes ***(12–14)***. However, they rely either on the presence of a well-resolved crystal or nuclear magnetic resonance structure, that is, a defined Cα trace, or on the availability of numerous homologous sequences, preferably from thermophilic sources. In addition, evolutionary approaches mimicking the Darwinian optimization cycle "mutagenesis, recombination, and selection" have been introduced to overcome the boundaries of current rational protein design ***(15,16)***.

1.1. Terminal Truncations

The importance of the terminal regions of proteins for structural or functional integrity varies from protein to protein. Some proteins readily tolerate the removal of terminal residues without any impairment of the native three-dimensional structure and conformational stability, which is in line with the observation that, in crystal structures, the terminal sections are frequently either poorly defined or not ordered at all. In contrast, some proteins are very sensitive to terminal shortening. In this case, the removal of several terminal residues can result in a large reduction in conformational stability, a looser structure ***(17)***, and an enhanced susceptibility to proteolysis or aggregation ***(18)***. This has been demonstrated for RNase A ***(19)***, RNase HI ***(20)***, Staphylococcal nuclease R ***(17)***, Rhodanese ***(21,22)***, the Stoffel fragment of *Taq* DNA polymerase ***(23)***, and chloramphenicol acetyl transferase ***(18)***. Although some proteins may not tolerate the removal of even a single terminal amino acid residue, for the majority of proteins, a threshold level of truncation seems to exist within which the expression level, the native conformation, and the stability are significantly affected although they are still compatible with protein function under certain circumstances (e.g., lowered reaction temperature).

Beyond this threshold, however, truncations lead to irreversible damaging of the protein structure and result in a complete loss of function. Therefore, by carefully selecting the correct degree of truncation, structural perturbations can be introduced easily in many protein structures without interfering with the folding process and a functional conformation.

1.2. Compensation of Structural Perturbations by Directed Evolution

Structural perturbations introduced into a protein by amino acid exchanges, insertions, or deletions can be reversed by single or multiple compensatory mutations, which often act as so-called global or second-site suppressors *(24)* because they suppress the phenotypes of various otherwise detrimental mutations at sites distant to the primary exchanges. This phenomenon follows from the high complexity and plasticity of protein structures and the intramolecular interaction networks governing protein stability. In the end, it is facilitated by the degenerate nature of protein structures, which are determined essentially by a limited number of key residues, whereas a large number of residues can be replaced without any phenotypic effect. Second-site suppressors of truncated versions of RNase HI have been created and identified by random mutagenesis followed by selection of functional variants *(20)*. However, the introduction of mutations by a random mutagenic process alone is likely to be accompanied by the accumulation of detrimental mutations reducing the number of functional proteins and, thus, the number of testable beneficial mutations. To overcome the limitations of random mutagenesis alone, a combinatorial process known as DNA shuffling has been introduced *(15,16)*, marking a cornerstone in the field of DNA-based protein manipulation. The resulting approach comprises repeated cycles of creation and jumbling of point mutations, and is referred to as in vitro or directed evolution. It is based on the same principles that govern natural evolution, namely, random mutagenesis, recombination, and screening or selection. Using this approach, many enzymes have since been improved considerably in terms of activity, selectivity, stability, and even function in organic solvents *(25)*. Directed evolution features many advantages over rational approaches because no previous knowledge of three-dimensional structure is required to improve enzyme properties. However, its applicability is strictly restricted to those proteins for which appropriate screening or, even better, selection procedures exist.

This chapter summarizes a manageable, widely applicable approach for the improvement of the stability of proteins. It is based on the introduction of structural perturbations by trimming the chain ends of a target protein at the DNA level (**Subheadings 3.2.** and **3.3.**). The detrimental phenotypes accompanying the truncations (**Subheading 3.4.**) are overcome by an evolutionary, genetic in vitro optimization procedure (**Subheading 3.5.**). The resulting truncation–optimization –elongation approach (**Fig. 1A**) stands out from other techniques by its simplicity because it relies neither on available structural

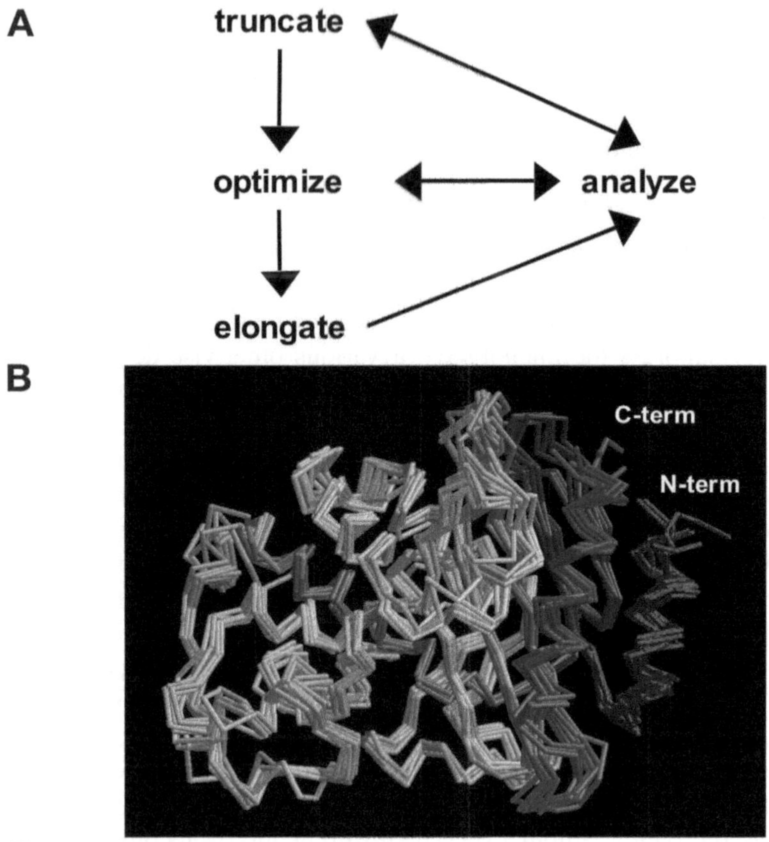

C

```
1BTL/        HPETLVKVKDAEDQL ~ MDERNRQIAEIGASLIKHW
1TEM
1SHV:A       spQPLEQIKLSESQL ~ MAERNQQIAGIGAALIEHWqr
1MBL:B       ktemkDDFAKLEEQF ~ ---
DDKLIAEATKVVMKalnmng
1BZA         anSVQQQLEALEKSS ~ R--RRDILAAAAKIVThgf
1BUL         ntkGIDEIKNLETDF ~ H--EDKVIAEASRIAIdnlk
1BSG   ghgsgsvsDAERRLAGLERAS ~ R--DDGLVADAARVLAETLg
1MFO         -APIDDQLAELERRD ~ N--LRPLIGELTALVLPSLl
1PIO:A       ----mKELNDLEKKY ~ P--NDKLISETAKSVMKEF--
```

Fig. 1. **(A)** Scheme of truncation–optimization–elongation. The enzyme is truncated, optimized by directed evolution, and, finally, elongated to the full-length protein. All steps information nor on the existence of known homologous proteins. However, if this kind of information can be obtained using the multitude of current databases, the truncation design process will be, of course, more straightforward, and, in some cases, more efficient. Finally, the described approach features an additional benefit because it allows the generation of thermostable enzyme variants with the selection being performed at physiological temperatures.

2. Materials

2.1. Plasmid Construction, DNA Shuffling, and Random Mutagenesis

1. Forward primer pr_sfi_pelB_DG_blawt: 5′-TTACTCACGGCCCAGCCGGCCATG-GCTGACGG **TCACCCAGAAACGCTGGTG**-3′; restriction sites are underlined and hybridizing regions are in bold.
2. Forward primer pr_sfi_pelB_DG_bladelHPE: 5′-TTACTCGCGGCCCAGCCG-GCCATGGCTGACGGT**ACGCTGGTGAAAGTAAAAGATG**-3′.
3. Forward primer pr_sfi_pelB_DG_bladelHPETL: 5′-TTACTCACGGCCCAGC-CGGCCATGGCTGACGGT**GTGAAAGTAAAAGATGCTGAAG** -3′.
4. Reverse primer pr_blawt_GG_his5_hind: 5′-TAGTCAAGCTTACTAGTGATG-GTGATGGTGGCCACC**CCAATGCTTAATCAGTGAGG**-3′.
5. Reverse primer pr_bladelW_GG_his5_hind: 5′-TAGTCAAGCTTACTAGTGAT GGTGATGGTGGCCACC**ATGCTTAATCAGTGAGGCACC**-3′.
6. Reverse primer pr_bla_delKHW_GG_his5_hind: 5′-TAGATCAAGCTTACTAG TGATGGTGATGGTGGCCACC**AAT CAGTGAGGCACCTATCTC**-3′.
7. Plasmid pKMENGRbla, a derivative of pAK400 ***(26)***.
8. Restriction enzymes *Sfi*I and *Hin*dIII (NEB).
9. Plasmid pKJE_Bla-NΔ5: expresses the TEM-1 β-lactamase with the first five N-terminal residues deleted (*see* **Subheading 3.3.**).
10. Forward primer pr_sfi_pelB_DG_shuffl: 5′-TTACTCACGGCCCAGCCGGC-CATGGCTGACGG-3′; restriction sites are underlined.

Fig. 1. *(Continued)* can be analyzed for activity and stability. **(B)** Superimposed Cα traces show various β-lactamases from different bacterial species using the CE algorithm ***(30)***. Contiguous secondary structural elements are in different shades of gray, using RasMol ***(47)***. **(C)** Structure-based sequence alignment of the N- and C-terminal parts of the β-lactamases displayed in (B), corresponding to the N- and C-terminal helix of the TEM-1 enzyme (PDB entries 1btl/1tem) are shown. Sequences are specified by given PDB codes. Residues that were ignored by the CE algorithm and are, therefore, not included in the superposition, are given in lowercase letters. Residues present in the SWISSPROT database, but missing in the corresponding X-ray structure, are indicated in italics. Omitted parts of the sequence are symbolized by ~. Selected β-lactamase genes showed identity ranging from 67.8 to 32.4%. Corresponding structures had root mean square deviation values ranging from 1.2 to 2.2 Å (with reference to the TEM-1 β-lactamase) and Z-scores greater than 7. The structural superposition demonstrates the structural variability of the terminal helices.

11. Reverse primer pr_GG_his5_hind_shuffl: 5′-TAGTCAAGCTTACTAGTGATG-GTGATGGTGGCCACC-3′.
12. *Taq* DNA polymerase (Sigma).
13. DNase I (Sigma).
14. DNase buffer: 50 m*M* Tris-HCl, pH 7.5, and 1 m*M* $MgCl_2$.
15. EDTA solution: 0.5 *M* EDTA.
16. Agarose gel: 1 to 2% (w/v) agarose in 0.5X Tris-base–boric acid–EDTA (TBE) buffer.
17. 10X TBE buffer: 1 *M* Tris-base, 1 *M* boric acid, and 20 m*M* EDTA, pH 8.0.
18. QiaexII gel extraction kit (Qiagen).
19. dNTP (Amersham).
20. Thermocycler (Eppendorf).
21. $MgCl_2$ stock solution: 25 to 100 m*M* $MgCl_2$.
22. $MnCl_2$ stock solution: 5 m*M* $MnCl_2$.
23. GFX polymerase chain reaction (PCR) DNA and gel band purification kit (Amersham).

2.2. Transformation, Protein Expression, and Purification

1. Butanol.
2. Electrocompetent *Escherichia coli* XL-1 Blue cells.
3. Electroporator (Bio-Rad).
4. 2YT: dissolve 16 g bacto-tryptone, 10 g yeast extract, and 5 g NaCl in 1 L H_2O and autoclave.
5. Transformation salt stock: 250 m*M* KCl and 1 *M* $MgCl_2$.
6. Ampicillin stock solution: dissolve 100 mg/mL ampicillin in water and filter through 0.22-μm, aliquot, and store at –20°C.
7. LB/Cm agar plates: 1% bacto-tryptone, 0.5% yeast extract, 0.5% NaCl, 1.5% agar, and 25 μg/mL chloramphenicol.
8. LB/Cm: 1% bacto-tryptone, 0.5% yeast extract, 0.5% NaCl, 1.5% agar, and 25 μg/mL chloramphenicol.
9. 2YT/Cm: 2YT medium supplemented with 25 μg/mL chloramphenicol.
10. High-speed centrifuge (Sorvall, with GS-3 and SS-34 rotor).
11. Resuspension buffer: 50 m*M* sodium phosphate and 500 m*M* NaCl, pH 7.0.
12. Benzonase (Sigma).
13. 0.45-μm polyethersulfone syringe filter.
14. 2-mL phenylboronate columns (MoBiTec).
15. Borate buffer: 0.5 *M* borate and 0.5 *M* NaCl, pH 7.0.
16. 4-mL Ni-nitrilotriacetic acid (NTA) column: 4 mL Ni-NTA superflow matrix (Qiagen) in a C10/10 column (Amersham-Pharmacia).
17. Imidazole buffer: 50 m*M* sodium phosphate, 0.25 *M* imidazol, and 50 m*M* NaCl, pH 7.0.
18. Phosphate buffer: 50 m*M* sodium phosphate and 150 m*M* NaCl, pH 7.2.
19. EDTA.
20. 1-mL phenylsuperose HR 5/5 column (Amersham-Pharmacia).

21. Tris-HCl buffer: 25 m*M* Tris-HCl and 25 m*M* NaCl, pH 8.0.
22. Mono Q HR 5/5 column (Amersham-Pharmacia).
23. Tris-HCl–NaCl buffer: 25 m*M* Tris-HCl and 0.5 *M* NaCl, pH 8.0.

2.3. Enzyme Assays and Urea-Induced Unfolding

1. Nitrocefin solution: 0.2 m*M* nitrocefin, 50 m*M* potassium phosphate and 0.5% dimethylsulfoxide, pH 7.0.
2. Photospectrometer that can at least measure one data point per second, e.g., Ultrospec 3000 (Amersham-Pharmacia) or V-550 (Jasco).
3. Spectrofluorometer, e.g., FluoroMax-2 (Jobin-Yvon) or FP-6500 (Jasco).
4. Phosphate buffer (*see* **Subheading 2.2., item 18**).
5. Urea: ultrapure, at least 99% purity (ICN).

3. Methods

This chapter presents a general method to increase the thermostability of proteins without the need of screening for activity at elevated temperatures. After an introduction of our model system, β-lactamase (**Subheading 3.1.**), we describe the design considerations for terminal truncations (**Subheadings 3.2.** and **3.3.**) and their effect on enzyme activity (**Subheading 3.4.**). Libraries are generated by directed evolution and error-prone PCR (**Subheading 3.5.**), and selected in vivo by applying different selection stringencies (**Subheading 3.6.**). The last part of this chapter discusses the re-elongation of evolved truncation mutants (**Subheading 3.7.**), expression strategies and purification methods (**Subheading 3.8.**), and enzymatic assays (**Subheading 3.9.**) and stability tests (**Subheading 3.10.**) for characterization.

3.1. β-Lactamase as a Model System

To demonstrate the applicability of the truncation–optimization–elongation approach (**Fig. 1A**) for the improvement of thermostability, TEM-1 β-lactamase was chosen as a model system. As a clinically relevant pathogenesis factor and important resistance marker, it belongs to a well-characterized family of proteins, for which several crystal structures are available. In addition, many mutagenic studies have been performed facilitating the study and interpretation of structure–function relationships and the rational design of truncations. Furthermore, β-lactamase is used for prodrug activation cancer therapy, making a stable β-lactamase a potential lead compound *(27,28)*.

Class A β-lactamases (EC 3.5.2.6) are bacterial periplasmic enzymes with relatively diverse amino acid sequences and stabilities but very similar tertiary structures. Although the polypeptide backbones of class A β-lactamases superimpose particularly well in the core region, they do so poorly at the ends of the terminal helices. The structural differences at the termini are accompanied by a high variability in length and amino acid composition of these regions, making these enzymes well-suited to test general ideas regarding the relation of sequence homology and functional importance.

3.2. Design of Terminal Truncations

The correct design of the deletions constitutes a central step in the approach described here: if the termini are shortened by too many residues, the resulting truncation variant is likely to be nonfunctional because of insufficient stability or defective folding; alternatively, if not enough residues are removed, it is unlikely that an impaired phenotype will arise and, thus, no selection pressure can be imposed on the gain of stabilizing exchanges.

If a crystal structure of the protein of interest exists, the Protein Data Bank (PDB; http://www.pdb.org/), SWISSPROT (http://www.expasy.org/sprot/ and http://www.ebi.ac.uk/swissprot/), and SCOP (http://scop.mrc-lmb.cam.ac.uk/scop/) databases in combination with structure analysis tools, such as the molecular modeling program WHATIF (http://swift.cmbi.kun.nl/WIWWWI/) and the CE algorithm (http://cl.sdsc.edu/ce.html) can be used for structural comparison and analysis. In particular, contact maps can be used to identify possible nonessential and essential interactions, which may help in the design process. If no high-resolution structure is available, other databases (e.g., HSSP) and the SWISS-MODEL homology-modeling server (http://swissmodel.expasy.org) can be applied, or, alternatively, sequential deletions in one or two amino acid steps can be performed to define the threshold level.

In this study, four different truncation variants were planned using information from mutagenic studies ***(29)***, a structure-based alignment of various homologous class A β-lactamases, and structural analysis using the CE algorithm ***(30)*** and the SWISS-PDB Viewer (http://www.expasy.org/spdbv) ***(31)***, respectively (**Fig. 1B,C**). The truncations were intended to introduce structural perturbations at physiological temperatures without rendering the protein completely nonfunctional. The first three residues at the N-terminus (His, Pro, and Glu) have high temperature factors (up to 23.8 $Å^2$; average: 13.1 $Å^2$) and solvent accessibilities (PDB code 1btl), and their deletion (yielding the mutant NΔ3) was expected to affect stability only marginally. The second N-terminal truncation (NΔ5) included the adjacent threonine and leucine, the first of which is nearly completely buried and forms a hydrogen bond to Ser285 of the C-terminal helix. At the C-terminal end, only one or three residues (mutants CΔ1 and CΔ3) were removed, because the distal tryptophan has been shown to be "essential" in a previous study ***(32)***.

3.3. Plasmid Design

The plasmids were designed to encode the mature wild-type (wt) β-lactamase or the respective deletion mutants as fusion proteins with an N-terminal pelB

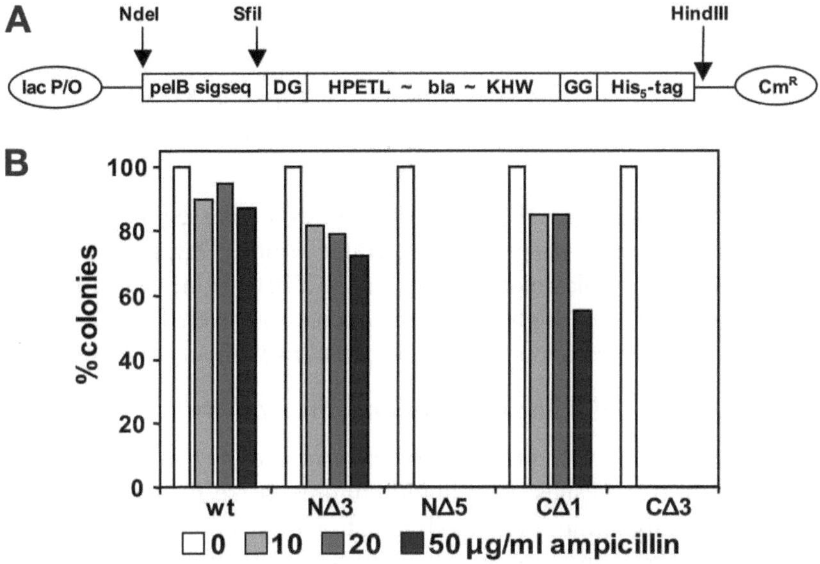

Fig. 2. (**A**) Expression vector design. A pelB signal sequence followed by aspartate and glycine was fused to the coding region of mature TEM β-lactamase, which was followed by two glycine residues and a His_5-tag to facilitate purification. Important endonuclease restriction sites are shown. For cytoplasmically expressed variants, the pelB signal sequences and aspartate/glycine residues were replaced by a single methionine residue. All plasmids contained a chloramphenicol resistance gene. (**B**) In vivo fitness of initially generated β-lactamase deletion mutants. Approximately 3500 freshly transformed XL-1 Blue cells harboring the respective plasmid were selected on LB agar plates containing chloramphenicol and 10 to 50 µg/mL ampicillin or chloramphenicol alone (100% clones). Plates were analyzed after 20 h incubation at 37°C. For the NΔ5-clone and CΔ3-clone, no colonies could be observed in the presence of ampicillin.

signal sequence *(33)* required for periplasmic targeting. A short aspartate–glycine linker was included to ensure an equally efficient processing rate for the variants that differ in their N-termini. Finally, a C-terminal His-tag was inserted to allow for immobilized metal ion affinity chromatography (IMAC), even of nonnative β-lactamase variants (**Fig. 2A**).

The plasmids were constructed as follows:

1. TEM-1 β-Lactamase deletion mutants, named NΔ3-clone, NΔ5-clone, CΔ1-clone, and CΔ3-clone, were generated by amplifying the *TEM*-1 gene from a pUC-derived plasmid *(34)*, using the respective forward and reverse primers. The forward primers provided the resulting PCR products with a 5′ *Sfi*I restriction site for cloning, the coding sequence for the C-terminal part of a pelB signal peptide, an aspartate–glycine linker, and the respective N-terminal modification (truncation) of the TEM-1 variants. At the 3′-end of the gene, the reverse primers introduced the

coding sequence for the modified TEM-1 C termini, a short spacer (two glycine residues), and a penta-histidine tag, two stop codons, and a *Hin*dIII restriction site.
2. The PCR products were digested with *Sfi*I/*Hin*dIII, isolated, and cloned into the *Sfi*I/*Hin*dIII fragment of pKMENGRbla (provided by K. M. M., unpublished), a derivative of the expression plasmid pAK400 ***(26)***, which carries a chloramphenicol resistance gene.

The resulting plasmids (pKJE_Bla-NΔ/CΔ series) expressed the various *TEM*-1 genes under control of a *lac* promoter/operator region. Despite the constitutive expression of *lac*I on the plasmids, a strong phage T7g10 Shine-Dalgarno sequence ***(35)*** resulted in a relatively high basal gene expression level. Therefore, all selection steps were performed under stringent conditions without overexpression.

3.4. Effect of Terminal Truncation on Enzyme Activity

The effect of the terminal truncations on enzyme activity can be estimated using growth assays on petri dishes or in liquid cultures in the presence of increasing concentrations of ampicillin. In general, selection on plates is more stringent than in liquid culture. Stringency can additionally be varied by looking at basal gene expression or after induction with isopropylthiogalactoside (IPTG), and also by incubating at different temperatures.

1. If necessary, transform the plasmids in BL21 cells using standard methods and plate on LB/Cm agar plates.
2. Grow overnight cultures started from glycerol stocks or from single colonies on plates (**step 1**) in LB/Cm for approx 16 h.
3. From the overnight culture, inoculate a preculture at an optical density (OD) at 600 nm (OD_{600}) of 0.1 and grow the culture to an OD_{600} of 1 (*see* **Note 1**).
4. Plate an aliquot of the preculture from **step 3** on LB/Cm agar plates with various amounts of ampicillin with or without IPTG. The amount needed for plating can be estimated by approximating that 2.5×10^8 cells/mL have an OD_{600} of 1.
5. Alternatively, inoculate a liquid LB/Cm culture with the preculture from **step 3** (starting $OD_{600} = 0.1$) with various amounts of ampicillin with or without IPTG.
6. Growth assays (**steps 4** and **5**) can be repeated at various incubation temperatures.

Without induction, colony numbers of cells producing any of the deletion mutants decreased significantly if the selective pressure increased from 0 to 50 μg/mL ampicillin on plates incubated at 37°C (**Fig. 2B**). For the mutants, NΔ5-clone and CΔ3-clone, no colonies were observed after 40 h incubation, suggesting an important role for the terminal residues. With induction, cells harboring the NΔ5-clone grew on plates with up to 50 μg/mL ampicillin at 25°C and in liquid culture with optimal aeration also at 37°C. This suggests that both truncations induced structural perturbations, but, at the same time, allowed for folding into active states. The NΔ5-clone conferred resistance to a much lower extent than the CΔ3-clone and was, therefore, chosen for subsequent optimization.

3.5. DNA Shuffling and Random Mutagenesis

The central step of every optimization procedure designed to reverse the structural perturbations brought about by terminal shortening or the introduction of deleterious mutations is the generation of a highly diverse library of variants at the genetic level. Random mutagenesis alone as well as in combination with DNA recombination has been harnessed to raise the required mutations.

DNA shuffling ***(15,16)*** is a widely used method for DNA recombination in the test tube. Because the jumbling of different homologous DNA sequences realized by DNA shuffling marks a crucial step of directed evolution, its central aspects should be depicted here in short: after the random generation of point mutations along the entire gene, primarily by error-prone PCR (*see* **step 7**) the resulting library of genes is subjected to DNA fragmentation to produce short, slightly different, interchangeable sequences of approx 50 to 200 basepairs length. Overlapping homologous sequences differing in one or more nucleotides are then assembled into larger fragments by a PCR procedure, starting with low annealing temperatures to allow mutual priming of the smallest fragments present in the mixture. With increasing cycle numbers, the annealing temperature is elevated steadily to select for longer overlaps favoring larger fragment sizes. Finally, the complete gene is fully assembled and amplified using a pair of flanking primers that overlap the terminal nucleotides of the gene and provide the resulting library with appropriate restriction sites for subsequent cloning.

The following section includes concrete instructions for how random mutagenesis and DNA shuffling can be performed to create a library of sufficient complexity to find some needles in a large haystack of sequence space. Note that the procedure described here is optimized with respect to our test system and might have to be modulated depending on the composition and length of the target gene. It should be taken into account that the usual order of reactions—first random mutagenesis, second recombination—was not kept here to determine the error rate of the shuffling process itself before the generation of point mutations by error-prone PCR (*see* **Note 2**).

1. Produce sufficient amounts (5–10 μg) of the gene of interest, either by PCR using primers homologous to constant regions flanking the target gene, or by the use of restriction enzymes cutting near to the ends of the gene. If PCR is used, *Taq* DNA polymerase should be used throughout to augment misincorporation of nucleotides. In our example, we amplified the truncated β-lactamase using the plasmid pKJE_Bla-NΔ5, and the primers pr_sfi_pelB_DG_shuffl and pr_GG_his5_hind_shuffl.
2. Digest approx 4 to 5 μg of the target gene using 0.2 U DNase I in a total volume of 50 μL DNase buffer for 12 min at room temperature (25°C).
3. Stop the endonucleolytic degradation process by adding 4 μL EDTA solution and storing on ice.

4. Analyze the resulting DNA fragments on a 2% (w/v) agarose gel and excise fragments of approx 50 to 150 basepairs length.
5. Purify the fragments using the QiaexII gel extraction kit (*see* **Note 3**).
6. Assemble approx 100 to 300 ng of the isolated fragments in a primer extension PCR in a total volume of 50 µL (*see* **Note 4**). The reaction mix contains 2.5 U *Taq* DNA polymerase, 0.4 m*M* of each dNTP, and 2 m*M* $MgCl_2$ in the supplied *Taq* buffer. Assembly PCR reactions are carried out, for example, in a thermocycler (Eppendorf) using the following program: 94°C for 3 min; 10 cycles of: 94°C for 1 min (denaturation), 45°C + 0.3°C/cycle for 1 min (annealing), 72°C for 1 min (elongation), and 72°C for 5 min after the last cycle; 15 cycles of: 94°C for 1 min, 50°C + 0.4°C/cycle for 1 min, 72°C for 1 min, 72°C for 5 min after the 15 cycles; 10 cycles of: 94°C for 1 min, 56°C + 0.5°C/cycle for 1 min, 72°C for 1 min, and 72°C for 5 min after the last cycle; 10 cycles of: 94°C for 1 min, 61°C + 0.5°C/cycle for 2 min, 72°C for 2 min, and 72°C for 5 min after the 10 cycles. Alternatively, a simplified program can be used consisting of: 94°C for 3 min; 35 cycles of: 92°C for 30 s, 30°C + 1°C/cycle for 1 min, 72°C for 1 min + 4 s/cycle, and 72°C for 5 min after the final cycle.
7. Amplify 1/5 volume of the assembly PCR reaction (no purification required) to ensure full length of the reconstituted genes and to add the appropriate restriction sites required for subsequent cloning. This step can be done at error-prone conditions to extend the library diversity. In addition to using *Taq* DNA polymerase for elongation, the error-rate of the amplification reaction can be increased by including 7 m*M* $MgCl_2$ and 0.5 m*M* $MnCl_2$ in the reaction mixture ***(36)*** (*see* **Note 5**). In our example, we used primers pr_sfi_pelB_DG_shuffl and pr_GG_his5_hind_shuffl at a final concentration of 0.5 µ*M* each. The reaction mixture included 7 m*M* $MgCl_2$, 0.5 m*M* $MnCl_2$, each dNTP at 0.4 m*M*, and 2.5 U *Taq*. The program used was 3 min at 94°C, 25 cycles of 1 min at 94°C, 1 min at 68°C, and 1 min at 72°C, with a final step of 7 min at 72°C.
8. Purify the PCR products using, e.g., the GFX PCR DNA and Gel Band Purification Kit, digest it with the appropriate enzymes (*Sfi*I and *Hin*dIII) and clone it into the expression vector (pKJE_Bla-ΔN5) treated with the same enzymes. Transformation is performed as described in **Subheading 3.6.**

We performed three rounds of directed evolution (S1 to S3) using the DNA shuffling and random mutagenesis procedures in combination with in vivo selection steps. Clones selected were pooled and served as templates for subsequent rounds. In the last round of directed evolution, the error-prone PCR step after DNA shuffling was replaced by a standard PCR procedure to prevent the introduction of deleterious mutations into the recombined clones.

3.6. Transformation and In Vivo Selection of Mutant Libraries

1. Desalt the plasmid mutant libraries (pKJE_Bla-NΔ5_Lib-S1 to -S3; S indicates shuffling rounds one to three, *see* **Subheading 3.5.**) by butanol precipitation. To 10- to 20-µL ligation mixtures, add double-distilled water to a volume of 50 µL. Mix thoroughly with 500 µL butanol and spin at maximal speed (~2000*g*) at room

temperature for 30 min. Remove supernatant, air-dry, and dissolve the pellet in 10 to 20 µL of water.

2. Transform desalted DNA into 100 µL of electrocompetent *E. coli* XL-1 Blue cells using 1.7 kV, 200 Ω, and 25 µF. Immediately after transformation, add 900 µL 2YT and 1/100 volume of transformation salt stock.
3. Transfer the cell suspensions in 10-mL glass test tubes and incubate for 60 to 70 min at 37°C with orbital shaking.
4. Assess the transformation efficiency by plating a dilution series of each transformation mixture on LB/Cm agar.
5. Use another aliquot of the transformed cells to estimate the level of selection pressure for the following round by plating on LB/Cm agar plates supplemented with various concentrations of ampicillin (*see* **Note 6**). Take the highest ampicillin concentration at which colonies can be observed as selection level for the subsequent round.
6. For selection, plate the transformed cells on LB/Cm agar plates with the ampicillin concentration determined in **step 5** (*see also* **Note 6**). We used ampicillin concentrations of 20 µg/mL, 100 µg/mL, and 200 µg/mL for the first, second, and third selection, respectively. The first selection round was performed in liquid medium, the other two rounds were selected on agar plates.

The impaired truncation mutant NΔ5-clone was revitalized by three rounds of directed evolution comprising random mutagenesis and DNA shuffling of a PCR product (as described in **Subheading 3.5.**). In three successive rounds, approx 19×10^3, 147×10^3, and 260×10^3 clones were generated and screened for survival on plates containing 20, 100, and 200 µg/mL ampicillin (*see* **step 6**), respectively, giving rise to approx 3, 600, and 7000 colonies.

Table 1 summarizes all sequenced clones isolated in the course of directed evolution sorted by shuffling round (S1–S3). Additionally, **Table 1** includes relative solvent accessibilities and average side chain atomic temperature factors (taken from PDB 1btl) ***(37)*** for each of the substituted wt residues. The five clones sequenced after the second optimization round shared two mutations, M182T and T265M (numbering according to Ambler et al., **ref.** ***38***), which were already present after the first round.

After the final optimization step, the library was pooled, and the maximum level of ampicillin resistance was determined on plates without IPTG, as described in **Subheading 3.4.** (**Fig. 3**). Normal colony development of a significant number of clones was seen up to 700 µg/mL, and, after prolonged incubation times of 40 h, clones could even be detected on 1000 µg/mL. In contrast, the wt construct grew only up to 500 µg/mL. Twenty-six clones were picked from the 800 to 1000 µg/mL plates and sequenced. The 26 sequences corresponded to 15 individual clones (**Table 1**). All clones shared two mutations, M182T and A224V. The M182T mutation has previously been demonstrated to compensate for folding defects ***(39)*** and stability losses ***(29,40)***. The second mutation (A224V) prevailed at the increased selective pressure of the third round. This

Table 1
Amino Acid Substitutions Selected in the Course of Directed Evolution of the NΔ5 Deletion Mutant

position	31	37	38	42	52	56	59	63	82	84	88	96	104	120	147	150	153	159	168	177	182	192	195	198	206	208	224	240	247	256	265	277
wt	V	E	D	A	N	I	S	E	S	I	Q	H	E	R	E	A	H	V	E	E	M	K	T	L	Q	I	A	E	I	K	T	R
% acc	63.3	15.1	55.7	18	84.3	45.8	28.7	66.3	24	19.1	59.9	68.2	66.3	47.4	50.1	40.1	51.1	25.9	54.2	48.1	20.3	36	45.5	40.7	25.6	18.5	27.5	56.5	0	73.1	4.7	60.8
B factor (sc)	8.3	12.2	19.9	11.6	16.7	8.7	11.0	25.5	8.8	13.8	33.5	20.7	34.1	25.0	20.6	8.3	16.5	6.5	21.1	23.5	7.9	12.1	14.6	25.7	14.0	13.8	8.2	26.1	5.4	26.0	7.0	30.9
S1/1																			A		T										M	
S1/2													G								T		S				V			R		
S1/3																	R												V			
S2/1																	R				T				H		V				M	
S2/2															G					K	T										M	
S2/3																					T										M	
S2/4																			A		T	E	S				V				M	
S2/5				G																	T										M	
S3/1 (×2)					S					V					G				V		T						V					
S3/2 (×4)														G			R	A			T						V					
S3/4 (×3)		D										Y									T						V				M	
S3/5 (×4)	A													G							T						V		V			
S3/6																	R				T		S				V					
S3/7 (×3)							G			V											T		F				V		V			
S3/17										V											T						V					
S3/18										V							R				T		S				V					
S3/19														G							T						V	G			M	K
S3/20	A		N	G					F	V											T		S	P			V		V	R	M	
S3/21	A																R				T		S				V					
S3/22									E		E						R		A		T						V			R		
S3/23					D			K		V						T	R				T		S	P			V					
S3/24						T				V							R				T					M	V					
S3/26																	R				T					M	V					
position	31	37	38	42	52	56	59	63	82	84	88	96	104	120	147	150	153	159	168	177	182	192	195	198	206	208	224	240	247	256	265	277

Amino acid substitutions of the selected deletion mutants are shown with regard to wt sequence. Residues are numbered according to Ambler et al. *(38)*. Mutants are grouped according to the round of directed evolution in which they were selected (S1–S3). The frequency of individual selected clones containing the same set of mutations is shown in brackets. The most prevalent amino acid substitutions of the third round mutants are shown in boldface type. % acc, relative solvent accessibility, the ratio between calculated and vacuum accessibility expressed as a percentage using the program What If *(48)*; B-factors corresponding to the average of side-chain atoms were taken from the Protein Data Bank (PDB entry: 1btl).

A

B

C

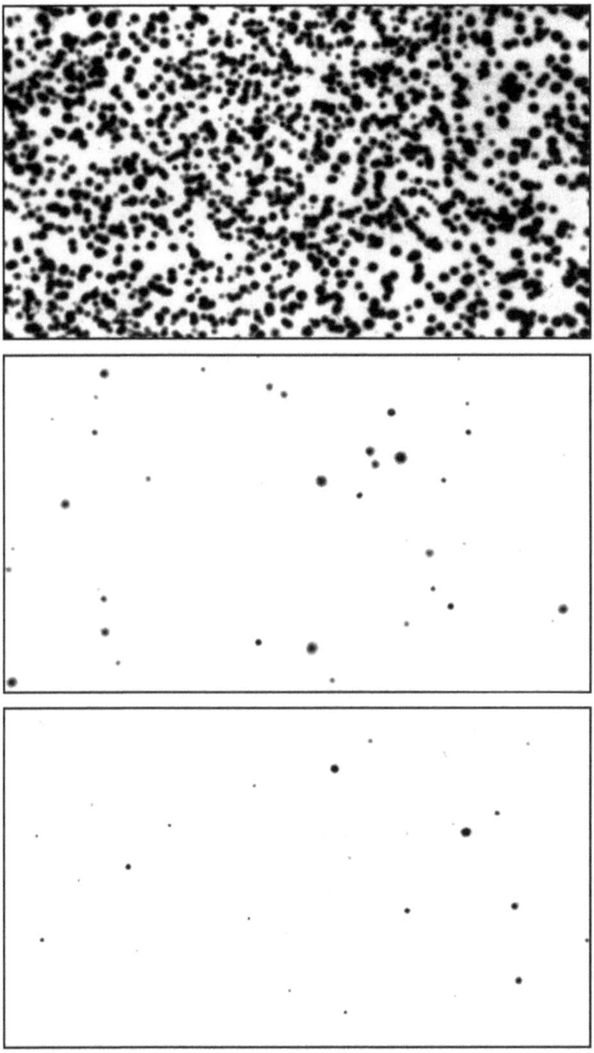

Fig. 3. Colonies of optimized NΔ5-β-lactamase deletion mutants on LB agar plates containing **(A)** no, **(B)** 350 μg/mL, or **(C)** 700 μg/mL ampicillin. Approximately 1500 cells of a regrown culture of pooled colonies isolated in the third round of directed evolution were plated. Plates are photographed after (A) 20 h or (B and C) 40 h incubation at 37°C.

mutation has been listed once, but no experiments were performed with this mutant ***(41)***.

Both mutations are at least 17 Å away from the site of deletion, indicating independent compensation of the structural interference imposed by deletion. **Figure 4** highlights the most-frequently mutated residues mapped to the known tertiary structure of TEM-1 β-lactamase. In general, the sets of mutations found

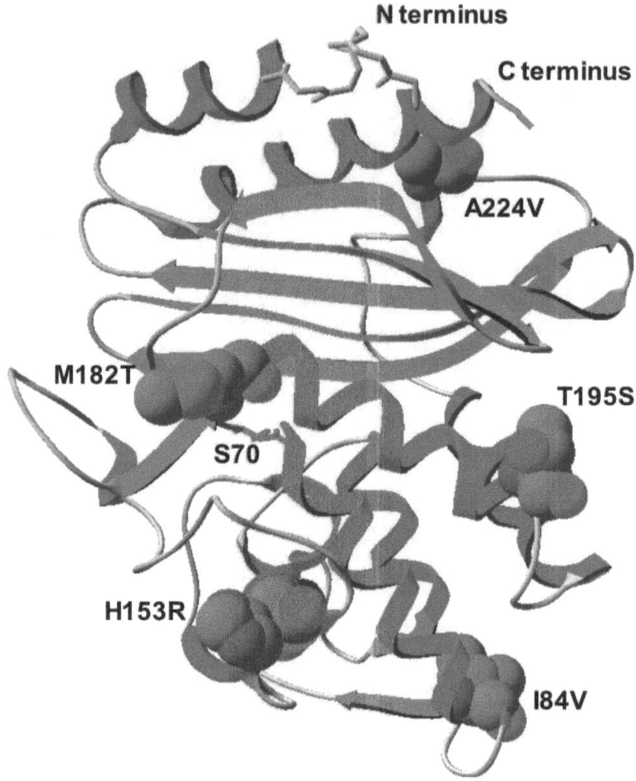

Fig. 4. Structure of wt TEM-1 β-lactamase (PDB identifier: 1btl). Residues frequently mutated are shown in full atoms (e.g., M182T), the catalytic serine residue (S70) and the N- and C-terminus are indicated. Residues are numbered in agreement with Ambler *(38)*. Most mutated amino acid residues are located near to the protein surface. The figure was drawn using SWISS-PDB Viewer ***(31)***.

in individual optimized variants were always scattered throughout the entire structure and were never clustered adjacent to the site of deletion.

3.7. Construction of Full-Length Mutants

Optimized clone NΔ5-S3/6 was re-elongated to elucidate to what extent the set of mutations compensating for protein destabilization after terminal truncation affect the complete enzyme. The full-length clone named FL-S3/6 was constructed by adding the missing residues according to the schematic representation shown in **Fig. 2A**. This variant was designed for translocation to the periplasm and was expressed in *Escherichia coli* XL-1 Blue cells to probe its functionality in vivo. Unexpectedly, the numbers of colonies on plates (20 h at 37°C) containing chloramphenicol and 100 μg/mL ampicillin was only 50% of the number of colonies grown on control plates without ampicillin. This was in contrast to NΔ5-S3/6, which grew equally well on both plates. To distinguish between protein function and protein translocation, enzymatic activities of crude whole-cell extracts of FL-S3/6 were tested relative to the nonextended mutant, NΔ5-S3/6. No significant differences in hydrolase activity with the chromogenic substrate nitrocefin were observed, hinting that enzyme translocation could be the main reason. In agreement with these data, decreased cleavage of the signal sequence of clone FL-S3/6 was also observed in the purification process (*see* **Subheading 3.8.**). Consequently, a cytosolically expressed variant was cloned (FL-S3/6-cyt) by replacing the signal sequence and the Asp–Gly tag by a methionine start codon. To ensure correct biophysical comparisons, the analogous construct for wt lactamase (wt-clone-cyt) was cloned.

3.8. Protein Expression and Purification

The generated β-lactamase variants (wt-clone periplasmatic, wt-clone-cyt, NΔ5-clone, optimized mutants NΔ5-S3/6 and NΔ5-S3/7, and FL-S3/6-cyt) were expressed with a His_5-tag in *E. coli* under control of the lac promoter (**Subheading 3.8.1.**). Primarily, two purification steps were applied, first, a substrate analog affinity chromatography using phenylboronate (**Subheading 3.8.2.**), and, second, an IMAC (**Subheading 3.8.3.**). This purification procedure worked very well for the deletion mutant, NΔ5-clone (**Fig. 5**), and the optimized deletion mutant, NΔ5-S3/7.

With some variants of β-lactamase (wt-clone, optimized deletion mutant NΔ5/S3-6, and re-elongated mutant FL-S3/6), purified native enzymes were highly contaminated (30–65%; *see* **Note 7**) with their respective unprocessed forms, most likely because of overexpression and/or rapid folding. Because even a periplasmic extraction aiming at a high product yield resulted in contaminants, a hydrophobic interaction chromatography (HIC) was chosen to separate the native from the much more hydrophobic unprocessed form (**Subheading 3.8.4.**).

Cytoplasmatically expressed β-lactamase variants wt-clone-cyt and FL-S3/6-cyt were purified by phenylboronate affinity chromatography and IMAC. To further increase purity for biophysical characterization, an additional anion exchange chromatography was carried out (**Subheading 3.8.5.**).

3.8.1. Protein Expression and Cell Disruption

1. Transform the vector in cells suitable for protein expression (e.g., BL21).
2. Inoculate 2YT/Cm overnight cultures from glycerol stocks and grow at 28°C for approx 16 to 18 h (*see* **Note 8**).
3. For the expression culture, inoculate four 1-L 2YT/Cm with the overnight culture to obtain a starting OD_{600} of 0.15, and grow at 24°C for the NΔ5 clone, and at 25 to 30°C for all other constructs (*see* **Note 9**).

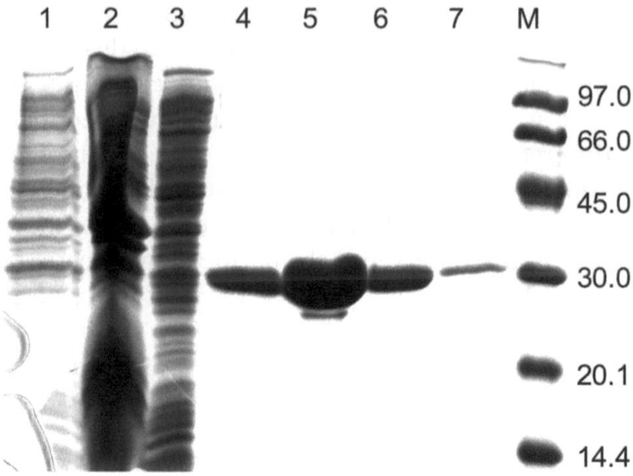

Fig. 5. Expression and purification of the initial β-lactamase deletion mutant (NΔ5-clone). A 12.5% SDS-PAGE, Coomassie stained, shows samples collected before and after purification by phenylboronate (PheBo) and IMAC, respectively. Lane 1, whole-cell lysate before induction; lane 2, whole cell lysate after induction; lane 3, disrupted cell supernatant; lanes 4–7, IMAC elution peak fractions; M, molecular weight marker with individual molecular weights (kDa) indicated.

4. Induce protein expression at OD_{600} of 0.7 with 0.5 m*M* IPTG. Forty minutes after induction, add 100 μg/mL ampicillin to select for β-lactamase-producing cells.
5. Harvest cells after 4 to 5 h (OD_{600} is ~4–5) at 6000*g* in a GS-3 rotor, and freeze the pellet at –80°C.
6. For purification, resuspend one pellet (from 1-L expression culture) in resuspension buffer (*see* **Note 10**) with 200 U benzonase.
7. Disrupt cells in a French press at approx 97 MPa (~14,000 psi) with 5 to 6 cycles in a prechilled cell. Hold samples on ice whenever possible.
8. Clarify crude cell extracts by centrifugation for 40 min at 41,000*g* in a SS-34 rotor at 4°C. Filtrate supernatant using 0.45-μm polyethersulfone syringe filters.

3.8.2. Phenylboronate Affinity Chromatography

1. Equilibrate a 2-mL phenylboronate column with resuspension buffer, load samples from **Subheading 3.8.1., step 8** and wash with resuspension buffer until absorbance at 280 nm approaches baseline.
2. Elute β-lactamase variants with borate buffer.
3. Assess purity by 12.5% sodium dodecyl sulfate (SDS)-polyacrylamide gel electrophoresis (PAGE) followed by Coomassie staining.

3.8.3. Immobilized Metal Ion Affinity Chromatography

1. Equilibrate a 4-mL Ni-NTA column with phosphate buffer with or without 5 m*M* imidazole, and load the samples from **Subheading 3.8.2., step 2**.

2. Wash and elute samples using a step gradient (5, 10, 16, and 100%) of imidazole buffer.
3. Assess purity by 12.5% SDS-PAGE followed by Coomassie staining.
4. For further characterization, dialyze (~8 h) protein samples against three times 1 L of phosphate buffer containing an additional 1 m*M* EDTA in the first dialysis.

3.8.4. Hydrophobic Interaction Chromatography

1. Equilibrate a 1-mL phenylsuperose column with 1 *M* $(NH_4)SO_4$.
2. Supplement the sample from **Subheading 3.8.3., step 2** or from **Subheading 3.8.2., step 2** with ammonium sulfate crystals to a final concentration of 1 *M* $(NH_4)SO_4$ (*see* **Note 11**), and clear the sample by centrifugation (10 min at 27,000*g* and 4°C; SS-34 rotor) and filtration (0.45-µm syringe filter).
3. Load sample onto the HIC column and apply a linear gradient (30 mL) from 0 to 100% of 0.5X phosphate buffer. Because the more hydrophobic unprocessed form should bind more tightly to the column than the processed form, both forms should elute as separate peaks.
4. If HIC chromatography is applied directly after phenylboronate affinity chromatography, the sample is subsequently purified via IMAC, as described in **Subheading 3.8.3.**

3.8.5. Anion Exchange Chromatography

1. Dialyze samples from **Subheading 3.8.3., step 2** three to four times in 1 to 1.5 L of Tris-HCI buffer.
2. Equilibrate a Mono Q anion exchange column with Tris-HCI buffer and load the sample.
3. Elute with a 30 mL linear gradient from 0 to 100% of Tris-HCI–NaCl buffer.
4. Dialyze the eluted protein as described in **Subheading 3.8.3., step 4**.

Amino acid compositions of all variants were confirmed by electrospray mass spectrometry within typical deviations from calculated masses. Enzymes were dialyzed as described in **Subheading 3.8.3., step 4** and characterized within 1 wk after purification. Before use, enzyme solutions were clarified by centrifugation (30 min at 15000*g* and 10°C). Protein concentrations were taken from absorbance spectra at 280 nm (*see* **Note 12**).

3.9. Enzyme Assays

The kinetic parameters of the β-lactamase variants were assayed photospectrometrically at 486 nm, using the chromogenic substrate nitrocefin ($\Delta\varepsilon_{486}$ = 16,000/M/cm).

1. Add 20 µL of enzyme dilution to 980 µL nitrocefin solution, mix, and immediately measure the change in absorbance at 486 nm for approx 1 min (*see* **Note 13**). We used a final enzyme concentration of 2.5 n*M* for the initial deletion mutant and 0.5 n*M* for all other variants.
2. Repeat measurements at least twice and calculate standard deviations.

Table 2
Kinetic Parameters

β-Lactamase variant	k_{cat} (1/s)	K_M (μ*M*)	k_{cat}/K_M (/M/s)
Wt-clone	780 ± 9	84.4	9.24×10^6
NΔ5-clone	123 ± 2	61	2.02×10^6
NΔ5-S3/6	728 ± 7	58.9	12.4×10^6
NΔ5-S3/7	732 ± 19	64.3	11.4×10^6
Wt-clone-cyt	602 ± 8	52.2	11.5×10^6
FL-S3/6-cyt	633 ± 9	70.6	8.97×10^6

Kinetic constants were determined at 25°C in 50 m*M* potassium phosphate buffer, 0.5% dimethylsulfoxide, pH 7.0, using the β-lactam compound nitrocefin as substrate.

3.9.1. Determination of Michaelis Constant and Turnover Number

1. To determine Michaelis constant (K_M) values, measure initial rates with substrate concentrations ranging from 10 to 500 μ*M* and enzyme concentrations of 2.7 to 5.9 n*M* for the NΔ5-clone and 0.5 n*M* for all other mutants.
2. Fit data to the Michaelis-Menten equation using, e.g., the Marquardt-Levenberg algorithm implemented in the program SigmaPlot (SPSS).

The K_M of NΔ5-clone approximated that of wt-clone, but the turnover number (k_{cat}) decreased approximately sixfold, implying an active site organization sufficient for binding but not for effective catalysis (**Table 2**).

3.9.2. Thermoactivity Profiles and Half-Life Time

To test and compare the temperature dependency of activity and the kinetics of heat-induced inactivation, a thermoactivity screen can be used. We assayed turnover rates at increasing temperatures (25–70°C) after specific incubation times (**Fig. 6**).

1. To assess thermoactivity, prepare a 292 n*M* solution for NΔ5-clone and a 25 n*M* solution for all other variants, and split these into three aliquots.
2. Incubate the first aliquot for 5 min at a given temperature (25–70°C) in a heated water bath and keep both of the remaining samples on ice.
3. Before measurement, mix the respective sample by brief vortexing.
4. Transfer 20 μL of the sample to 980 μL of nitrocefin solution preheated to the respective assay temperature, and determine the initial reaction rate between 5 and 25 s in a heated spectrophotometer.
5. Preheat the second aliquot for 30 s at the respective temperature, and assay as detailed in **steps 3** and **4**.
6. Assay the third aliquot after 10 min incubation on ice.
7. Repeat the entire procedure for a different temperature.

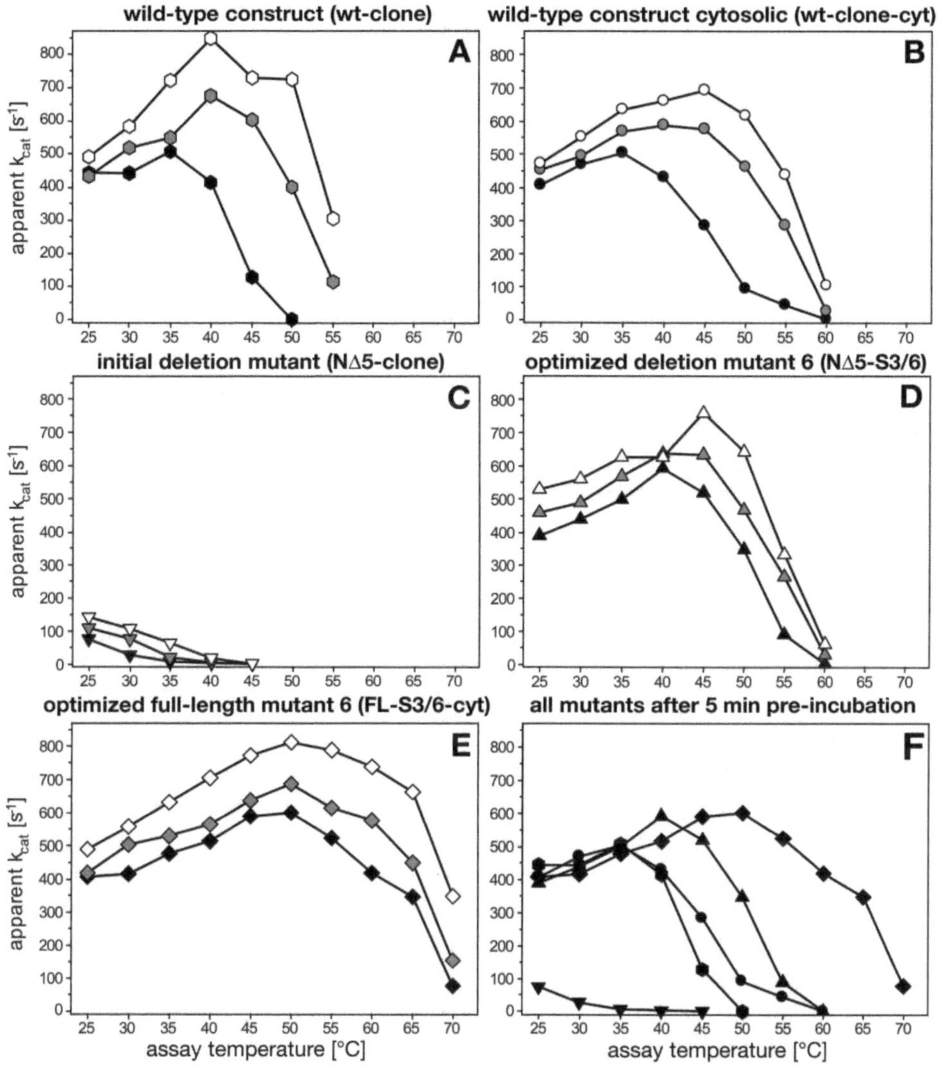

Fig. 6. Temperature-dependent activity profiles of investigated β-lactamase variants. **(A–E)** apparent k_{cat} values were determined photospectrometrically from 25 to 70°C, using nitrocefin as the substrate. Enzyme solutions are diluted (initial deletion mutant: 292 n*M*, all other mutants: 25 n*M*), split into three fractions, and subjected to different pretreatments: incubation for 5 min or 30 s at assay temperature (black and grey symbols, respectively), or storage on ice for 10 min (white symbols). Assay buffer (980 μL) was prewarmed in a heated water bath before the reactions were started by adding 20 μL of enzyme solution. Final enzyme concentrations of 5.9 n*M* for the initial deletion mutant and 0.5 n*M* for all other variants were used. **(F)** Summary of thermoactivity profiles obtained after 5-min preincubation. ⬢, wt-clone (periplasmatic); ●, wt-clone (cytosolic); ▼, NΔ5-clone; ▲, NΔ5-S3/6; ◆, FL-S3/6-cyt.

The thermoactivity profiles of the wt-clone varied clearly depending on the pretreatment (**Fig. 6A**). The temperature optimum after ice incubation and 30 s preheating at the assay temperature was 40°C. However, when the enzyme was preheated for 5 min, the maximum shifted down to 35°C, indicating commencing thermal unfolding.

The truncated NΔ5-clone maintained its highest activity at 25°C, the lowest temperature examined (**Fig. 6C**). With increasing temperatures, enzyme activity dropped rapidly and approached zero at 40°C. A detailed analysis of the reaction rate at 40°C after incubation on ice showed that this mutant was heat inactivated during the measurement, within seconds after addition to the preheated assay mixture (**Fig. 7A**). The observed decrease of reaction rate over time could be fitted by an exponential decay equation (*see* **Note 14**), yielding a half-life time of 7 s at 40°C (**Fig. 7B**).

The 0- and 30-s temperature–activity profiles of the optimized mutant NΔ5-S3/6 (**Fig. 6D**) largely resembled those of the wt-clone. However, the 5-min preheating profile differed significantly. The temperature–activity curve of this optimized mutant was shifted to higher temperatures by approx 8°C compared with the wt-clone (**Fig. 6A,F**).

The full-length optimized mutant, FL-S3/6-cyt (**Fig. 6E**), exhibited thermostability features superior to those of the optimized deletion mutant, NΔ5-S3/6, and the corresponding wt-clone-cyt (**Fig. 6B,F**). The catalytic activities of these clones were nearly identical at assay temperatures up to 40°C, irrespective of pretreatment. At 50°C and greater, FL-S3/6-cyt retained significant higher activities, especially at conditions of prolonged heat stress of 5-min preincubation. The maximum catalytic activity of wt-clone-cyt was at 45°C after ice incubation, but decreased to 35°C after 5 min of preincubation. In contrast, the temperature optimum of FL-S3/6-cyt (50°C) remained unchanged. Only the reaction rate decreased after heat incubation. Interestingly, the alterations between wt-clone and wt-clone-cyt at the N terminus (AspGly replaced by Met) influenced stability.

Comparison of the 5-min preincubation thermoactivity profiles of all truncated and full-length variants (**Fig. 6F**) illustrates how terminal truncation diminished activity at 35 to 40°C. Compensating amino acid substitutions restored activity and even improved stability. Re-extension of optimized mutant S3/6 further increased thermostability as well as thermoactivity.

3.10. Urea-Induced Unfolding and Data Analysis

Unfolding of the β-lactamase variants was evaluated by fluorescence spectroscopy. The red-shift of the intrinsic tryptophan fluorescence emission maximum was monitored as a function of urea concentration (**Fig. 8**, **Table 3**). Data analysis was performed according to a three-state model.

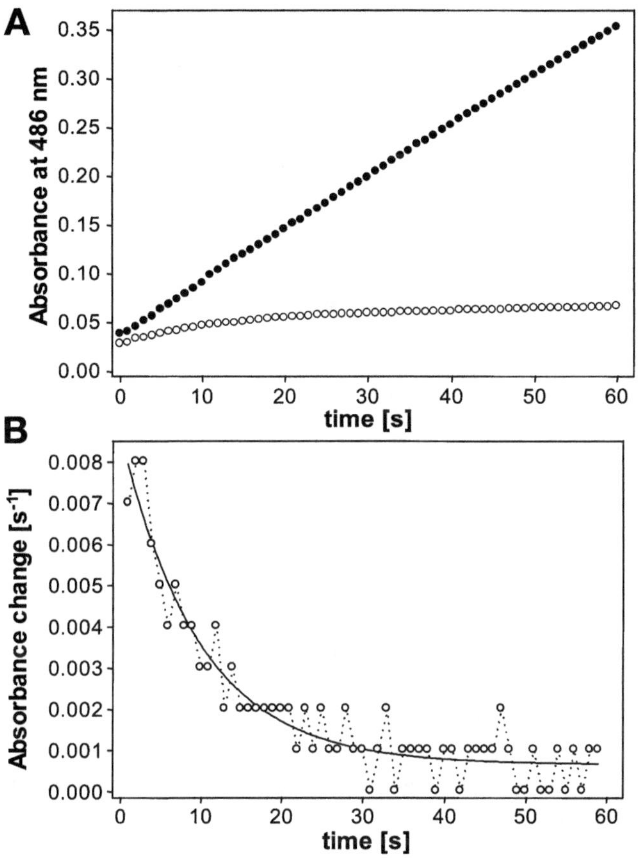

Fig. 7. Enzyme kinetics and exponential decay of catalytic activity of NΔ5-clone and stable kinetics of NΔ5-S3/6 at 40°C. **(A)** Spectrophotometric analysis of product formation using the chromogenic substrate nitrocefin. The final enzyme concentration was 2.9 n*M* for NΔ5-clone (○) and 0.5 n*M* for NΔ5-S3/6 clone (●). Twenty microliters of a concentrated enzyme solution was mixed with 980 µL of prewarmed assay buffer. **(B)** Exponential decay fit of enzymatic activity of NΔ5-clone using the change of absorbance per second (ΔA/Δt) from the data shown in (A). A three-parameter exponential decay fit revealed a half-life of 7 s at 40°C (using *SigmaPlot*).

1. Equilibrate 300 to 400 n*M* enzyme solutions from **Subheading 3.8.3., step 4** containing 0.25 to 8 *M* urea for 18 to 20 h at 19°C.
2. Record fluorescence emission spectra from 320 to 380 nm at 20 to 23°C while exciting at 280 nm. For each data point, average four scans.
3. If necessary, correct fluorescence spectra for the background fluorescence of the solution (phosphate buffer plus denaturant).

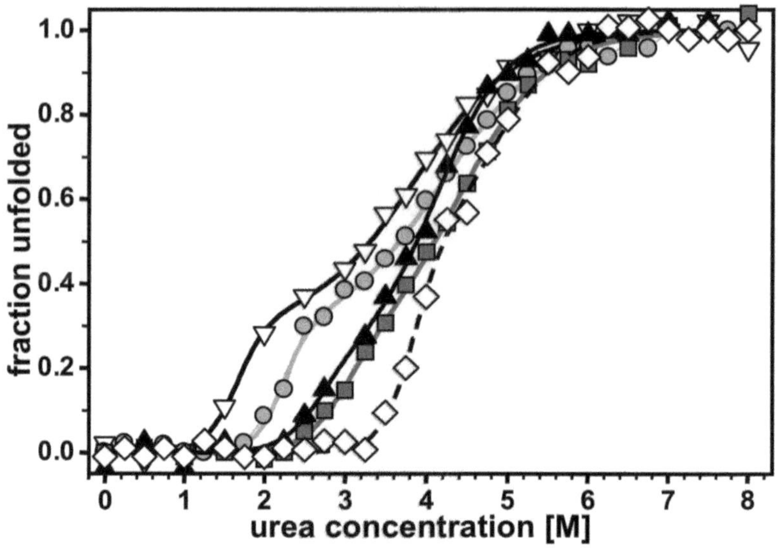

Fig. 8. Unfolding of β-lactamase variants in urea. Denaturation is followed using the red-shift of intrinsic fluorescence emission maxima. Enzyme samples (0.3–0.4 m*M* in 50 m*M* sodium phosphate and 150 m*M* NaCl, pH 7.2) containing 0.25 to 8 *M* urea were equilibrated for 18 to 20 h at 19°C and measured at 20°C to 23°C. Data are fitted assuming a three-state unfolding mechanism. NΔ5-clone, inverted triangles (white); wt-clone-cyt, circles (light grey); optimized deletion mutant NΔ5-S3/7, squares (dark grey); mutant NΔ5-S3/6, triangles (black); and optimized full-length mutant FL-S3/6-cyt, diamonds (white, dashed line).

Table 3
Thermodynamic Parameters

Variant	$\Delta G^0_{NI, H_2O}$ (kJ/mol)	m_{NI} (kJ/mol/*M*)	$D^{1/2}_{NI}$ (/*M*)	$\Delta G^0_{IU, H_2O}$ (kJ/mol)	m_{IU} (kJ/mol/*M*)	$D^{1/2}_{IU}$ (/*M*)
Wt-clone-cyt	30.4 ± 4.5	13.7 ± 2.1	2.9	15.8 ± 1.1	3.8 ± 0.2	4.2
NΔ5-clone	20.2 ± 5.5	12.3 ± 3.7	1.6	15.3 ± 1.7	3.9 ± 0.4	3.9
NΔ5-S3/7	27.1 ± 6.4	9.3 ± 2.6	2.9	19.0 ± 2.6	4.3 ± 0.5	4.4
NΔ5-S3/6	26.2 ± 10.1	9.8 ± 4.3	2.7	23.4 ± 3.2	5.7 ± 0.6	4.1
FL-S3/6-cyt	45.8 ± 7.7	11.9 ± 2.2	3.8	27.2 ± 7.0	5.6 ± 1.2	4.9

Data were analyzed using the linear extrapolation method ***(49)*** assuming a biphasic unfolding transition (native, intermediate, and unfolded). The values for half denaturation were obtained according to **Eq. 2** (**Subheading 3.10., step 7**).

4. Smooth the fluorescence intensity spectra to obtain the associated λ_{max} values. We used a tricube weighting function (Loess; sampling proportion, 0.25; polynomial degree, 3) implemented in the program Sigma Plot (**Note 15**).
5. Analyze the data assuming an appropriate model for the unfolding reaction.

6. For a three-state model, N I U (where N resembles the native, I the intermediate, and U the unfolded state), with two equilibrium constants, K_{NI} and K_{IU}, fit the thermodynamic parameters $\Delta G^0_{NI,H_2O}$, $\Delta G^0_{IU,H_2O}$, m_{NI}, m_{IU}, and y_I to the fluorescent data using the following equation (*see* **Note 16**):

$$y_{obs} = \frac{y_N + y_I \times e^{\left(\frac{-\Delta G^0_{NI,H_2O} + m_{NI} \times [D]}{R \times T}\right)} + y_U \times e^{\left(\frac{-\Delta G^0_{NI,H_2O} + m_{NI} \times [D]}{R \times T}\right)} \times e^{\left(\frac{-\Delta G^0_{IU,H_2O} + m_{IU} \times [D]}{R \times T}\right)}}{1 + e^{\left(\frac{-\Delta G^0_{NI,H_2O} + m_{NI} \times [D]}{R \times T}\right)} + e^{\left(\frac{-\Delta G^0_{NI,H_2O} + m_{NI} \times [D]}{R \times T}\right)} \times e^{\left(\frac{-\Delta G^0_{IU,H_2O} + m_{IU} \times [D]}{R \times T}\right)}} \qquad (1)$$

Set y_N and y_U to the average of the first and last data points, respectively. A slope in the baselines was not seen in our case.

7. The midpoints of transitions are given by:

$$[D]_{1/2} = \frac{\Delta G^0_{H_2O}}{m} \qquad (2)$$

The order of denaturation (**Fig. 8**, *see* **Note 17**) corresponded to the order seen in the thermoactivity assays (**Fig. 6** and **Subheading 3.9.2.**). The NΔ5-clone was the least stable, followed by the wt-clone-cyt. The optimized mutants, NΔ5-S3/6 and NΔ5-S3/7, started to unfold at approximately the same urea concentration, and the elongated FL-S3/6 was the last one to unfold.

The unfolding showed a clear three-state behavior in the case of the wt-clone-cyt and the NΔ5-clone. Taking a closer look at the FL/S3-6-cyt data and comparing two- and three-state fits also strongly indicated a three-state behavior. The data of the optimized mutants, NΔ5-S3/6 and NΔ5-S3/7, could be explained either by a two- or three-state unfolding. For comparison, all data were fitted using the three-state model, N I U with **Eq. 1** (*see* **step 6** and **Table 3**). This approach is supported by previously described intermediate folding states for TEM-type β-lactamases ***(42–44)***. Enzyme activity tests with urea-denatured proteins showed decreasing activity after first denaturation. The extracted thermodynamic values aid in the comparison and explanation of the obtained mutants, but require careful interpretation because of underlying assumptions. The deletion of five N-terminal amino acids primarily affected the first transition, reducing the stability by approx 10 kJ/mol and had little effect on the second transition. Optimizing for catalytic activity in vivo yielded clones displaying stabilization effects for both transitions, whereby the achieved stability of the first transition is below and the second is above the wt-clone. Elongation (clone FL-3/6-cyt) considerably stabilized the first phase (19.6 kJ/mol compared with NΔ5-S3/6), but had little effect on the second transition (3.8 kJ/mol compared with NΔ5-S3/6). This is in agreement with the view that elongation compensated truncation independent of the introduced mutations. Comparing FL-S3/6-cyt with wt-clone-cyt revealed a stabilization of 15 kJ/mol for the first transition, and of 11.4 kJ/mol for the second transition. It is important to note that the optimized mutants denature at higher urea concentrations than the wt-clone, confirming the stability ranking observed in the thermoactivity assays.

4. Notes

1. To ensure that growth results can be compared, a preculture is required to provide consistent starting conditions.
2. The error rate of the shuffling process was determined to be 0.83 by analysis of 2529 bases.
3. Elution of bound the DNA fragment is increased by elongated incubation (15 min) at elevated temperatures, e.g., 50°C.
4. No extra primers are added at this point to facilitate mutual priming and shuffling of the isolated DNA fragments.
5. Increased Mg^{2+} concentrations stabilize noncomplementary basepairs, whereas the presence of 0.5 m*M* manganese ions diminishes the template specificity of the polymerase. Error rates can be controlled over a wide range, for example, by varying the number of template molecules, the cycle number, the period of extension, the source and amount of polymerase, the concentrations and ratios of dNTPs, and the type of dNTP analogs used. The error rate also depends on the nucleotide composition of the target gene. Thus, the use of an identical protocol may give quite different results in different trials.
6. If the enzyme activity is very low in the first round, selection can alternatively be performed in liquid medium (LB/Cm with different amounts of ampicillin), where the selection stringency is lower than on plates.
7. The percent of unprocessed protein was quantified from scanned Coomassie-stained gels using the image analysis software, Scion/NIH image.
8. If expression is performed at temperatures significantly below 37°C, it is best to also lower the temperature of the overnight culture to avoid a lag phase when starting the expression culture.
9. The optimal expression temperature needs to be adjusted for every mutant by performing a small-scale growth test at various temperatures.
10. Two pH values (7.0 and 7.2) were tested, with equal results.
11. 1 *M* $(NH_4)SO_4$ is desirable to ensure proper binding, but, if protein precipitation is a problem, the concentration of $(NH_4)SO_4$ can be reduced. Because samples with wt β-lactamase started to precipitate after addition of 0.5 *M* $(NH_4)SO_4$, a final concentration of only 0.65 *M* was used. In this case, the native form of the enzyme was in the flow-through, whereas the unprocessed form was retained on the column.
12. Absorbance spectra were measured from 350 to 220 nm, and the absorption at 280 nm was corrected for background absorbance by extrapolating the absorbance between 320 and 350 nm linearly to 280 nm. Molar extinction coefficients and molecular masses were calculated according to Gill and Hippel ***(45)***.

13. Because the first seconds of the reaction are especially important for enzymes with very short half-lifes, it is essential to start the measurement immediately after adding the enzyme. In addition, instruments should be used that record at least one data point per second.
14. The enzyme decay was fitted using the three-parameter exponential decay fit implemented in SigmaPlot: $\frac{\Delta A}{\Delta t} = a + \left[\frac{\Delta A}{\Delta t}\right]_0 e^{-\lambda \times t}$ whereby a is a parameter accounting for basal and accumulated absorption of the nitrocefin solution and $T_{1/2} = \ln 2/\lambda$
15. Depending on the number of data points per nanometer acquired, the parameters for smoothing need to be adjusted. Alternatively, a running mean average can be used for smoothing. In addition to determining λ_{max}, we also tested the shift of center of mass from 330 to 370 nm. Despite the use of more data points, this method did not necessarily give better results.
16. The emission maximum (y_{obs}) as a function of denaturant concentration, [D], was deconvoluted in the constituting signals of the three conformational states (y_N, y_I, y_U) according to their fraction present (f_N, f_I, f_U), which is described by $y_{obs} = y_N \times f_N + y_I \times f_I + y_U \times f_U$. The law of mass action, $K_{NI} = [I]/[N]$, $K_{IU} = [U]/[I]$, was combined with mass conservation, $[N]_0 = [N] + [I] + [U]$, to calculate the fractions, $f_N = 1/(1 + K_{NI} + K_{NI} \times K_{IU})$, $f_I = K_{NI}/(1 + K_{NI} + K_{NI} \times K_{IU})$, and $f_U = K_{NI} \times K_{IU}/(1 + K_{NI} + K_{NI} \times K_{IU})$. Thermodynamic parameters were introduced using the linear extrapolation method based on $\Delta G^0 = -RT\ln K = \Delta G^0_{H_2O} - m \times [D]$, assuming a linear dependence for all states ***(46)***.
17. To account for small changes between the absolute maxima measured with the two fluorometers, the normalized fraction unfolded, $f_{unfold} = (y_{obs} - y_F)/(y_U - y_F)$, is given in the plot.

References

1. Jaenicke, R. and Böhm, G. (1998) The stability of proteins in extreme environments. *Curr. Opin. Struct. Biol.* **8,** 738–748.
2. Ladenstein, R. and Antranikian, G. (1998) Proteins from hyperthermophiles: stability and enzymatic catalysis close to the boiling point of water. *Adv. Biochem. Eng. Biotechnol.* **61,** 37–85.
3. Querol, E., Perez-Pons, J. A., and Mozo-Villarias, A. (1996) Analysis of protein conformational characteristics related to thermostability. *Protein Eng.* **9,** 265–271.
4. Vieille, C. and Zeikus, G. J. (2001) Hyperthermophilic enzymes: sources, uses, and molecular mechanisms for thermostability. *Microbiol. Mol. Biol. Rev.* **65,** 1–43.
5. Pace, C. N., Grimsley, G. R., Thomson, J. A., and Barnett, B. J. (1988) Conformational stability and activity of ribonuclease T1 with zero, one, and two intact disulfide bonds. *J. Biol. Chem.* **263,** 11,820–11,825.
6. Mason, J. M., Gibbs, N., Sessions, R. B., and Clarke, A. R. (2002) The influence of intramolecular bridges on the dynamics of a protein folding reaction. *Biochemistry* **41,** 12,093–12,099.

7. Thompson, M. J. and Eisenberg, D. (1999) Transproteomic evidence of a loop-deletion mechanism for enhancing protein thermostability. *J. Mol. Biol.* **290,** 595–604.
8. Vogt, G., Woell, S., and Argos, P. (1997) Protein thermal stability, hydrogen bonds, and ion pairs. *J. Mol. Biol.* **269,** 631–643.
9. Szilagyi, A. and Zavodszky, P. (2000) Structural differences between mesophilic, moderately thermophilic and extremely thermophilic protein subunits: results of a comprehensive survey. *Structure Fold Des* **8,** 493–504.
10. Karshikoff, A. and Ladenstein, R. (2001) Ion pairs and the thermotolerance of proteins from hyperthermophiles: a "traffic rule" for hot roads. *Trends Biochem. Sci.* **26,** 550–556.
11. Petsko, G. A. (2001) Structural basis of thermostability in hyperthermophilic proteins, or "there's more than one way to skin a cat." *Methods Enzymol.* **334,** 469–478.
12. Malakauskas, S. M. and Mayo, S. L. (1998) Design, structure and stability of a hyperthermophilic protein variant. *Nat. Struct. Biol.* **5,** 470–475.
13. Filikov, A. V., Hayes, R. J., Luo, P., et al. (2002) Computational stabilization of human growth hormone. *Protein Sci.* **11,** 1452–1461.
14. Lehmann, M., Loch, C., Middendorf, A., et al. (2002) The consensus concept for thermostability engineering of proteins: further proof of concept. *Protein Eng.* **15,** 403–411.
15. Stemmer, W. P. (1994) Rapid evolution of a protein in vitro by DNA shuffling. *Nature* **370,** 389–391.
16. Stemmer, W. P. (1994) DNA shuffling by random fragmentation and reassembly: in vitro recombination for molecular evolution. *Proc. Natl. Acad. Sci. USA* **91,** 10,747–10,751.
17. Yang, F., Cheng, Y., Peng, J., Zhou, J., and Jing, G. (2001) Probing the conformational state of a truncated staphylococcal nuclease R using time of flight mass spectrometry with limited proteolysis. *Eur. J. Biochem.* **268,** 4227–4232.
18. Van der Schueren, J., Robben, J., and Volckaert, G. (1998) Misfolding of chloramphenicol acetyltransferase due to carboxy-terminal truncation can be corrected by second-site mutations. *Protein Eng.* **11,** 1211–1217.
19. Sherwood, L. M. and Potts, J. T., Jr. (1965) Conformational studies of pancreatic ribonuclease and its subtilisin-produced derivatives. *J. Biol. Chem.* **240,** 3799–3805.
20. Haruki, M., Noguchi, E., Akasako, A., Oobatake, M., Itaya, M., and Kanaya, S. (1994) A novel strategy for stabilization of Escherichia coli ribonuclease HI involving a screen for an intragenic suppressor of carboxyl-terminal deletions. *J. Biol. Chem.* **269,** 26,904–26,911.
21. Trevino, R. J., Tsalkova, T., Kramer, G., Hardesty, B., Chirgwin, J. M., and Horowitz, P. M. (1998) Truncations at the NH2 terminus of rhodanese destabilize the enzyme and decrease its heterologous expression. *J. Biol. Chem.* **273,** 27,841–27,847.
22. Trevino, R. J., Gliubich, F., Berni, R., et al. (1999) NH2-terminal sequence truncation decreases the stability of bovine rhodanese, minimally perturbs its crystal structure, and enhances interaction with GroEL under native conditions. *J. Biol. Chem.* **274,** 13,938–13,947.

23. Vainshtein, I., Atrazhev, A., Eom, S. H., Elliott, J. F., Wishart, D. S., and Malcolm, B. A. (1996) Peptide rescue of an N-terminal truncation of the Stoffel fragment of taq DNA polymerase. *Protein Sci.* **5,** 1785–1792.
24. Shortle, D. and Lin, B. (1985) Genetic analysis of staphylococcal nuclease: identification of three intragenic "global" suppressors of nuclease-minus mutations. *Genetics* **110,** 539–555.
25. Petrounia, I. P. and Arnold, F. H. (2000) Designed evolution of enzymatic properties. *Curr. Opin. Biotechnol.* **11,** 325–330.
26. Krebber, A., Bornhauser, S., Burmester, J., et al. (1997) Reliable cloning of functional antibody variable domains from hybridomas and spleen cell repertoires employing a reengineered phage display system. *J. Immunol. Methods* **201,** 35–55.
27. Orencia, M. C., Yoon, J. S., Ness, J. E., Stemmer, W. P., and Stevens, R. C. (2001) Predicting the emergence of antibiotic resistance by directed evolution and structural analysis. *Nat. Struct. Biol.* **8,** 238–242.
28. Rodrigues, M. L., Presta, L. G., Kotts, C. E., et al. (1995) Development of a humanized disulfide-stabilized anti-p185HER2 Fv-beta-lactamase fusion protein for activation of a cephalosporin doxorubicin prodrug. *Cancer Res.* **55,** 63–70.
29. Huang, W. and Palzkill, T. (1997) A natural polymorphism in beta-lactamase is a global suppressor. *Proc. Natl. Acad. Sci. USA* **94,** 8801–8806.
30. Shindyalov, I. N. and Bourne, P. E. (1998) Protein structure alignment by incremental combinatorial extension (CE) of the optimal path. *Protein Eng.* **11,** 739–747.
31. Guex, N. and Peitsch, M. C. (1997) SWISS-MODEL and the Swiss-PdbViewer: an environment for comparative protein modeling. *Electrophoresis* **18,** 2714–2723.
32. Huang, W., Petrosino, J., Hirsch, M., Shenkin, P. S., and Palzkill, T. (1996) Amino acid sequence determinants of beta-lactamase structure and activity. *J. Mol. Biol.* **258,** 688–703.
33. Lei, S. P., Lin, H. C., Wang, S. S., Callaway, J., and Wilcox, G. (1987) Characterization of the Erwinia carotovora pelB gene and its product pectate lyase. *J. Bacteriol.* **169,** 4379–4383.
34. Yanisch-Perron, C., Vieira, J., and Messing, J. (1985) Improved M13 phage cloning vectors and host strains: nucleotide sequences of the M13mp18 and pUC19 vectors. *Gene* **33,** 103–119.
35. Olins, P. O., Devine, C. S., Rangwala, S. H., and Kavka, K. S. (1988) The T7 phage gene 10 leader RNA, a ribosome-binding site that dramatically enhances the expression of foreign genes in Escherichia coli. *Gene* **73,** 227–235.
36. Cadwell, R. C. and Joyce, G. F. (1994) Mutagenic PCR. *PCR Methods Appl.* **3,** S136–S140.
37. Jelsch, C., Mourey, L., Masson, J. M., and Samama, J. P. (1993) Crystal structure of Escherichia coli TEM1 beta-lactamase at 1.8 A resolution. *Proteins* **16,** 364–383.
38. Ambler, R. P., Coulson, A. F., Frere, J. M., et al. (1991) A standard numbering scheme for the class A beta-lactamases. *Biochem. J.* **276 (Pt 1),** 269, 270.
39. Sideraki, V., Huang, W., Palzkill, T., and Gilbert, H. F. (2001) A secondary drug resistance mutation of TEM-1 beta-lactamase that suppresses misfolding and aggregation. *Proc. Natl. Acad. Sci. USA* **98,** 283–288.

40. Wang, X., Minasov, G., and Shoichet, B. K. (2002) Evolution of an antibiotic resistance enzyme constrained by stability and activity trade-offs. *J. Mol. Biol.* **320,** 85–95.
41. Osuna, J., Perez-Blancas, A., and Soberon, X. (2002) Improving a circularly permuted TEM-1 beta-lactamase by directed evolution. *Protein Eng.* **15,** 463–470.
42. Vanhove, M., Raquet, X., and Frere, J. M. (1995) Investigation of the folding pathway of the TEM-1 beta-lactamase. *Proteins* **22,** 110–118.
43. Zahn, R., Axmann, S. E., Rucknagel, K. P., Jaeger, E., Laminet, A. A., and Plückthun, A. (1994) Thermodynamic partitioning model for hydrophobic binding of polypeptides by GroEL. I. GroEL recognizes the signal sequences of beta-lactamase precursor. *J. Mol. Biol.* **242,** 150–164.
44. Frech, C., Wunderlich, M., Glockshuber, R., and Schmid, F. X. (1996) Competition between DsbA-mediated oxidation and conformational folding of RTEM1 beta-lactamase. *Biochemistry* **35,** 11,386–11,395.
45. Gill, S. and von Hippel, G. (1989) Calculation of protein extinction coefficients from amino acid sequence data. *Anal. Biochem.* **182,** 319–326.
46. Pace, C. N. (1986) Determination and analysis of urea and guanidine hydrochloride denaturation curves. *Methods Enzymol.* **131,** 266–280.
47. Sayle, R. A. and Milner-White, E. J. (1995) RASMOL: biomolecular graphics for all. *Trends Biochem. Sci.* **20,** 374.
48. Vriend, G. (1990) WHAT IF: a molecular modeling and drug design program. *J. Mol. Graph.* **8,** 52–56.
49. Santoro, M. M. and Bolen, D. W. (1988) Unfolding free energy changes determined by the linear extrapolation method. 1. Unfolding of phenylmethanesulfonyl alpha-chymotrypsin using different denaturants. *Biochemistry* **27,** 8063–8068.

Index

C

D

H

M

N

MIX
Papier aus verantwortungsvollen Quellen
Paper from responsible sources
FSC® C105338

If you have any concerns about our products,
you can contact us on
ProductSafety@springernature.com

In case Publisher is established outside the EU,
the EU authorized representative is:
Springer Nature Customer Service Center GmbH
Europaplatz 3, 69115 Heidelberg, Germany

Printed by Libri Plureos GmbH
in Hamburg, Germany